Brainmedia

Thinking|Media

Series Editors

Bernd Herzogenrath

Patricia Pisters

Brainmedia

One Hundred Years of Performing Live Brains, 1920–2020

Flora Lysen

BLOOMSBURY ACADEMIC
NEW YORK • LONDON • OXFORD • NEW DELHI • SYDNEY

BLOOMSBURY ACADEMIC
Bloomsbury Publishing Inc
1385 Broadway, New York, NY 10018, USA
50 Bedford Square, London, WC1B 3DP, UK
29 Earlsfort Terrace, Dublin 2, Ireland

BLOOMSBURY, BLOOMSBURY ACADEMIC and the Diana logo are trademarks of Bloomsbury Publishing Plc

First published in the United States of America 2022
Paperback edition published 2024

Copyright © Flora Lysen, 2022, 2024

For legal purposes the Acknowledgments on pp. x–xii constitute an extension of this copyright page.

Cover design: Daniel Benneworth-Gray
Cover image © Paolo Sanfilippo

All rights reserved. No part of this publication may be reproduced or transmitted in any form or by any means, electronic or mechanical, including photocopying, recording, or any information storage or retrieval system, without prior permission in writing from the publishers.

Bloomsbury Publishing Inc does not have any control over, or responsibility for, any third-party websites referred to or in this book. All internet addresses given in this book were correct at the time of going to press. The author and publisher regret any inconvenience caused if addresses have changed or sites have ceased to exist, but can accept no responsibility for any such changes.

A catalog record for this book is available from the Library of Congress.

ISBN: HB: 978-1-5013-7875-1
PB: 978-1-5013-7872-0
ePDF: 978-1-5013-7873-7
eBook: 978-1-5013-7874-4

Series: Thinking Media

Typeset by Newgen KnowledgeWorks Pvt. Ltd., Chennai, India

To find out more about our authors and books visit www.bloomsbury.com and sign up for our newsletters.

Contents

List of Figures	viii
Acknowledgments	x
Introduction	1
Prelude: Live Brain Demonstrations in 1934 and 2014	1
Studying Performances of Live Brains: An Introduction	2
A Critical History of Brain Science, or: Is the Medium the Message?	4
From Metaphors to Brainmedia	9
A Media-Historical Approach to the Media Enmeshed with Brain Science	13
Understanding Scientific Practices and Communications through "Performing Knowledges"	16
Forms of Liveness: Watching the Brain at Work	20
Critical Histories and the Neuro-Enchanted Present	24
1 The Birth of the Live Brain, 1820–1920	29
Nervous Subject, Modern Sensorium	31
Foucault's "Apparatus of Neurological Capture"	38
Examining and Imagining the Living Brain	43
Conclusions	46
2 Displaying Dynamic Brains: Illuminated Brain Models and the Enchanted Loom, 1928–38	49
A Feverish Image of the Brain Gone Mad: Electro-Brains and a Crisis of Representation	52
A Glow-in-the-Dark Brain from Vienna	57
Animated Brains for Modern Citizens	61
Searching for a Dynamic Image: Between *Gesamtbild* and *Gehirnwahrheiten*	63
Streaming Headlines and the Motograph Brain	68

	Tele-visual and Televisual Neurophysiology	72
	Brains at Work in Office and Factory: Living Diagrams and a Logic of Direct Display	75
	Conclusions	79
3	Demonstrating Brainwaves beyond the Laboratory: EEG as White Magic and Dark Media, 1934–41	85
	Brainwave Imaginaries and Popularizing Science	90
	Framing EEG in Print: Vivid Demonstrations and "the Stuff That Dreams Are Made On"	96
	The "White Magic" of Science: EEG at the 1937 *Paris International Exhibition*	102
	Dark Brain Media in Hollywood	110
	Conclusions	115
4	Broadcasting Live Brains: The Brain on Television and as Television, 1949–57	119
	Broadcasting Science and a "Television of Attractions"	123
	The Brain "*En direct*" and the Epistemological Seductions of Television	126
	Toward the Brain as TV: The Cybernetic Living Brain	133
	Cortical (Television) Scanning, the "Ineluctable Inference"	138
	Too Close to the Screen: Television as Trigger and Mirror	142
	The Toposcope: The Brain Displays Itself	147
	Conclusions	151
5	Interfacing the Real-Time Brain: EEG Feedback in Art and Science, 1964–77	155
	The Groovy Science of Alpha	158
	"Broadening Horizons" and "The Golden Age of Man"	162
	"Disproportionate Excitement" and the Alpha Fad	167
	Circuited Selves, Media Environments, and Radical Software	171
	Techno-Sensory Interface Projects and New Modes of Communication	178

Contents vii

    New Micro-Temporalities of the Brain in Real Time   182
    Conclusions   192

6   Synchronizing Two Dynamic Brains: Art–Science Experiments and Neuroscience in the Wild, 2013–19   195
    Investigating New Forms of Neuroscientific Life   200
    A Real-World Neuroscience with Hyper-Stakes   204
    When Works of Art Become Scientific Papers: Neurocentrism Revisited   211
    The Allure of Synchronization: Toward a Critical Media History of Being on the Same Wavelength   215
    Conclusions   224

Conclusion   227
    Understanding Contemporary Live Brains   229
    Engaging Live Brains Today   231

List of Sources of Figures   235
Bibliography   239
Index   279

Figures

1.1	Diagram showing the four chief association centers of the human brain, *c.* 1919	30
1.2	Pierre Marie's scheme for interconnected left hemisphere cortical centers involved in oral and written language, 1888	36
2.1	Fritz Kahn, *Die Lichtwarhnehmung*, 1929	54
2.2	Dr. Edith Klemperer and Dr. Robert Exner, luminous brain model from Vienna, *c.* 1931	58
2.3	Edith Klemperer, patent of anatomical model, 1931/4	59
2.4	Push-button brain diagram, *c.* 1930	63
2.5	Constantin Von Economo's plaster models of cytoarchitecture, *c.* 1927	66
2.6	Fritz Lang, *Die Spione*, 1928	77
2.7	Science's futile attempt to build a perfect mechanical brain, 1934	82
3.1	"Why radio may have uncovered a sixth sense! Science Now investigating cases of broadcast programs being picked up, unaided, by the human nervous system," 1926	94
3.2	Electroencephalography booth in the exhibition *La Biologie: Exposition Internationale de Paris* at the Palais de la Découverte as part of the *Paris International Exposition*, 1937	106
3.3	Edward Dmytryk, *The Devil Commands*, 1941	113
4.1.1	Catching a brain wave, 1954	120
4.1.2	Catching a brain wave, 1954	121
4.2.1	*En direct du Cerveau Humain*, 1956	127
4.2.2	*En direct du Cerveau Humain*, 1956	128
4.2.3	*En direct du Cerveau Humain*, 1956	129
4.2.4	*En direct du Cerveau Humain*, 1956	130
4.3	Illustration of an analogy between a profile-scanning device and the human brain, 1953	137
4.4	Illustration of a trigger circuit, 1949	145

Figures

4.5	Toposcope display visible in EEG-set-up at the Burden Neurological Institute in Bristol c. 1954, with researcher Vivian Walter	147
4.6	Toposcope display, c. 1957	148
5.1	Alvin Lucier (left) and John Cage (right) preparing a performance of *Music for Solo Performer* at the festival "John Cage at Wesleyan," 1988	156
5.2	Joe Kamiya with EEG feedback setup and subject, c. 1968	166
5.3.1	Marvin Karlins and Lewis M. Andrews, *Biofeedback: Turning on the Power of Your Mind*, 1973	170
5.3.2	Larry Kettelkamp, *A Partnership of Mind and Body, Biofeedback*, 1976	170
5.3.3	Anthony A. Zaffuto, *Alphagenics; How to Use Your Brain Waves to Improve Your Life*, 1974	170
5.3.4	Barbara B. Brown, *New Mind, New Body; Bio-Feedback: New Directions for the Mind*, 1974	170
5.4	David Rosenboom, *Ecology of the Skin*, 1972	173
5.5	Nina Sobell and Michael Trivich, interactive electroencephalographic video drawings, 1972–4	174
5.6	Marc Bjorlund, no title, 1971	175
5.7	Richard Lowenberg, *Environetic Synthesis*, 1970	176
5.8	Feedback training with alpha train in Barbara Brown's lab, c. 1974	180
5.9.1	Edmond Dewan, brain-controlled lamp setup, 1964	181
5.9.2	Edmond Dewan, brain-controlled lamp setup, 1964	181
6.1	Suzanne Dikker and Matthias Oostrik, *Mutual Wave Machine*, 2014	196
6.2	Jean Livet et al., *Brainbow* image of dentate gyrus, 2007	198
6.3	Guillaume Dumas, illustration of two-body neuroscience, *Face-to-phases*, 2013	219
6.4	Two museum attendees participating in David Rosenboom's Vancouver piece at the Vancouver Art Gallery, 1973	222

Acknowledgments

This book is the product of a long journey through many academic and non-academic institutions. Along the way, marvelous colleagues have contributed to my interdisciplinary approach. I would like to acknowledge my various institutional "homes" for hosting and supporting me: the Amsterdam School of Historical Studies (ASH), the Visual and Environmental Studies Department at Harvard University, the Amsterdam School of Cultural Analysis (ASCA), and the Media Studies Department of the University of Amsterdam.

I felt privileged to be part of ASCA's scholarly network, which continues to be a vibrant, interdisciplinary, and engaged research space. Thank you former ASCA colleagues: Annelies Kleinherenbrink, Blandine Joret, Daniel de Zeeuw, Erin La Cour, Florian Göttke, Halbe Kuipers, Irene Villaescusa, Lonnie van Brummelen, Lucy van de Wiel, Moosje Goossen, Noortje de Ley, Nur Ozgenalp, Paula Albuquerque, Peyman Amiri, Rik Spanjers, Selçuk Balamir, Simon Ferdinand, Thijs Witty, Tim Yaczo, and many others. The ASCA neurocultures and neuroaesthetics research group has been an important scholarly sounding board to me throughout the period of working on this study. Organizing the Worlding the Brain conferences and workshops together has been a very enjoyable and instructive experience. Thank you, fellow core team: Julian Kiverstein, Machiel Keestra, Nim Goede, Patricia Pisters, and Stephan Besser for co-creating this hybrid space welcoming junior and senior academics, artists, and other collaborators, all with a similar enthusiasm for the strange world of the "neuro."

The University of Amsterdam's Mediastudies department was a vital academic home base for many years. Many thanks to my smart and kind office mates: Amanda Paz, Arnoud Arps, Christian Olesen, David Duindam, David Gauthier, Eva Sancho Rodriguez, Judith Naeff, Miriam Meissner, Natalia Sánchez Querubín, Niels Kerssens, Pedram Dibazar, Sal Hagen, Penn Ip, and Simone Kalkman. Over the years, Amsterdam's media scholars have increasingly influenced and inspired me. Thank you, Maryn Wilkinson and Daisy van de Zande, my first friends in acade-media, for welcoming me and for laughs. Thank you, Sebastian Scholz, for our discussions on the intersections between histories of science and media. Thank you, Abe Geil, for our conversations about academia and about the world beyond academia. I am beholden to my thoughtful and persistent mentors Rob Zwijnenberg at Leiden University, and Patricia Pisters at the University of Amsterdam, who have been advising this study from its inception and have helped me find my own way of studying histories of science and media. Thank you, Rob, for your insistence on finding a more argumentative voice. I can hardly imagine bringing this project to a successful completion without the warm commitment and trust of Patricia, who shows it is possible to be a feminist scholar, manager, and teacher, and to be a caring and creative person at the same time.

Researching one hundred years of science and media history would have been impossible without the wonderful support of library and archival staff at the University of Amsterdam and many other institutions. Thank you for copying, mailing, and acquiring; for (digital) access to archives; for renewals; and, most of all, for keeping things. I am also very grateful to Florian Duijsens for his excellent editorial work on an earlier version of the manuscript for this study.

Versions of parts of chapters in this study have previously appeared in the following publications:

Chapter 1: "It Blinks, It Thinks? Luminous Brains and a Visual Culture of Electric Display, Circa 1930." *Nuncius* 32, no. 2 (2017): 412–39.

Chapter 5: "The Interface Is the (Art)Work: EEG-Feedback, Circuited Selves and the Rise of Real-Time Brainmedia (1964–1977)." In *Brain Art: Brain–Computer Interfaces for Artistic Expression*, edited by Anton Nijholt, 33–63. Cham: Springer International Publishing, 2019.

During the final period of working on this study, I combined my research with a new job as first program coordinator for the Amsterdam Research Institute of the Arts and Sciences. I thank the ingenious and inspiring artists and scholars I met during this period, who dared to experiment with different forms of research. A special thanks to Suzanne Dikker and Matthias Oostrik, whose art-science work I discuss in Chapter 6, for taking the time to introduce me to their experimental investigation of brains and behavior and for their open discussion of methodologies-in-the-making. Many thanks also to the artist Antye Guenther for exploring the world of art-science and cerebral imaginaries together, as well as for her friendship.

Working on this book overlapped with building a home and living a communal life in de Lepelkruisstraat. This alternate universe has meant the world to me. Thank you for infinite amounts of coffee, giggles, meetings, dishes, activism, relaxation, relativism, and reassurance: Anna Galenkamp, Eef Hengst, Flint Hignett, Hannah Murray, Hans Wijnberg, Jenny Pannenbecker, John Visser, Leontien Simis, Marten van den Berge, Matthias Post-Uiterweer, Rik Sibbing, Rosa van de Griendt, Rutger Post-Uiterweer, Ryuko Reid, Serhat Ozcelik, Sev van Dorst, Shanti Straub, and Staf Post-Uiterweer. Beyond this home, so many other kind friends have been important in the course of writing this book. Thank you, Adinda Post-Uiterweer, Anna Grasskamp, Anne de Ley, Carlos Vaquero, Caroline von Courten, Danielle Voestermans, Eelco Jacobs, Erik Rietveld, Hanneke Mooren, Hein Meijer, Ida Lysen, Joeri Scholtens, Johannes Koeppel, Kim Bruggeman, Lotte Batelaan, Marianne van Asperen, Marta Zarzycka, Neeltje Bollen, Olga Boerwinkel, Puck van Dijk, Rik Peters, Ruben Jacobs, Rutger Kaput, Rose Nijssen, Sarah Beeks, and Tomm Velthuis. To Robert Owsiejko: a hug, thank you so much. To Alieke van Saarloos, Bas Voorwinde, and Charlotte Brugman: you are incredible friends.

With my sister, Thalia Lysen, it only takes a millisecond to synchronize: thank you for being my best friend. My parents, Jojet Lysen and Chris Leonards, provided unwavering support and warm encouragement, thank you for teaching me to explore with ambition. I owe tremendous gratitude to Chris, who patiently assisted in figuring out different reference tools and traced the copyright holders of the images in this book. Thank you, my dear Nora Rausch for your uplifting hugs and for your sticky notes with brain smileys. Finally, there are not enough words to thank Christoph Rausch, who, together with Nora, continues to show me what it means to care. Picking your brain is wonderful. Loving you even more.

Introduction

Prelude: Live Brain Demonstrations in 1934 and 2014

On May 12, 1934, members of the British Physiological Society gathered in a large Cambridge University lecture hall to witness the first UK demonstration of a technology measuring the electrical activity of the human brain. To prove that the brain displayed varying patterns of electrical activity and that this could be measured from the scalp, neurophysiologists Bryan Matthews and Edgar Adrian used a new technical setup that included an amplifier and an oscillograph, all visible on stage.[1] During the demonstration, Matthews would measure Adrian's brain activity, recorded by a pen zigzagging on a moving piece of paper. Simultaneously, the audience of invited scientists could see a magnified projected image of the moving arm of the ink writer on a large projection screen above the stage.[2] This direct projection allowed spectators to see the recorded activity change when Adrian closed his eyes or solved a math problem. One scientist who attended wrote in a later account how "those who were privileged to be present will never forget … the dramatic effect of the first demonstration of the Berger [now called EEG] rhythm. I suppose that, to anyone present at that demonstration, it offered a first glimpse of great new possibilities. Adrian and Matthews's demonstration … will be recognized as an event of epoch-making importance!"[3]

On November 1, 2014, the city of Amsterdam was unusually brimming with brains. As part of a celebration of the "European Year of the Brain,"

---

[1] Theories and techniques involved in this demonstration are explained in E. D. Adrian and B. H. C. Matthews, "The Berger Rhythm: Potential Changes from the Occipital Lobes in Man," *Brain* 57 (1934).

[2] This demonstration took place on May 12, 1934, with the help of an epidiascope (a type of image projector). See Adrian's account in Edgar Douglas Adrian, "The Discovery of Berger," in *Handbook of Electroencephalography and Clinical Neurophysiology*, ed. Antoine Rémond (Amsterdam: Elsevier, 1971). Adrian cannot recall whether flicker lights were used on stage to produce distinct alpha waves. For another account of the demonstration, see W. Grey Walter, "Thought and Brain: A Cambridge Experiment," *Spectator Archive*, October 4, 1934.

[3] Henry Dale (source and year unspecified) cited in David Millett, "Wiring the Brain: From the Excitable Cortex to the EEG, 1870–1940" (PhD dissertation, University of Chicago Press, 2001), 346.

visitors of the Amsterdam Museum Night could, for example, witness a dissection of a brain in the university library, walk into a giant brain at the NEMO Science Center, and measure their brain activity at the former anatomical theater (De Waag). For the latter event, titled *Brain Waves*, volunteers had signed up through an open call on Facebook, agreeing to participate in an experiment in which their brain activity would be measured on stage and directly shown to an audience. Three participants at a time were fitted with EEG headsets and asked to take a seat in the middle of a circular room. Scientists dressed in white lab coats encouraged the subjects to stay focused while being confronted with a range of stimuli, such as a burlesque dancer, loud electronic music, and a brain teaser. Colorful visualizations of the participants' corresponding brain activities were projected on three large screens lining the walls of the space. An adjacent explanatory legend helped visitors to interpret and compare the differing brainwave states on view. An unexpectedly overwhelming number of people lined up to participate in *Brain Waves*, perhaps because, as one participant exclaimed, "It just seemed so cool to see my own brain activity in real-time!"[4]

Studying Performances of Live Brains: An Introduction

Adrian and Matthew's 1934 demonstration and the 2014 *Brain Waves* event are examples of spectacular appearances of the active brain outside the scientific laboratory. Both demonstrations are instances of the deliberate ways scientists and science educators perform a direct view of the living brain, invoking a sense of seeing the brain "in action," of perceiving it live, as it were. This "live brain" is at the center of this study. It interests me as a practice, idea, myth, and imaginary.

This study analyzes original historical material with a transdisciplinary scope. First, I synthesize scholarship in the history of science and media studies, introducing analytical concepts to better understand emerging practices of imagining, conceptualizing, and performing the brain in action. To that end, my analysis uses three key concepts—*brainmedia, performing knowledges*, and *forms of liveness*—that I introduce in this chapter. Second, based on my research of many not previously accessed historical documents (focused on the United States and Europe), I present five case studies of

---

[4] "Brainwaves during Museum Night," Waag Society, https://waag.org/en/blog/brainwaves-during-museum-night.

live brains from the 1920s until today. Introducing these five case studies, I first locate the emergence of the live brain in the early decades of the twentieth century, embedding it in a broader historical context with special attention to Michel Foucault's writings on the correlative constitution of the neurological gaze.

My study traces one hundred years of performing live brains (characterized by shifting forms of "liveness") through five case studies: Chapter 2 tackles the creation of illuminated brain models around 1930; Chapter 3 covers staged recordings of brainwaves from the 1930s to the 1940s; Chapter 4 focuses on live brains *on* television and conceptions of brains *as* television in the 1940s and 1950s; Chapter 5 addresses real-time EEG feedback circuits around 1970; and Chapter 6 discusses "brain-to-brain" art–science installation works between 2013 and 2019. To analyze these five case studies, I examine practices and imaginaries that shaped and enacted the idea of perceiving the live brain. Particularly, I inquire into the way new approaches to technical and broadcast media and new ideas of mediation undergird different imaginaries of seeing the brain in action.

Describing the emergence of a new conception of the living brain in the twentieth century, my argument is that it was specific *assemblages* of brains and media—brainmedia—that allowed for the formation of what I call the live brain. My main question is how conceptions of the "brain at work" were practically produced and negotiated in specific historical situations. Asking what is required to create a live brain, I am interested in the enmeshments of screens, nerve preparations, experimental subjects, and computer displays, but also rhetorical tropes, science on television, or art–science installations that together imagine and conceive the brain in action. I particularly inquire into the role of media and mediation in such assemblages, asking how various media (including both technical and broadcast media) *and* changing historical ideas of mediation shape the conception and examination of the active brain. Throughout, I show how new conceptions of media and mediation allow the human brain to be thought differently—as a brain in action—and have opened new ways to perform and enact this live brain.

Turning back to this chapter's preludes from two very disparate moments in time prompts a series of questions: What was needed to produce a live brain in 1934 and in 2014? What repertoires of scientific performance could demonstrate the brain in action in a scientific lecture hall in Cambridge and on a platform for "public science" in Amsterdam? How were audience members prompted to understand this performance, and how do their accounts reflect historical understandings of mediation, as well as previous experiences and instances of seeing the living brain and body at work? How were brainmedia assembled across a projector apparatus, a screen, but also in newspaper accounts (in 1934) and online (in 2014)? Which expertise and what kinds of dramatization did these knowledge performances invoke? What type of live brain was enacted, and what type of interventions into our lives did this live brain allow?

A Critical History of Brain Science, or: Is the Medium the Message?

By introducing the concept of brainmedia to approach the past and the present of seeing the brain at work, I aim to nuance two key strands in critical humanities scholarship of the brain sciences. The first is the dominant narrative in the historiography of the brain and mind sciences in the twentieth century that attributes a special historical status to the rise of functional imaging in the 1990s, viewing their emergence as a watershed moment in the history of the brain at work, as creating a live brain that was persuasively transparent and heralded powerful popular acclaim for the neurosciences. With this study, I present alternative histories of the functioning, living brain—inside and outside the laboratory—before and after the 1990s, thus opening up different lineages and historical perspectives. Second, critical literature about the rise of a new neurocentered culture since the 1990s tends to depict this culture as one in which the scientific and the popular are rather rigidly demarcated. I, however, claim that there is no such clear demarcation, rather a back and forth between different realms of knowledge production.

It is by now commonplace to start an analytical account of the brain and mind sciences with the so-called Decade of the Brain in the 1990s, which is said to mark the start of the omnipresence of the brain and the brain sciences in cultural discourses (omnipresent, that is, understood from the viewpoint of affluent Western societies where the majority of brain studies are conducted and circulated).[5] This historical flagpole then gives way to a characterization of the present as a "brain culture" or "neuroculture": a present in which the brain makes striking appearances in popular media, in which brains are increasingly equated with the "self," and in which the brain sciences are attributed a remarkable explanatory power for complex phenomena ranging from adolescent behavior and gender differences to consumption, criminality, and depression.[6] Already during the 1990s and in

---

[5] About the "Decade of the Brain" initiative and other neuroscience awareness initiatives, see Dolores Modic and Maryann P. Feldman, "Mapping the Human Brain: Comparing the US and EU Grand Challenges," *Science and Public Policy* 44, no. 3 (2017).

[6] "Brain Culture" in Davi Johnson Thornton, *Brain Culture: Neuroscience and Popular Media* (New Brunswick, NJ: Rutgers University Press, 2011). G. Frazzetto and S. Anker,

the early 2000s a number of cultural and sociological studies critiqued this hegemonic status of the neuro-disciplines. Their ascent is often located in the emergence of functional brain imaging in the 1990s (PET and fMRI), which produced beautiful, colorful images of brains "lighting up" with varying mental behaviors, leading to speculation about, and promises of, mapping cognitive functions that ran far ahead of theoretical understandings and scientific proof.[7]

Numerous scholars have remarked on the seductive allure of these new images of the brain in action and critiqued image-oriented brain research (pejoratively called "blobology").[8] Perhaps, some authors have argued, it is because these imaging methods mesh morphological and functional data that they make it seem, in historian Michael Hagner's words, as if one "observes the mind at work just as one observes the blushing or the mimicry of a face. Neuroimaging, then, is a kind of physiognomy turned inward."[9] Around 2000, a number of ethnographers and social scientists examining new PET, MRI, and fMRI practices analyzed the theories and assumptions that are built into visualizations of active brains and demonstrated that these imaging practices' claims to truth depend heavily on the technologies created to construct them.[10] In 1996, Hagner argued that functional brain images had caused a shift in

---

"Neuroculture," *Nature Reviews Neuroscience* 10, no. 11 (2009). Francisco Ortega and Fernando Vidal, eds., *Neurocultures: Glimpses into an Expanding Universe* (Frankfurt am Main: Peter Lang, 2011). For a summary of the scholarly backlash against the neuro-turn, see Melissa M. Littlefield, "'A Mind Plague on Both Your Houses': Imagining the Impact of the Neuro-Turn on the Neurosciences," in *The Human Sciences after the Decade of the Brain*, ed. Jon Leefmann and Elisabeth Hildt (London: Academic Press, 2017).

[7] Anne Beaulieu, "Images Are Not the (Only) Truth: Brain Mapping, Visual Knowledge, and Iconoclasm," *Science, Technology & Human Values* 27, no. 1 (2002); Adina L. Roskies, "Neuroimaging and Inferential Distance: The Perils of Pictures," in *Foundational Issues in Human Brain Mapping*, ed. Stephen José Hanson and Martin Bunzl (Cambridge, MA: MIT Press, 2010); Letitia Meynell, "The Politics of Pictured Reality: Locating the Object from Nowhere in FMRI," in *Neurofeminism: Issues at the Intersection of Feminist Theory and Cognitive Science*, ed. Robyn Bluhm, Anne Jaap Jacobson, and Heidi Lene Maibom (London: Palgrave Macmillan, 2012); Fitsch, Hannah. *Dem Gehirn beim Denken zusehen?: Sicht- und Sagbarkeiten in der funktionellen Magnetresonanztomographie*. Bielefeld: Transcript Verlag, 2014.

[8] "Seductive Allure" in D. S. Weisberg et al., "The Seductive Allure of Neuroscience Explanations," *Journal of Cognitive Neuroscience* 20, no. 3 (2008). "Blobology" in Cordelia Fine, *Delusions of Gender: How Our Minds, Society, and Neurosexism Create Difference* (New York: W.W. Norton, 2011), 153.

[9] Michael Hagner, "The Mind at Work: The Visual Representation of Cerebral Processes," in *The Body Within: Art, Medicine and Visualization*, ed. Renée van de Vall and Robert Zwijnenberg (Leiden: Brill, 2009), 87.

[10] Anne Beaulieu, "The Space Inside the Skull: Digital Representations, Brain Mapping and Cognitive Neuroscience in the Decade of the Brain" (Dissertation, University of Amsterdam, 2000); Joseph Dumit, *Picturing Personhood: Brain Scans and Biomedical Identity* (Princeton, NJ: Princeton University Press, 2004); Andreas Roepstorff, "Brains in Scanners: An Umwelt of Cognitive Neuroscience," *Semiotica* 134, nos. 1/4 (2001),

conceptualizing the brain when the abstract information processing model of cognitive science ("the brain as computer") was replaced by visual images of brain activation that offered a remarkable concreteness.[11] These images seemed to show the mind at work, thus visually strengthening the idea that cognitive functions could be decisively mapped and hence that these cognitive functions ("brain states") were real.[12] Because these images buttressed the idea that the living, working brain had become transparent, various longue-durée narratives frame the advent of functional imaging in the 1990s as a decisive turning point in the cultural history of the brain sciences.

For some cultural historians, notably Cornelius Borck and Roger Smith, the Decade of the Brain and the advent of functional imaging not only suggest a critical juncture but also seem to present an *endpoint* with regard to the history of cultural imaginaries of the brain at work. In this account, the 1990s birth of the colorfully imaged brain constitutes the conclusion of a long history of models of, and metaphors for, the brain's mechanisms, replacing it "with the immediacy of an artificially real brain image," as Cornelius Borck has argued, now "the brain has become the medium and the message."[13] Similarly, historian Roger Smith argues that these imaging practices constitute a major turning point in the history of psychology and psychophysiology: "It is tempting to think that in brain science, Marshall McLuhan was right: the medium is the message—the technology of representation determines the content of knowledge."[14]

Neither Borck nor Smith would adhere to a strong version of the claim that there are no more metaphors in the brain sciences. They simply mean to underline a quantitative shift in the brain sciences, a change in the relation of technology (medium) vis-à-vis theorization (message). In fact, histories of functional imaging (such as Borck's work on the history of EEG, and Smith's work on the history of the mind–brain problem) show that promises of transparent brains and the possibility to trace mental states through physiological measurements have much longer and varied histories of cerebral imaginaries tied to technological media. A well-known historical account,

---

and "Transforming Subjects into Objectivity: An Ethnography of Knowledge in a Brain Imaging Laboratory," *FOLK, Journal of the Danish Ethnographic Society* 44 (2002).

[11] Michael Hagner, "Der Geist Bei Der Arbeit: Überlegungen Zur Visuellen Repräsentation Cerebraler Prozesse," in *Anatomien Medizinischen Wissens: Medizin- Macht- Moleküle*, ed. Cornelius Borck and Susan M. DiGiacomo (Frankfurt am Main: Fischer-Taschenbuch-Verlag, 1996).

[12] Ibid.

[13] Cornelius Borck, "Toys Are Us: Models and Metaphors in Brain Research," in *Critical Neuroscience: A Handbook of the Social and Cultural Contexts of Neuroscience*, ed. Suparna Choudhury and Jan Slaby (Oxford: Wiley Blackwell, 2011), 129.

[14] Roger Smith, *Between Mind and Nature: A History of Psychology* (London: Reaktion Books, 2013), 259.

in this respect, is Gerd Gigerenzer's observation on a shift from theory to instruments (what he calls a "tools-to-theory heuristics") that took place in the domain of cognitive psychology, where he claims that the empirical method of inferential statistics (emerging with the ubiquity of computers in psychological and psychophysiological laboratories in the 1960s) positioned "the mind as an intuitive statistician."[15] Yet, even though the move from theories to tools has a long history, there seems to be something qualitatively different, according to various authors, about image making in the 1990s. As Hagner has argued, while the old questions pertaining to the relationship between mind, experience, and brain formulated in the nineteenth century essentially remain the same, brain science is now more than ever determined by "newly enabled views and grasps by new technology."[16] Something happened in the wake of the Decade of the Brain that allowed these images to gain great power.

This attribution of an exceptional position, that is, the particularly persuasive power of the 1990s blinking brain, is problematic, as it is made possible by associating this persuasion to a new, ever-expanding popular realm for neuroscience, a so-called neuroculture. Yet, this focus on a pervasive neuroculture has the faulty tendency to suggest a division between imaging practices in the laboratory and the circulation of the "active brain" outside the laboratory. When scholars describe visual and public manifestations of neuroculture (most often in popular science journals or newspapers), they invoke a general public of laypeople who are easily fascinated by the beautiful and authoritative appearance of brain images, and in turn just as easily convinced by the claims the images are asked to support. Brain images, with their seductive reality effects and promise of objectivity, can cause the "apparent materialization," as Maurizio Meloni puts it, of particular emotions, affects, and behaviors at the level of observed processes in the brain, and they seem to do so especially in popular media and public discourse.[17] Functional brain images are, as Joseph Dumit puts it, "expert images" that exert "undue persuasiveness" over nonexperts because of their "apparent legibility."[18]

[15] Gerd Gigerenzer, "From Tools to Theories: A Heuristic of Discovery in Cognitive Psychology," *Psychological Review* 98, no. 2 (1991): 255.
[16] Michael Hagner, *Der Geist Bei Der Arbeit: Historische Untersuchungen Zur Hirnforschung* (Göttingen: Wallstein, 2006), 36.
[17] Maurizio Meloni, "Philosophical Implications of Neuroscience: The Space for a Critique," *Subjectivity* 4, no. 3 (2011): 310.
[18] Joseph Dumit, "Objective Brains, Prejudicial Images," *Science in Context* 12, no. 1 (1999): 175. While Joseph Dumit admits that scientists are equally complicit in "neurorealism" by their presentation of brain scans, he is especially attentive to ways in which brain images "overflow" their status as "expert images" when they enter arenas beyond the laboratory, such as popular science magazines, doctor's waiting rooms, Hollywood movies, discourses of mental illness, or courtrooms. See also ibid.

Laypeople's perception of the spectacular-looking functional brain images and performances (such as the *Brain Waves* example) is hence positioned as an important driver of the promissory culture of neuroscience. When an overly credulous public is exposed to expert or appropriated brain images, it is all too eager to adopt neurocentral views of phenomena such as love, voting preferences, or criminal behavior. According to critical neuroscience scholar Jan Slaby, this cultural support is due to a "hunger" for self-objectification or, more generally, "people's natural affinity for reductionistic explanations of cognitive phenomena," as David McCabe and Alan Castel claim in a much-cited article on the interpretation of brain images.[19] Similarly, Matthew Crawford, talking about "the limits of Neurotalk," refers to brain imaging as "that fast-acting solvent of critical faculties," and Steven Poole writes about the "rise of popular Neurobollocks," including brain images that "like religious icons, inspire uncritical devotion."[20] Some scholars emphasize these processes of deceitful seduction are not the work of neuroscientists, but of popular science writers. There is a distinction, as Jan MacVarish and colleagues claim, between proper neuroscience and "neuroscientism," that is, a difference "between legitimate findings emerging from … research on neurological functioning and the activities of those who appropriate the authority of scientific objectivity to pursue moral, political or commercial agendas in the public sphere."[21]

My analysis reveals that by creating stark bifurcations between "legitimate" science and those who "appropriate" science in the public sphere, these critical narratives foreclose a deeper understanding of the circulation between these two realms, as well as the ways in which certain scientific practices of conceptualizing and mediating research afford particular appropriations over others. Here, I challenge the seemingly straightforward dichotomy between "genuine" and "appropriated" science, that is, between the sincere intentions of brain scientists and the wrongful, spectacularizing, grandiose

---

[19] Jan Slaby, "Steps towards a Critical Neuroscience," *Phenomenology and the Cognitive Sciences* 9, no. 3 (2010): 398; David P. McCabe and Alan D. Castel, "Seeing Is Believing: The Effect of Brain Images on Judgments of Scientific Reasoning," *Cognition* 107, no. 1 (2008): 344.

[20] Matthew B. Crawford, "The Limits of Neurotalk," *New Atlantis* 19 (2008): 65; S. Poole, "Your Brain on Pseudoscience: The Rise of Popular Neurobollocks," *New Statesman*, September 6, 2012, par. 18. Both cited in Martha J. Farah, "Brain Images, Babies, and Bathwater: Critiquing Critiques of Functional Neuroimaging," *Hastings Center Report* 44, no. s2 (2014).

[21] Jan MacVarish, Ellie Lee, and Pam Lowe, "Understanding the Rise of 'Neuroparenting,'" in *We Need to Talk about Family: Essays on Neoliberalism, the Family and Popular Culture*, ed. Roberta Garrett, Tracey Jensen, and Angie Voela (Newcastle: Cambridge Scholars, 2016), 100.

claims of brain science communicators. I also do not presume the public to have any reductionist affinity, but instead I aim to historicize how notions of the popular and the public, as well as notions of (legitimate) science popularization and science expertise, are invoked in different situations. Going beyond both neuro-enthusiasm and neuro-skepticism, I investigate the way live brains are performed in the interplay between scientists, science communicators, and spectators. If, within this interplay in brain science, "the medium is the message," I ask, what *was* the medium?

From Metaphors to Brainmedia

The central concept of this study is "brainmedia," a term I coined to denote the material-discursive assemblages that produce and demonstrate the live brain. "Brain" in brainmedia stands for the cerebral organs as they have been (and are) shaped and enacted by the historically variable concepts and practices used to examine them. "Media" refers to three different kinds of media: the technological apparatuses neuroscientists use in laboratories; an environment understood as a medium that affected modern brains; and the new broadcast—and other forms of recording—and presentational media that brain scientists use to communicate and think about their research of the "brain in action." Throughout my case studies, I show how brains are constituted, through brainmedia, as "live brains."

With my conceptualization of brainmedia, I build on—but also depart from—what could be called a history of metaphors of the brain. By analyzing brainmedia as assemblages, I can relate to these metaphors and analogies in a different way—dive deeper, so to say, into the tropes' tissue. I do so by moving beyond an exclusively linguistic or discursively oriented approach to metaphor. This is not to dismiss the long and important tradition in the study of cerebral metaphors and analogies (about a battle of nerve cells or human nerves as an internet network, for example).[22] There is a vast and important body of work that describes and uncovers the effects of the figurative language employed to describe processes of brain activity, thinking,

---

[22] "Battle of nerve cells" in Stephan Besser, "From the Neuron to the World and Back: The Poetics of the Neuromolecular Gaze in Bart Koubaa's 'Het Gebied Van Nevski' and James Cameron's 'Avatar,'" *Journal of Dutch Literature* 4, no. 2 (2013). For a critical view on the brain as network metaphor versus a view of the brain as today's (actual) complicated material internet network, see Philip Haueis and Jan Slaby, "Connectomes as Constitutively Epistemic Objects: Critical Perspectives on Modeling in Current Neuroanatomy," *Progress in Brain Research* 233 (2017).

memory, and emotion.²³ As historian Douwe Draaisma noted, the domain of mental processes is particularly abundant with metaphor, because "the literal description of mental processes seems to be fundamentally excluded … the problem with much figurative usage in psychology is that no literal alternative is available."²⁴ The brain provokes trope-heavy language because it performs complex and important functions, yet it does not straightforwardly reveal the internal mechanisms at work in its intriguing gray matter.

Metaphors are a central conceptual device in knowledge production; as Laura Otis puts it: "Metaphors do not 'express' scientists' ideas; they *are* the ideas."²⁵ Indeed, brain research discourse is rife with metaphors and tropes, especially analogies between brains and technology, including, since the nineteenth century, what we may call media technology (phonograph, telegraph, telephone, computing, internet). Responding to this longstanding metaphorical language, many scientists have commented on the inaptitude of such analogies (the ever-nearing "end of the computational metaphor") and the call for new, more imaginative analogies.²⁶ My aim in this book, however, is different. I show how a sole focus on the discursive translations and rhetorical figurations between technological media and biological entities cannot do justice to the imploded and recursive qualities of their technoscientific entanglement. The history of tropes for the brain, the nervous system, and the neurological subject reveals a material-discursive fabric in which we can never untangle the neural and its "other"—whether technology, tissue, or text.²⁷ Synaptic gaps, cerebral hemispheres, neural

---

[23] Juliana Goschler, "Metaphors in Cognitive and Neurosciences: Which Impact Have Metaphors on Scientific Theories and Models?" *Metaphorik* 12 (2007), http://www.metaphorik.de/12/goschler.pdf; Borck, "Toys Are Us."

[24] Douwe Draaisma, *Metaphors of Memory: A History of Ideas about the Mind* (Cambridge: Cambridge University Press, 2000), 11. Draaisma observes a shift to mechanical analogies for the mind starting in the seventeenth century and a second shift in the nineteenth century to new technologies of recording (photography, phonograph, cinematography).

[25] Laura Otis, *Networking: Communicating with Bodies and Machines in the Nineteenth Century* (Ann Arbor: University of Michigan Press, 2001), 48, citing N. Katherine Hayles, *Unthought: The Power of the Cognitive Nonconscious* (Chicago: University of Chicago Press, 2017), 120.

[26] Matthew Cobb, *The Idea of the Brain: The Past and Future of Neuroscience* (New York: Basic Books, 2020), 373. For a different twentieth-century genealogy focusing on the influence of physics, mathematics, and engineering on models of brain activity and behavior, see Grace Lindsay, *Models of the Mind: How Physics, Engineering and Mathematics Have Shaped Our Understanding of the Brain* (London: Bloomsbury Sigma, 2021).

[27] Laura Otis's work on the historical relation between metaphors of networks and the study of nerve cells points to the limits of a simple metaphorical analysis, showing how similar metaphorical referents (networks, for example) could be used to make differing arguments. Laura Otis, "The Metaphoric Circuit: Organic and Technological

networks, nervous transmissions, and brainwaves are always already built by analogy, always already part of infinitely recursive loops—what Laura Otis has helpfully called "metaphorical circuits."[28] The danger of a study of machine-organism metaphors is that it takes up a too "easy-going metaphorics," as Jussi Parikka notes, placing phenomena "on an explanatory grid that has already stabilized the relations of nodes."[29] On the contrary, as I emphasize with this study, technology does not "mediate" biology in the sense of a metaphorical model "translating" bodies through representation. Instead, media modulate the world through what Pasi Väliaho calls "transposition"—bodies and technologies form new assemblages in which nothing was ever "natural" or "other."[30]

It is with this transposed perspective in mind that I coin the term brainmedia. I conceptualize brainmedia as organic/technical/media/cultural assemblages, that is, as the material-discursive assemblages that configure the "live brain." I mean for the term "material-discursive" to resonate with Donna Haraway's study of technoscience. Haraway speaks of the "collapse of metaphor and materiality" to advocate a study of science not as "a question of ideology but of modes of practice among humans and nonhumans that configure the world materially and semiotically—in terms of some objects and boundaries and not others."[31] Her perspective makes it possible to describe the "naturecultures" that we both inhabit and are part of, worlds in which materiality and semiotics cannot be separated but have to be "imploded" (considered as ever more densely knotted), to use Haraway's vocabulary.[32] My focus on the imploded, the material-discursive, aims to bridge what James Bono has described as the false rift between studying the materialities of scientific practice (the assumed focus of science studies) and the discursive analysis of science (the assumed focus of the cultural or literary analysis of science).[33]

Communication in the Nineteenth Century," *Journal of the History of Ideas* 63, no. 1 (2002): 107.

[28] Ibid.
[29] Jussi Parikka, *Insect Media: An Archaeology of Animals and Technology* (Minneapolis: University of Minnesota Press, 2010).
[30] Pasi Väliaho, "Bodies Outside In: On Cinematic Organ Projection," *Parallax* 14, no. 2 (2008): 17.
[31] Donna Jeanne Haraway, *Modest−Witness@Second−Millennium. Femaleman−Meets−Oncomouse: Feminism and Technoscience* (New York: Routledge, 1997), 97. Haraway uses the term "material-semiotic" to speak of sites where metaphors and materialities collapse, which she variably calls "material-semiotic apparatuses," "articulations," "fields," "worlds," or "webs." See also ibid.
[32] Ibid., 97. "Naturecultures," in *The Companion Species Manifesto: Dogs, People, and Significant Otherness*, vol. 1 (Chicago: Prickly Paradigm Press, 2003), 12.
[33] Formulating a material-discursive approach, James Bono has proposed the terms "metaphorics of science," which insists on the embodied, performative, and material

The hundred years of brainmedia that I present in this study—an era in which media have often been understood as ever more intimately present—warrants a new specification of Haraway's naturecultures. New forms of media are not only always enmeshed in the material world themselves, but are particularly potent forces in enmeshing other entities we may have previously viewed as separate bodies, machines, or things.[34] Hence, we need articulations of material-discursive assemblages, of naturecultures, that pay special attention to the position of media and mediation in these entanglements, invoking what Rosi Braidotti has called "medianaturecultures."[35] In the past decade, media studies scholars have coined various hybrid notions with different inflections to approach such an analytical project: "medianatures" (Jussi Parikka), "biomedia" (Eugene Thacker), "mediabodies" (Marie-Louise Angerer), or the "biomediated body" (Patricia Clough).[36] Recently, N. Katherine Hayles has offered the term "cognitive assemblages" to describe the particular assembling that happens between humans and new technologies (what she dubs technical cognizers), such as novel types of sensors that exteriorize some of the microtemporal processes happening below human consciousness.[37] Working in resonance with these articulations of medianaturecultures, I offer my conception of brainmedia as a viable means to analyze the proliferation of the live brain.

For me, studying historical assemblages of brainmedia means both studying how brains, nerves, minds, and selves are articulated as

dimensions of metaphor, what he calls "embodied metaphors-in-action," "material metaphors," or "performative metaphors"; they are "part of the material world and *instruments* for our knowing and manipulating it." James J. Bono, "Why Metaphor? Toward a Metaphorics of Scientific Practice," in *Science Studies. Probing the Dynamics of Scientific Knowledge*, ed. Sabine Maasen and Matthias Winterhager (Bielefeld: Transcript Verlag, 2003), 226; "Making Knowledge: History, Literature, and the Poetics of Science," *Isis* 101, no. 3 (2010).

[34] Richard Grusin, "Radical Mediation," *Critical Inquiry* 42, no. 1 (2015): 146; Sarah Kember, "Doing Technoscience as ('New') Media," in *Media and Cultural Theory*, ed. James Curran and David Morley (London: Routledge, 2007).

[35] Rosi Braidotti, "The Critical Posthumanities; or, Is Medianatures to Naturecultures as Zoe Is to Bios?" *Cultural Politics* 12, no. 3 (2016).

[36] Jussi Parikka, "New Materialism as Media Theory: Medianatures and Dirty Matter," *Communication and Critical/Cultural Studies* 9, no. 1 (2012); Eugene Thacker, *Biomedia* (Minneapolis: University of Minnesota Press, 2004); Marie-Luise Angerer, "Medienkörper: Zur Materialität Des Medialen Und Der Medialität Der Körper," in *Kultur—Medien—Macht: Cultural Studies Und Medienanalyse*, ed. Andreas Hepp and Rainer Winter (Wiesbaden: VS Verlag für Sozialwissenschaften/Springer, 1997); N. Katherine Hayles, "Unfinished Work: From Cyborg to Cognisphere," *Theory, Culture & Society* 23, nos. 7–8 (2006).

[37] Hayles, *Unthought*.

interwoven with our (media-)technological environment and examining how notions of mediation and technological media have changed how the brain could be thought. As part of brainmedia, the brain is always already a medianatureculture, always already entangled with media technologies, social practices, and metaphors; and media denote the historically variable processes of mediation, conceptions of mediating, and technical media through which knowledge about the active brain emerges.

By analyzing brainmedia as assemblages, I also enact these brainmedia. Describing, theorizing, negotiating, and criticizing assemblages in this study brings those assemblages into being, makes them emerge more clearly. This is what John Law has described as the performance of writing, "telling stories about the world also helps to perform that world ... what is being performed is thereby rendered more obdurate, more solid, more real than it might otherwise have been."[38]

A Media-Historical Approach to the Media Enmeshed with Brain Science

Writing about one hundred years of brainmedia, from the 1920s until 2020, I methodologically align myself with a heterogeneous body of historical media studies that comprise both media-archaeological and media-genealogical approaches. I sidestep ongoing debates about Foucauldian concepts here, as neither approach refers to rigidly circumscribed methodologies, but both circumscribe a set of concerns and analytical emphasis in investigating the historical development of media apparatuses and practices, and media's relation to power/knowledge formations.[39] I use the term "media-historical analysis" to emphasize this combined approach, though at times I refer to (media) genealogy to signal an interest in a "history of the (neurocultural) present," redrawing the development of the relation between brains and media in the twentieth century.

A media-historical analysis is sensitive to media as an emergent part of the fabric of knowledge formation. This means that a combined archeological

---

[38] John Law, *Aircraft Stories: Decentering the Object in Technoscience* (Durham, NC: Duke University Press, 2002), 6.
[39] See the discussion on media archeology and media genealogy in Erkki Huhtamo and Jussi Parikka, "Introduction: An Archaeology of Media Archaeology," in *Media Archaeology: Approaches, Applications, and Implications*, ed. Erkki Huhtamo and Jussi Parikka (Berkeley: University of California Press, 2011).

and genealogical study of researching the active brain is also always a study of the formation of the (brain)media that co-constitute this active brain. Recently, we can see media-sensitive approaches being taken up by historians of science. Philip Sarasin, for example, has rallied for attention to media in the history of knowledge (*Wissensgeschichte*), emphasizing the constitutive role of media: "Knowledge is shaped by the logic of the media because it always has to be formatted"; "media function as filters that select, emphasize and repress knowledge, change it and connect it with other elements of knowledge."[40]

Yet, a more radical media-historical approach takes media's "formatting" role in knowledge production and cultural formation a step further. Media, in the media-historical sense that I propose here, are as much constitutive *of* culture as they are constituted *by* culture; media have world-making capabilities. This also means that what counts as a medium is itself always in flux. Media are not fixed but always in a state of "becoming media," as Joseph Vogl has emphasized, transforming "apparatuses, codes, symbolic systems, forms of knowledge, specific practices, and aesthetic experiences" into media.[41]

I believe that when media are studied in their emergence, it is imperative to investigate the "gray zones" in their development when new forms of mediation have not yet been delineated as a medium. This is why I also study the imaginary media and media imaginaries related to brainmedia: the longed-for machines and fantastic ways to access the active brain, as well as the imaginaries of mediation that surpass conventional spatiotemporal boundaries of outer world and inner brain, living and dead, visible and invisible.[42] In Chapter 3, for example, I analyze a fictional brain wave reader that can supposedly communicate with the dead, while Chapter 6 features a contemporary media installation that invokes the possibility of getting feedback on "brain-to-brain synchronization." I discuss such gray zones, media imaginaries, and imaginary media as they relate to brain research not to trace the origins of particular

---

[40] Philipp Sarasin, "Was Ist Wissensgeschichte?" *Internationales Archiv für Sozialgeschichte der deutschen Literatur* 36, no. 1 (2011): 168.

[41] Joseph Vogl, "Becoming-Media: Galileo's Telescope," *Grey Room* 29 (2007): 16.

[42] On "imaginary media" and "media imaginaries," see Jussi Parikka, "Imaginary Media: Mapping Weird Objects," in *What Is Media Archaeology*, ed. Jussi Parikka (Cambridge, MA: Polity Press, 2012); Simone Natale and Gabriele Balbi, "Media and the Imaginary in History: The Role of the Fantastic in Different Stages of Media Change," *Media History* 20, no. 2 (2014); Eric Kluitenberg, "On the Archeology of Imaginary Media," in *Media Archaeology: Approaches, Applications, and Implications*, ed. Erkki Huhtamo and Jussi Parikka (Berkeley: University of California Press, 2011).

(brain)media assemblages but rather to emphasize contingencies and describe their rise and demise.[43]

My media-historical analysis examines how media, as Alexander Monea and Jeremy Packer put it, "allowed certain problems to come to light, be investigated, and chosen for elimination and how media aided in the various solutions that have been enacted."[44] For me, studying phenomena media-historically means positioning media assemblages within their historical discourse and examining the practices and thoughts that enable media to come into being and in turn also influence this field of possibilities. In my case studies of brainmedia, a genealogy of live brains requires studying specific media assemblages and remaining attentive to media as a constitutive part of knowledge formation and to shifts from one discursive formation to another. My media-historical analysis thus uncovers the required conditions for legitimate statements about the active brain and brain activity. Different at different points in time, such articulations of active brains were fueled by—and at the same time opened up—the possibility for (normative) categorization and governmentality through (discourses about) brain science. Note, for example, that an EEG exhibition in 1937 (discussed in Chapter 2) was part of a broader, government-steered, normative public discourse that also included personality and intelligence testing and an emphasis on biometrics. Or note that new devices developed to show spatiotemporal brain activity in the 1950s (discussed in Chapter 3) fortified a project to develop categorizations of personalities and mental (ab)normalities. Conditions of knowledge (the subject of archaeology) are always conjoined with networks of power that provoke knowledge formation (the subject of genealogy).

Significantly then, my analysis of configurations of knowledge about the live brain shows how much brains (as minds, selves, and subjects) are articulated as part of the fabric of a modern media landscape. This placement of the subject within media opens the subject with a live brain up—makes it conducive—to new networks of knowledge/power. It is with an eye to studying the networks of knowledge/power informing contemporary assemblages of the live brain in the twenty-first century that I present historical accounts of brainmedia.

---

[43] On the concern for the nonteleological, nonlinear, and contingent in media-historical studies, see Jussi Parikka, ed., *What Is Media Archaeology?* (Cambridge, MA: Polity Press, 2012); Timothy Druckery, "Foreword," in *Deep Time of the Media: Toward an Archaeology of Hearing and Seeing by Technical Means*, ed. Siegfried Zielinski (Cambridge, MA: MIT Press, 2006).

[44] Alexander Monea and Jeremy Packer, "Media Genealogy: Technological and Historical Engagements of Power—Introduction," *International Journal of Communication* 10 (2016): 3154.

Understanding Scientific Practices and Communications through "Performing Knowledges"

Next to the concept of brainmedia, a second important analytical dimension is my consideration of scientific practices as *performing knowledges*. I approach knowledge production about live brains as taking place across different spheres, both inside and beyond the academic laboratory. The concept of performing knowledges allows me to attend to many actors and sites where live brains are produced and move away from the dichotomy between "genuine" (serious, academic, institutional) brain science versus "popular" views about the brain and the implications of brain science. I thus adhere to a more inclusive and recursive conception of science as it is performed, mediated, and configured within and beyond the inner circles of the science establishment.

In the past decades, historians of science, scholars in science studies, and historians of knowledge have proposed less dichotomous conceptual frameworks to understand scientific practices; for example, James Secord's notion of "knowledge in transit" and the idea of "circulating knowledge" (as developed by a number of authors).[45] To study circulating knowledge means acknowledging the fact that the nature and status of knowledge change during moments of transition and zooming in on the particularity of such changes and mutations instead of forging easy distinctions between "genuine" science and popular, para-, or pseudoscience, or between the history of science and the history of science popularization.[46] The popular must itself be historicized as a changing category, as part of the transformation of public knowledge.[47]

To study circulating knowledge as being performed means moving away from a diffusionist model of science popularization—the idea that authentic science is only dramatized, staged, or performed when it is mediated to lay

---

[45] James A. Secord, "Knowledge in Transit," *Isis* 95, no. 4 (2004). On "circulation," see Lynn K. Nyhart, "Historiography of the History of Science," in *A Companion to the History of Science*, ed. Bernard Lightman (Chichester: John Wiley, 2016), 12–16; Johan Östling et al., "The History of Knowledge and the Circulation of Knowledge: An Introduction," in *Circulation of Knowledge: Explorations in the History of Knowledge*, ed. Johan Östling et al. (Lund: Nordic Academic Press, 2018).

[46] Bernadette Bensaude-Vincent, "A Historical Perspective on Science and Its 'Others,'" *Isis* 100, no. 2 (2009); Jonathan R. Topham, "Rethinking the History of Science Popularization/Popular Science," in *Popularizing Science and Technology in the European Periphery, 1800–2000*, ed. Agustí Nieto-Galan, Enrique Perdiguero, and Faidra Papanelopoulou (Farnham: Ashgate, 2009).

[47] Andreas W. Daum, "Varieties of Popular Science and the Transformations of Public Knowledge: Some Historical Reflections," *Isis* 100, no. 2 (2009): 320.

audiences—and also steering clear of the notion of a "legitimate" sphere of science that becomes mitigated or diluted through circulation.[48] Instead, following the example of scholars in cultural studies of scientific knowledge, it is important to understand this distinction between the scientific and the popular as a dichotomy that is itself continuously enacted.[49] At every stage, it takes effort and rhetoric to separate the supposedly scientific from the popular, and this boundary work deserves particular investigation.[50]

My study of brainmedia examines *how* knowledge about the brain circulates across scenes, scenarios, and spheres of knowledge production, thus also uncovering the materiality and mediality of circulating knowledges. Examining live brains as assembled across different sites (lectures, public demonstrations, magazine articles) allows me to examine the emergence of new knowledges shaped by new stages, spaces, and repertoires for performing these knowledges. I borrow the term from Mary Dupree and Sean Franzel, who, in their edited volume *Performing Knowledge, 1750–1850*, highlight "the specific physicality, materiality and temporality of the performance situation," as well as the emergence of new types of knowledge performers and audiences.[51]

In this study, I use "performativity," "performing," and "performances" for two main reasons: to flag the potential performativity of presenting knowledge, and to highlight the potential dramatic structure of performances of knowledge. First, studying brainmedia as heterogeneous assemblages requires analyzing their reality-producing (performative) dimension—that is, their ability to enact (to call into being, to bring forth, to hail) that which it has set out to disclose. Media, as used in performances of knowledge, have special performative or reality-producing qualities, what media theorist Sybille Kramer has called *das in-Szene-setzende Wahrnehmbarmachen*, meaning that media also create and stage the things they make perceptible.[52]

---

[48] Roger Cooter and Stephen Pumfrey, "Separate Spheres and Public Places: Reflections on the History of Science Popularization and Science in Popular Culture," *History of Science* 32, no. 3 (1994): 249; Stephen Hilgartner, "The Dominant View of Popularization: Conceptual Problems, Political Uses," *Social Studies of Science* 20, no. 3 (1990).
[49] Joseph Rouse, "What Are Cultural Studies of Scientific Knowledge?" *Configurations* 1, no. 1 (1993): 13.
[50] Thomas Gieryn, "Boundary-Work and the Demarcation of Science from Non-Science: Strains and Interests in Professional Ideologies of Scientists," *American Sociological Review* 48, no. 6 (1983).
[51] Mary Helen Dupree and Sean B. Franzel, "Introduction: Performing Knowledge, 1750–1850," in *Performing Knowledge, 1750–1850*, ed. Mary Helen Dupree and Sean B. Franzel (Berlin: Walter de Gruyter GmbH, 2015), 9.
[52] Sybille Krämer, "Was Hat 'Performativität' Und 'Medialität' Miteinander Zu Tun? Plädoyer Für Eine in Der 'Aisthetisierung' Gründende Konzeption Des Performativen. Zu Einführung in Diesen Band," in *Performativität Und Medialität*, ed. Sybille Krämer (München: Fink, 2004), 25.

Media thus "phenomenalize," in Kramer's words, enacting what they mediate while they mediate, and this means that "which they embody is not a more or less stabile entity, but only exists in the fluid and processual presence of media action."[53]

Aside from performativity, I speak of "performing" brainmedia because that makes it clear that knowledge production (about the live brain) is a practice carried out for an (actual or imagined) public that reconfigures, by rhetorical and material means, what has been performed; it transforms knowledge. In this sense, these performances are related to (theatrical) plays or acts that have a persuasive objective—making something look truthful, beautiful, or cool, for example. These intersecting meanings of performativity and performance/ing resonate with the work of key historians and philosophers of science. Well known in this respect are Donna Haraway, Bruno Latour, Andrew Pickering, and Karen Barad, who analyze (with different emphases and vocabularies) the performativity of reality-producing agents assembled in lab experiments and other scientific practices.[54] Alternatively, scholars such as Stephen Hilgartner, José van Dijck, and Peter Weingart have focused on "staging" and "dramatization" within and beyond scientific laboratories, for example, "science on stage" in popular news media, science exhibitions, or policy debates.[55] Performativity and performance intermingle in these studies, as scholars use concepts such as "choreography of truth," "experimental scenography," "theater of proof," "demonstration assemblage," "material performativity," or "scenarios of knowledge" for analytical approaches that point to reality-producing actions and agents at the basis of material-discursive assemblages, with special attention to staging, both in the lab and in the public sphere.[56]

---

[53] Ibid.

[54] Bruno Latour, *Science in Action: How to Follow Scientists and Engineers through Society*, rev. ed. (Cambridge, MA: Harvard University Press, 1988); Haraway, *Modest−Witness@ Second−Millennium.Femaleman−Meets−Oncomouse*; Andrew Pickering, *The Mangle of Practice: Time, Agency, and Science* (Chicago: University of Chicago Press, 1995); Karen Barad, "Posthumanist Performativity: Toward an Understanding of How Matter Comes to Matter," *Signs* 28, no. 3 (2003).

[55] Stephen Hilgartner, *Science on Stage: Expert Advice as Public Drama* (Stanford, CA: Stanford University Press, 2000); José van Dijck, *Imagenation: Popular Images of Genetics* (Basingstoke: Macmillan, 1998); Peter Weingart, *Die Wissenschaft Der Öffentlichkeit: Essays Zum Verhältnis Von Wissenschaft, Medien Und Öffentlichkeit* (Weilerswist: Velbrück, 2005).

[56] "Choreography of truth," in Iwan Rhys Morus, "Placing Performance," *Isis* 101, no. 4 (2010): 777. "Experimental scenography," in Bruno Latour, "From Fabrication to Reality. Pasteur and His Lactic Acid Ferment," in *Pandora's Hope: Essays on the Reality of Science Studies*, ed. Bruno Latour (Cambridge, MA: Harvard University Press, 1999). "Theatre of proof," in *The Pasteurization of France* (Cambridge, MA: Harvard University Press, 1993), 86. "Demonstration assemblage," in Henning Schmidgen, "Pictures,

Studying the performing of scientific knowledge implies that science is rarely only about persuading an audience of a finding's truthfulness, of the scientists' trustworthiness, or of the authority of the instrumental setup and the scientific locale. Performances of scientific knowledge are never merely about "making real"; they always interface with other important aspects such as the interpellation of particular types of audiences and the specific ways in which science could be sensed, felt, and appreciated. Historian Henning Schmidgen gives the example of late nineteenth-century physiologists who were eager to connect their recording devices to projection machines that would allow the collective experience of seeing their immediate inscriptions.[57] Scientific practices were thus significantly influenced by this goal of making public, as Jimena Canales notes: "The history of these technologies thus intersects with the larger history of spectacle, from the classroom to the movie theater."[58] Scientists produce their performances of knowledge as part of, and with an eye to, broader structures and functioning of visuality. Hence, it is important to investigate the affective dimensions of performing knowledges—the fears, hopes, and desires they enact—as well as the forms of storytelling, imaging, and imagining by which they are produced.[59] As Ian Morus says, "Looking at performances should also get us thinking about science in terms of doing rather than writing, aesthetic pleasure rather than hard reason."[60]

For me, analyzing the emergence of the active brain as a live brain means to move from a confined analysis of scientific images to an analysis of performing knowledges that, in their circulation, entail a variety of processes of articulation, conceptualization, visualization, and demonstration. With this study I expand the analytical frame by looking at complex mediations in which brain models, metaphors, or installations are performed, that is, conceptualized, developed, demonstrated, and exhibited in public, across the boundaries of established scientific discourses and science.

Preparations, and Living Processes: The Production of Immediate Visual Perception (Anschauung) in Late-19th-Century Physiology," *Journal of the History of Biology* 37, no. 3 (2004): 481. "Material performativity," in Pickering, *The Mangle of Practice*, 16. "Scenarios of knowledge," in Victoria Tkaczyk, "The Making of Acoustics around 1800, or How to Do Science with Words," in *Performing Knowledge, 1750–1850*, ed. Mary Helen Dupree and Sean B. Franzel (Berlin: Walter de Gruyter GmbH, 2015), 36.

[57] Henning Schmidgen, "1900—The Spectatorium: On Biology's Audiovisual Archive," *Grey Room* 43 (2011).

[58] Jimena Canales, "Recording Devices," in *A Companion to the History of Science*, ed. Bernard Lightman (Chichester: John Wiley, 2016), 509.

[59] Maureen McNeil et al., "Conceptualizing Imaginaries of Science, Technology and Society," in *The Handbook of Science and Technology Studies*, ed. Ulrike Felt et al. (Cambridge, MA: MIT Press, 2016), 457.

[60] Morus, "Placing Performance."

Forms of Liveness:
Watching the Brain at Work

Why is it so cool to see our own brain activity in real time?—to repeat the question of a participant of the *Brain Waves* installation (mentioned in the prelude above) at Amsterdam's Anatomical Theater. How can we understand such a fascination with the view of the living brain? In the brainmedia case studies in this study, I analyze these as varying instances of "live" brains. My conception of liveness is as a category of spatiotemporal operativity that is differently configured in different media-historical constellations. "Liveness" is not solely a technical, nor a primarily experiential term, but denotes a more dynamic concept covering different configurations of directness, nearness, hereness, aliveness, liveliness, and nowness. These different configurations of liveness are what I call the specific *forms of liveness* that matter in my histories of brainmedia.

According to one of its meanings in the *Oxford English Dictionary*, the adjective "live" denotes a performance or event that is "heard or watched at the time of its occurrence … not pre-recorded."[61] Yet, by speaking of different forms of liveness in relation to brain research, I expand that notion (usually more narrowly associated with the realm of radio or television) to cast a specific media-analytical perspective on brainmedia in the twentieth century. In my case studies of live brains, I study the forms of liveness enacted and assembled in brainmedia up-close. Examining situations in which scientific measurements of the living brain are performed and presented in different ways, I note that they are never only about invoking immediacy, instantaneity, or transparency, as they are also always paired with other associated notions— tied to forms of liveness—such as authority, attraction, aliveness, liveliness, or intimacy. In fact, by studying various historical brainmedia with special attention to their forms of liveness, I present an alternative to the dominant scholarly focus on the apparent immediacy and transparency of situations in which the active brain is performed (the interpretations of which are often rhetorically assigned to a naïve spectator who imagines a direct window into the brain).

Various scholars have described the puzzling nature of scientific mediation practices (making visible the invisible) by speaking of the "paradox of transparency" (Cornelius Borck), "myth of total transparency" (José van Dijck), and "a 'dream' of media-free immediacy" (Florian Sprenger), on the one hand, positing the fact of mediation, the mediated nature of processes

---

[61] "Live, Adj.1, N., and Adv.," in *Oxford English Dictionary* (Oxford University Press).

of scientific transcription, and, on the other hand, the attribution, on a discursive or experiential level, of direct access to these newly visualized entities, hence erasing any mediation processes.[62] Though it is important to describe the tension between these two poles—a rhetorical opposition that has spurred important media-theoretical investigations—the notion of the paradox also sets up a too general a rift, one that cannot fully describe what happens in particular historical performances of knowledge. In contrast, my aim is to historicize attributions of (spatial) nearness and (temporal) speed in histories of performing the "live brain." My case studies demonstrate that performances of live brains are never merely about producing an immediate view inside the living brain; they are never only about producing a supposed encounter with nature. Media may conjure sensations of intimacy, liveliness, or aliveness that cannot be equated with just a desire for transparency: to become live means that mediation can become both transparent and present.

By emphasizing liveness over immediacy, I do not mean to dismiss the importance of examining the ideal of immediacy or transparency in studying the human brain. The rhetoric of transparency is a structuring narrative in the twentieth-century ambition to access the living brain. Indeed, in this study, this rhetoric is omnipresent in varying historical imaginaries of direct entry into the brain at work; my chapters present historical desires to make the skull translucent, imaginations of direct access to the interior world, and the wish to view the activities of the living brain as they happen. Yet, I am not tracing the twentieth-century discourse of increasing cerebral translucency, nor am I merely exposing the perils and pitfalls of the myth of the transparent brain at various moments in history. Instead, I demonstrate how historically situated conceptions of media and mediation are imbricated in the formation of the live brain.

My attention to varying forms of liveness contributes to scholarship in historical epistemology. The work of Lorraine Daston and Peter Galison has already enabled us to look at the way different historical time periods engender different "objectivities," connected to what they call shifting "epistemic virtues," that is, the changing ways in which scientists are

---

[62] "Paradox of transparency," in Cornelius Borck, "Die Unhintergehbarkeit Des Bildschirms," in *Mit Dem Auge Denken: Strategien Der Sichtbarmachung in Wissenschaftlichen Und Virtuellen Welten*, ed. Bettina Heintz, Arnold O. Benz, and Jörg Huber (Zürich: Ed. Voldemeer, 2001), 388. "Ideal of transparency," in José van Dijck, *The Transparent Body: A Cultural Analysis of Medical Imaging* (Seattle: University of Washington Press, 2005), 15. "'Dream' of Media-Free Immediacy," in Florian Sprenger, *Medien Des Immediaten: Elektrizität, Telegraphie, Mcluhan* (Berlin: Kadmos, 2012), 5. Sprenger in fact discusses the problem of this analytical frame and calls for a historization of narratives of immediacy.

supposed to behave in order to be regarded as good scientific practitioners (being restrained, creative, or trustworthy, for example). My analysis of performances of knowledges is especially attentive to the epistemic virtues attributed to (emerging) media, and I show how aesthetic, artistic, and performance-oriented modes of knowledge production arise in varying "virtuous" relations envisioned between scientists and the technical and broadcast media they use.

In developing the notion of *forms of liveness*, I argue that we can extend the term "live" from its narrow association with a presumed origin in 1934 (the liveness of radio) to function as a broader investigative concept that sensitizes the historian to discern the situated fabric of different temporal-spatial configurations and sensations that depends on existing technologies, affective positions, and situated knowledges. In relation to brainmedia, "liveness" is a particularly rich analytical concept, enabling us to see that producing spatiotemporal configurations of active brains—determining what counts as alive, direct, objective, engaging, aesthetic, automatic, or comprehensible, for example—is always contingent and relational: in different historical situations, experiential, technical, and rhetorical elements are intricately and variably intertwined. The prerecorded may be experienced as current; something brought from far away may be sensed as (co)present. If, as Philip Auslander states, media "make claims on us" about liveness, the question is how such a claim can emerge: liveness depends on the situated interpretation of mediation practices and technologies.[63]

Understood in this way, liveness is generated by more than technical advances in relay speed. This is not to downplay the importance of liveness' technological dimension, which is foregrounded in the scholarship of media theorist Wolfgang Ernst, who speaks of the media-specific, "proper" temporal dimensions (*Eigenzeit*) of technological media, what he calls their specific temporalities and temporealities, "tempor(e)alities" in short.[64] Ernst's study of "chronopoetics" highlights the technological processes and microtemporal configurations by which "technical objects embody complex temporalities," for example, the calculated windows of delay time in what is

---

[63] Philip Auslander, "Digital Liveness: A Historico-Philosophical Perspective," *PAJ: A Journal of Performance and Art* 34, no. 3 (2012): 8. Paddy Scannell has also pointed to this situated performativity of live events (their "event-character"); they enact liveness but can only "work" as live in appropriate situations. Paddy Scannell, *Television and the Meaning of "Live": An Enquiry into the Human Situation* (Cambridge, MA: Polity Press, 2014), 105.

[64] Also important is Ernst's observation that different media-induced tempor(e)alities (such as "real time") do not progress linearly alongside the development of different media technologies, but can be intermingled and connected in different time periods and within technological ensembles.

called "real time."[65] Employing Ernst's approach to a study of brainmedia, we can start to see that when a network of nerves is compared to a display device (in Chapter 2), or when a cathode ray oscillograph is used to both visualize groups of active nerve cells and provide visual stimuli (Chapter 4), this means that particular technical temporal configurations (embodying specific chronopoetics) were assembled and enacted to show the brain in action. While Ernst provides significant insights into tempor(e)alities as a technically oriented media archeologist, my emphasis in this study is on broader assemblages, on discourses on liveness, and on technological devices involved in performing the living, working brain. For me, studying brainmedia assemblages that perform the live brain means not only studying the forms of liveness embodied by technical objects but also analyzing the way "technical and living beings" assembled together in brainmedia are always "instantiating and embodying complex temporalities," to speak in N. Katherine Hayles's terms.[66]

Hence, the argument substantiated in my five case studies is that liveness emerges through socially configured, phenomenally experienced, and technologically mediated assemblages of (spatial) copresence and (temporal) simultaneity, or varying instantiations thereof. What I call forms of liveness are produced by different (configurations of) *technologies* that produce specific spatiotemporalities (direct, live, real time), by particular cultural *discourses* tied to mediation, and by the invocation and interpellation of particular *sensations*, that is, an "experience" of liveness for audience members. What is more, forms of liveness are also part of what I call the *politics* of liveness. This means studying the institutional structures creating particular claims to liveness through relations of center and periphery, current or past, dominant or subservient.[67]

---

[65] Wolfgang Ernst, *Chronopoetics: The Temporal Being and Operativity of Technological Media* (London: Rowman & Littlefield, 2016), citing N. Katherine Hayles, "Komplexe Zeitstrukturen Lebendiger Und Technischer Wesen," in *Die Technologische Bedingung. Beiträge Zur Beschreibung Der Technischen Welt*, ed. Erich Hörl (Frankfurt am Main: Suhrkamp, 2011), 217. As Lisa Gitelman has pointed out, Ernst hardly attends to the importance of sociohistorical context for the emergence of new media technologies. Lisa Gitelman, *Always Already New: Media, History, and the Data of Culture* (Cambridge, MA: MIT Press, 2006), 10.

[66] N. Katherine Hayles, *How We Think: Digital Media and Contemporary Technogenesis* (Chicago: University of Chicago Press, 2012), 106.

[67] As Nick Couldry has argued in this context, liveness is not a descriptive term but a category that depends on its "place within a wider system or structured pattern of values, which work to reproduce our belief in, and assent to, something wider than the description carried by the term itself." Nick Couldry, "Liveness, 'Reality,' and the Mediated Habitus from Television to the Mobile Phone," *Communication Review* 7, no. 4 (2004): 354.

Critical Histories and the Neuro-Enchanted Present

Historians have a role to play in critically evaluating the current hegemonic position of the neurosciences. This was one of many suggestions in the inaugural proposal of an initiative called "Critical Neuroscience," launched in Berlin in 2008, which urgently called for critical analyses of the brain sciences because it viewed neuroscience as a domain that often presents itself as apolitical despite its increasing complicity in normative and economic agendas of, for example, self-optimization and economic productivity.[68] Critical neuroscientists have issued an important caution against a problematic neurocentrism in today's culture: we should be wary of such neurocentrism because it does not do justice to the embodied, enculturated, enactive, and affective dimension of human behavior, and because it may feed into what has been called "neuro-governmentality," brain science that prescribes how people live their lives, builds norms and politics into brain facts, and makes the social look natural.[69]

While the Critical Neuroscience initiative no longer exists as a physical or institutional research group, its agenda is still important, not least because of its call for the need for critical histories of the neurocultural present. A number of historians of brain and mind sciences have employed this longer perspective to study histories of, for example, the double brain, the elite brain, ideas of brain localization, phrenology, and emotion research, asking us with these studies to reconsider the inevitability of current neuroscience, that is, "not to take today's solutions as the final answers."[70] One important

---

[68] See Suparna Choudhury and Jan Slaby, "Introduction," in *Critical Neuroscience: A Handbook of the Social and Cultural Contexts of Neuroscience*, ed. Suparna Choudhury and Jan Slaby (Oxford: Wiley Blackwell, 2011). For a recap, see Des Fitzgerald et al., "What's So Critical about Critical Neuroscience? Rethinking Experiment, Enacting Critique," *Frontiers in Human Neuroscience* 8 (2014). Cornelius Borck, "Auf Der Suche Nach Der Verlorenen Kultur: Vom Neuroimaging Über Critical Neuroscience Zu Cultural Neuroscience—Und Zurück Zur Kritik," *Berichte zur Wissenschaftsgeschichte* 41, no. 3 (2018).

[69] Meloni, "Philosophical Implications of Neuroscience." On "neuro-governmentality," see particularly Nikolas Rose and Joelle Abi-Rached, "Governing through the Brain: Neuropolitics, Neuroscience and Subjectivity," *Cambridge Journal of Anthropology* 32, no. 1 (2014).

[70] Choudhury and Slaby, "Introduction," 15. A number of important cultural and conceptual "histories of the brain" have been published in the past decades, tracing histories of ideas in the brain and mind sciences and examining the social, economic, political, and cultural situatedness of brain facts and cerebral models. Important examples are: Anne Harrington, *Medicine, Mind, and the Double Brain: A Study in Nineteenth-Century Thought* (Princeton, NJ: Princeton University Press, 1987); Cornelius Borck, *Hirnströme: Eine Kulturgeschichte Der Elektroenzephalographie* (Göttingen: Wallstein

observation from these historical accounts concerns the heterogeneity of the brain and mind sciences. The term "neuroscience," for example, only emerged in the 1960s and circumscribes a very heterogenous conglomerate of practices.[71] Framing historical narratives as histories of the brain, the mind, or the neurosciences may give the false impression that these have been constant, dominant, or meaningful objects of (scientific) research. In fact, the history of mind and brain sciences is, as Stephen Clark and Delia Gavrus emphasized, "a patchwork of loosely held together fragments," which means that any longue-durée account should avoid conjuring a false epistemological coherence of the many disciplines involved in the study of brains, minds, selves, bodies, psyches, mental pathologies, psychologies, behaviors, nerves, and humans.[72]

Writing big histories of the brain is risky, as, by conjuring historical narratives that "lead up" to the omnipresence of the brain today, historians may help to fortify the brain sciences' current hegemonic position.[73] Max Stadler thus argues that we need to "defamiliarize" ourselves from our neuroscientific past.[74] Other historians of brain and mind sciences have also emphasized the importance of unsettling established historical narratives by writing "different" and "marginal" histories.[75] Yet, what a marginal history

Verlag, 2005); Michael Hagner, *Geniale Gehirne: Zur Geschichte Der Elitegehirnforschung* (Göttingen: Wallstein Verlag, 2004); Roger Smith, *Inhibition: History and Meaning in the Sciences of Mind and Brain* (Los Angeles: University of California Press, 1992); Katja M. Guenther, *Localization and Its Discontents: A Genealogy of Psychoanalysis and the Neuro Disciplines* (Chicago: University of Chicago Press, 2015); Laura Salisbury and Andrew Shail, eds., *Neurology and Modernity: A Cultural History of Nervous Systems, 1800–1950* (London: Palgrave Macmillan, 2010).

[71] Max Stadler argues that since the 1990s, a new history of the neurosciences emerged with a "culturalist orientation" that is guilty of grafting the history of neuroscience onto a history of the brain and reimagining the history of the brain through the lens of "modern neuroscience." He also argues we should not overemphasize the brain as a unit of analysis in histories prior to the 1950s; less spectacular histories of research on nerves and muscles, often in animal models, were ubiquitous. Max Stadler, "The Neuromance of Cerebral History," in *Critical Neuroscience*, ed. Suparna Choudhury and Jan Slaby (Hoboken, NJ: Wiley Blackwell, 2011).

[72] In parallel, Roger Smith remarks there is no comprehensive history of the mind–body debate for the twentieth century; it is an unavoidably patchy narrative—psychologists, physiologists, and philosophers use different idioms and often do not speak to the same problems. Roger Smith, "Representations of Mind: C. S. Sherrington and Scientific Opinion, c.1930–1950," *Science in Context* 14, no. 4 (2001): 537.

[73] Roger Cooter, "Neural Veils and the Will to Historical Critique: Why Historians of Science Need to Take the Neuro-Turn Seriously," *Isis* 105, no. 1 (2014).

[74] Max Stadler, "Circuits, Algae and Whipped Cream: The Biophysics of Nerve, ca. 1930," in *The History of the Brain and Mind Sciences: Technique, Technology, Therapy*, ed. Stephen T. Casper and Delia Gavrus (Rochester: University of Rochester Press, 2017).

[75] "Marginal" histories in Stephen T. Casper and Delia Gavrus, eds., "Introduction. Technique, Technology, and Therapy in the Brain and Mind Sciences," in *The History of the Brain and Mind Sciences: Technique, Technology, Therapy* (Rochester: University of

of the mind and brain is always needs to be historicized in relation to established narratives about the past, as Casper and Gavrus note.[76] With this study, I aim to contribute to this by moving between histories of brain science in laboratories and scientific writing and histories of public performances of brain, and between cultural histories of the brain in relation to histories of media and mediation.

Still, writing big histories of the brain might be too risky. Historian Roger Cooter is convinced that historians will inevitably be coopted by what he calls the "neuro-turn."[77] By writing longue-durée histories that trace the formation of neurocentrism and neurobiologization, such histories may help to naturalize and sustain these tendencies and thus turn into technologies of power for, instead of critical reflections on, the new regime of truth. Pessimistic, Cooter writes, "we stand largely on the inside of the turn's enchantment, capable of offering some criticisms but not much by way of historical critique of the sort that entails both looking beyond the object of inquiry itself to the conditions of possibility for it and standing back to reflect on our own position as historians in relation to it."[78]

I am more optimistic about the potential of tracing some of the historical conditions of neurosciences' lofty position today. My work emerges from "inside the neuro-turn's enchantment," and it is from this vantage point in a fully media-enchanted world that I see the importance of studying the historical emergence of the notion of the living brain as the live brain. I believe that the public history of the brain and the brain sciences in the twentieth century has not received enough attention. With respect to the nineteenth century, there are ample studies of the presence and practice of mind and brain science beyond scientific circles (think of the study of phrenology, nervousness, and neurasthenia), yet there is considerably less research on the way knowledge of the brain was circulated between laboratories and lay audiences since the beginning of the twentieth century.[79] Additionally, by

---

Rochester Press, 2017). "Different" histories in Mattia Della Rocca and Claudio Pogliano, "Different Histories from 20th Century Neuroscience," *Nuncius* 32, no. 2 (2017).

[76] Casper and Gavrus, "Introduction. Technique, Technology, and Therapy in the Brain and Mind Sciences."

[77] Cooter, "Neural Veils and the Will to Historical Critique," 146.

[78] Ibid.

[79] For a good overview of nineteenth-century neuroscience in relation to popular audiences, see Stephen T. Casper, "History and Neuroscience: An Integrative Legacy," *Isis* 105, no. 1 (2014): 123–32. For the early twentieth century, Roger Smith is one of the few sources that note how popular literature on the brain offered "empowerment through materialist representations of mind, for example, by linking personality and the energies of the brain." Smith, "Representations of Mind," 514. Borck also remarks on the scarcity of research on the relation between publicity and science in the twentieth century. Cornelius Borck,

focusing on the imbrication of media in various historical case studies, I also contribute to an understanding of the way scientific techniques and practices do not emerge instantly, but always need to be negotiated and performed in specific historical contexts.

Summarizing the three conceptual tools introduced in this chapter, I analyze a genealogy of brainmedia in five case studies through the frame of *brainmedia as assemblages*, approached as a practice of *performing knowledges* that generate new *forms of liveness* in conceptualizing and imagining the active brain. Brainmedia configure different forms of liveness, that is, as assemblages, they conjure varying spatiotemporalities and spatio-tempo-realities. Brainmedia assemble new technical configurations as well as novel experiences of the directness, nearness, hereness, presence, aliveness, togetherness, and nowness of perceiving the brain in action—the live brain.

Brainwaves: A Cultural History of Electroencephalography (London: Routledge, 2018), 120 ff, 116. An exception in this respect is Melissa M. Littlefield, *The Lying Brain: Lie Detection in Science and Science Fiction* (Ann Arbor: University of Michigan Press, 2011).

1

The Birth of the Live Brain, 1820–1920

My story of the "live brain" begins with an image of a woman, writing (Figure 1.1).[1] She has the bobbed haircut and striped blouse fashionable for female white-collar workers in American cities around 1920—and though we might expect her to use a typewriter, she is using a quill pen and ink. Pasted over her head is a drawing of the cerebral organ in its simplified outlines. Hand-drawn arrows label four little blots inside the brain as the "centers" for hearing, speech, sight, and motion. From these, dotted lines run to the corresponding senses: the woman's ear, mouth, eyes, and writing hand. The view inside her head—not dissected, but simply made transparent through superimposition—gestures toward the idea of seeing a living brain at work, perceiving, as it were, that active cerebral substances mirror this writer's activities. Even more than an image of a woman at work, this is an image of a brain at work.

It is no coincidence that this is a woman's brain. Lisa Cartwright has emphasized the female body's frequent appearances in discourses of, and advertisements for, modern visualization technologies (such as X-rays) that frame them as opaque entities: ideal subjects to be rendered transparent by a superior, technological-scientific, male gaze, an act that was implicitly or even explicitly equated with sexual looking.[2] From the dawn of the brain and mind sciences until this day, women feature as exemplary subjects in photographs of laboratory work as well as in advertising for neuro-technologies.[3]

I zoom in on this specific image, inscribed with longstanding traditions of gendered scientific depiction, from around 1920, to signal a historical convergence. The image allows us to discuss both the nervous *past* extending into the 1920s and the 1920s *present* extending into the future. I will show how this image evinces both the persistence of nineteenth-century conceptions in the mind and brain sciences and a particular promise toward a future understanding. The early decades of the twentieth century were a time of

---

[1] Complete source information of figures is provided toward the end of this study.
[2] Lisa Cartwright, *Screening the Body: Tracing Medicine's Visual Culture* (Minneapolis: University of Minnesota Press, 1995).
[3] Paula Gardner and Britt Wray, "From Lab to Living Room: Transhumanist Imaginaries of Consumer Brain Wave Monitors," *ADA. Journal of Gender, New Media and Technology*, no. 3 (2013), https://adanewmedia.org/2013/11/issue3-gardnerwray/.

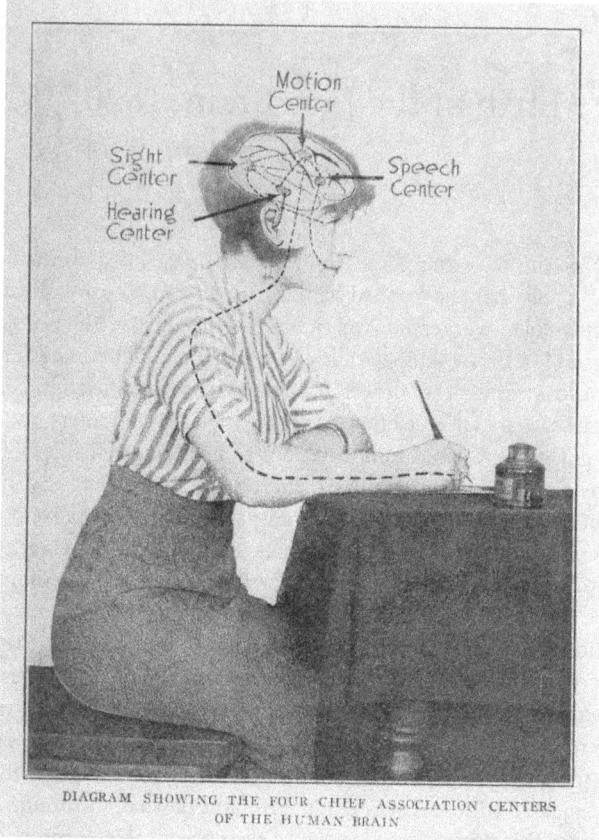

Figure 1.1 Diagram showing the four chief association centers of the human brain, c. 1919 (illustration).

ambiguity about how the brain and human nervous system could be thought. In broad strokes, and with the help of the writings of Michel Foucault, this chapter's historical account sketches the way conceptions of media and mediation were articulated in relation to the human nervous system and the modern human at this time. I argue it is in this historical context that a new way of thinking "the brain at work" emerges—the live brain—and my case studies of brainmedia in Chapters 2 through 6 should be read against the backdrop of this argument.

Beginning with this "minor" image—uncanonical, forgotten, idiosyncratic—this chapter proceeds in four steps. First, I describe how the image builds on longstanding ideas of the human sensorium, still present in 1920, that used new media technologies to articulate it as intimately part of the modern environment. This allows me to use the image as an example of the new structures and forms through which scientific knowledge circulated. Second, I turn to the image's blots and dotted lines to analyze what was left of nineteenth-century physiology and anatomy. Particularly important in this respect is to understand the changing ways in which the relation between brains and living bodies was actively configured. To do so, I turn to Foucault's reflections on the "anatomoclinical" and "neurological" gaze, zooming in on his narrative of the "disappearance of the great neurological hope": the nineteenth century's disappointment about the failure to systematically correlate brain lesions and (abnormal) human behaviors. Third, drawing on Foucault's observations on the "correlative constitution" of the neurological body before the "apparatus of neurological capture" (*les dispositifs de capture neurologique*), I argue for the importance of analyzing the role of media/mediation as integral to this correlative constitution.[4] Finally, I provide early twentieth-century examples of studying the human brain—both imagined and actual practices—in which new media played an important part. It is in this moment of convergence, when different mediated correlations accumulated, I contend, that we discern the nascent understanding of the active, living brain as a "live brain."

Nervous Subject, Modern Sensorium

Printed without a source reference in 1920, the image of the writing woman and her brain served as an illustration in Warren Hilton's motivational self-help guide *Applied Psychology: Making Your Own World*, part of a series of no less than twelve volumes on "the Applications of Psychology to the Problems of Personal and Business Efficiency," published between 1919 and 1920 by the Society of Applied Psychology under the auspices of the American publisher The Literary Digest (also founded by Hilton).[5] The volumes promised to teach their readers new "mental methods" for an "efficient use of the mind" in the modern

---

[4] "dans la constitution corrélative d'un corps neurologique en face de ce regard et de ce dispositif de capture." Michel Foucault, *Le Pouvoir Psychiatrique: Cours Au Collège De France, 1973-1974*, Hautes Études (EHESS/Gallimard/Seuil, 2003), 301.

[5] Warren Hilton, *Applied Psychology: Making Your Own World: Being the Second of a Series of Twelve Volumes on the Applications of Psychology to the Problems of Personal and Business Efficiency* (New York: The Literary Digest for the Society of Applied Psychology, 1920), 16. (While I cannot ascertain the print run of these twelve volumes (of over 1,100

work environment.[6] Hilton's books were advertised as applied psychology (his work was particularly influenced by "New Thought," American motivational literature from around 1900), a heterogeneous discipline that maintained only loose connections to other branches of psychology such as experimental psychology.[7] Although the author briefly invoked "facts ... by experiments from psychological physiology," there was little mention of neurophysiological or neuroanatomical science.[8] Applied psychology itself was a contested field. In 1920, as historian Michael Sokal notes, one prominent applied psychologist, Walter Van Dyke Bingham, worried about pseudoscientific branches such as the abovementioned Society of Applied Psychology, "whose president, Warren Hilton, was acquainted with the science of psychology only through one undergraduate course in the area at Harvard."[9] (Hilton indeed credited Harvard scholar Hugo Munsterberg as one of his sources of inspiration.) In Hilton's book, the image of the brain functioned as a gesture toward scientific authority and signaled some biological facts supporting his advice for mind improvement. On the relation between brain or nerve cell activity and mind, he remained rather agnostic, simply stating that mixing the two registers was "incompatible with scientific methods," and that his readers should be "investigating the mind, not the body."[10]

pages), it is likely that the series was widely distributed and ran through many editions. They were advertised at least until 1930. In fact, they are still available in reprints today.)
[6] Ibid., 4.
[7] As Nadine Weidman has pointed out, when applied psychology (a hybrid field of psychologists advising advertising firms and working on vocational tests and worker efficiency) rapidly developed as a separate branch, "a coalescence of psychology and neurology mattered less and less," and "applications could work without any understanding of or attention to the biological basis of behavior." Nadine M. Weidman, *Constructing Scientific Psychology: Karl Lashley's Mind-Brain Debates* (Cambridge: Cambridge University Press, 1999), 14.
[8] Hilton, *Making Your Own World*, 17. In the introduction to *Applied Psychology*, volume I, Hilton printed a list of more than a dozen scholars who influenced his thinking (but did not reference them directly anywhere else), including psychotechnical researcher Hugo Munsterberg, experimental psychologist Joseph Jastrow, psychologist Pierre Janet, and other self-help authors, applied psychologists, psychical researchers, intelligence testers, management consultants, "New Thought" thinkers, experimental psychologists, and therapists. Warren Hilton, *Applied Psychology: Psychology and Achievement. Being the First of a Series of Twelve Volumes on the Applications of Psychology to the Problems of Personal and Business Efficiency* (New York: The Literary Digest for the Society of Applied Psychology, 1919), prefatory note.
[9] Michael M. Sokal, "The Origins of the Psychological Corporation," *Journal of the History of the Behavioral Sciences* 17, no. 1 (1981): 57.
[10] Warren Hilton, *Applied Psychology: Mind Mechanism. Being the Eighth of a Series of Twelve Volumes on the Applications of Psychology to the Problems of Personal and Business Efficiency* (New York: The Literary Digest for the Society of Applied Psychology, 1920), 27.

Hilton's work is evidence of the persistent importance of a heterogeneous domain of mental sciences or psychoanalytics in the early decades of the twentieth century (dispersed and appropriated in fields of psychical research, psychotherapy, and management consulting, for example) that continued to conceptualize thought in terms of "nerve force," "mental power," and "nervous exhaustion." This conception of the nervous system in terms of a battery, of a "nervous economy" of force that could be depleted or reenergized, had been popularized by George Beard's study *American Nervousness* in 1881 and spurred different versions of the idea of "neurasthenia" (nervous exhaustion) in local research cultures across the globe.[11] Beard's neurasthenia was the preeminent condition of (often male) modern city dwellers, "brain workers" equipped with a human sensorium (a nervous system extended into the world by the senses) that was constantly exposed to—as well as networked with— the intense stimuli coming from new (media)-technologies so prevalent in their modern surroundings.[12]

By the 1920s, neurasthenia had gradually waned as a diagnostic label in scientific communities in a number of countries, viewed as too much of a blanket category that did not take biological and psychological facts into consideration.[13] And yet, elements of neurasthenic thinking remained: a relational conception of the modern experience, of the intimate and vulnerable connection between sensorium and environment. Within this conception, modern technologies offered analogies for the human nervous system (for its speed, communication, and movement) and the other way around: new conceptions of nervous matter (as a network and as transporting signals by means of energy) also shaped the way the relation between nervous subjects and their social world could be thought.[14]

This neurasthenic conception of a media-networked modern nervous subject was still very much present in *Making Your Own World* in 1920. Like previous motivational business literature, Hilton sketched a very contemporary image of the working nervous system in its modern environment: it was a world brimming with activity, where "at the busy corner

---

[11] Marijke Gijswijt-Hofstra and Roy Porter, *Cultures of Neurasthenia from Beard to the First World War* (Amsterdam: Rodopi, 2001).
[12] Andreas Killen, *Berlin Electropolis: Shock, Nerves, and German Modernity* (Berkeley: University of California Press, 2006), 42; Jeffrey Sconce, *The Technical Delusion: Electronics, Power, Insanity* (Durham, NC: Duke University Press, 2019), 195.
[13] Tom Lutz, "Varieties of Medical Experience: Doctors and Patients, Psyche and Soma in America," in *Cultures of Neurasthenia from Beard to the First World War*, ed. Marijke Gijswijt-Hofstra and Roy Porter (Amsterdam: Rodopi, 2001), 59.
[14] Laura Salisbury and Andrew Shail, *Neurology and Modernity: A Cultural History of Nervous Systems, 1800-1950* (London: Palgrave Macmillan, 2010), 10.

of a city street," "light, sound and tactual vibrations press upon you from every side," always "titillating the unsleeping nerve-ends of the sensorium."[15] The brain is portrayed as a "central office" that could only be reached after messages passed through an infinitesimal number of nerve cells.[16] The nervous system worked like the transmitter of a "human telephone" that connected to the world, and the mind functioned as "the receiving apparatus of the wireless telegraph," picking from the air only those vibrations "to which it is attuned."[17] The modern (male) worker could best navigate this busy world by being selective, the book claimed. Selection was the key to "making your own world": conscious mental acts were necessary to attune the mind to only those parts of the environment conducive to productivity. Consequently, judgment and conduct would be "made up" of specifically selected mental pictures of the environment only; they were part and parcel of your own mind: "*Your environment is within you.*"[18]

The functioning of the brain or the nerves was thus not discussed in anatomical terms; it was inscribed in a broader metaphorical circuit that connected everyday media technologies and sensory experience. The modern human subject—equipped with a nervous system connected to the outside world via the senses—was understood as always already caught up in a media-saturated environment. In fact, the term "media" must be understood as having its own historicity: it did not always mean a technological "medium" (an instrument or system such as a radio or television, as we use the term today; in fact, it did not have this meaning in 1920). A particularly important meaning—present in the work of a number of theorists at the time—is that of media as the material-technical environment in and through which perception takes place and is itself altered (what Walter Benjamin had started envisioning in the 1920s and later discussed as the "medium of present-day perception" and the "apparatus of apperception").[19] The biological senses of the modern citizen were viewed as literally changed by the omnipresence

---

[15] Hilton, *Making Your Own World*, 21.
[16] Ibid., 14.
[17] Ibid. While Hilton never credits New Thought writer William Walker Atkinson, this passage on the wireless receiver clearly seems copied from William Walker Atkinson, *Nuggets of the New Thought; Several Things That Have Helped People* (Chicago: The Psychic Research Company, 1902). Cited in Sconce, *The Technical Delusion*, 208.
[18] Hilton, *Making Your Own World*, 59.
[19] Walter Benjamin, "The Work of Art in the Age of Its Technological Reproducibility (First Version)," *Grey Room*, no. 39 ([1935] 2010): 11–38, 15, 33. On changing conceptions of "medium" and "media," see Antonio Somaini, "Walter Benjamin's Media Theory: The Medium and the Apparat," *Grey Room* 62 (2016): 62; John Guillory, "Genesis of the Media Concept," *Critical Inquiry* 36, no. 2 (2010). On "mediation," see Raymond Williams, *Keywords: A Vocabulary of Culture and Society* (New York: Oxford University Press, 1976), 204–7.

of media technologies, becoming part of a network of mediation, and by this process the nervous system and the senses were newly articulated and conceptualized vis-à-vis such modern media technologies.

Combining an altered photograph with drawn imagery, the figure of the writing woman is characteristic of the cut-and-paste methods of early twentieth-century photomontage, a technique that was favored by emerging popular science journals and magazines in the United States and Europe.[20] Generally, the early decades of the twentieth century were characterized, as Cornelius Borck has noted, by a conjunction of the increasing "scientification" of everyday phenomena (scientific research into quotidian phenomena like work environments, diet, clothing, sleeping habits, and mental testing), as well as the emergence of new public sites and media where science was experienced, negotiated, and also challenged by both scientists and laypeople (newspapers, radio, popular magazines, and science exhibitions, for example).[21] Hilton's book, for example, noted that "through the writings of lay-men the popular mind has become befuddled with vague and speculative explanations of the facts."[22] Research about the brain—and the status and authority of such research—was conveyed and disputed within new contexts (and through new forms) of popular mediation.

Looking closer at the image of the writing woman also reveals what was left of more than fifty years of studying the relation between living humans and their brains: blots and lines. The dashed lines stemmed from reaction-time experiments, started in the nineteenth century by psychophysiologists measuring the time it took from stimulating a nerve or a subject to its response, while the blots hailed from nineteenth-century *Zentrenlehre* localizing cerebral functions spatially in the brain.[23] In fact, this photomontage was

---

[20] Cornelius Borck, "Communicating the Modern Body: Fritz Kahn's Popular Images of Human Physiology as an Industrialized World," *Canadian Journal of Communication* 32, no. 3 (2007); Matthew Biro, "The New Man as Cyborg: Figures of Technology in Weimar Visual Culture," *New German Critique*, no. 62 (1994).

[21] Borck describes this as a shift from the nineteenth-century "experimentalization of life" (the transformation of life's processes into phenomena that could be experimentally observed in laboratories) to the "experimentalization of everyday life" in the twentieth century. Borck, *Brainwaves*, 119; Hilton, *Applied Psychology: Mind Mechanism*, 38.

[22] The book cautioned that various present-day authorities had falsely equated mind-facts and brain-facts, "explanations that may actually be true, but are in the very nature of things incapable of proof and are utterly out of place in a scientific study of the subject. They are excursions into the dream forest of mysticism, occultism and religion." *Applied Psychology: Mind Mechanism*, 37.

[23] On the history of psychophysiology, see Henning Schmidgen, *Hirn Und Zeit: Die Geschichte Eines Experiments 1800–1950* (Berlin: Matthes and Seitz Verlag, 2013); "Die Geschwindigkeit Von Gefühlen Und Gedanken," *NTM International Journal of History & Ethics of Natural Sciences, Technology & Medicine* 12, no. 2 (2004). On the history of the *Zentrenlehre*, see Guenther, *Localization and Its Discontents*.

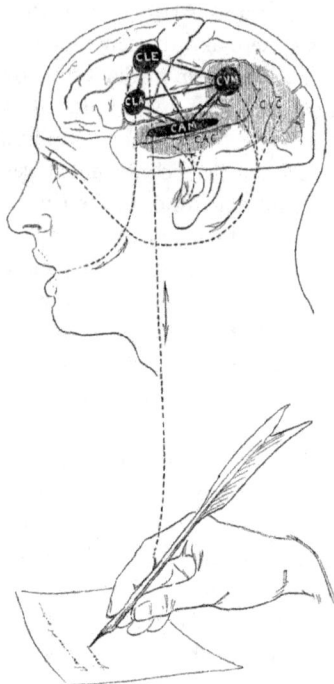

Figure 1.2 Pierre Marie's scheme for interconnected left hemisphere cortical centers involved in oral and written language, 1888 (illustration).

an appropriation of an earlier cross-section drawing of a writer's brain: that writer was holding a feather quill (hence that outmoded quill in the hand of the woman) (Figure 1.2). Created in 1888 by neurologist Pierre Marie, the drawing was based on an earlier diagram by Marie's famous teacher Jean-Marie Charcot, part of a project that correlated behavioral abnormalities (such as speaking difficulties in aphasic patients) with postmortem brain observations, which resulted in a diagram of four different cerebral centers involved in the production of oral and written language.[24]

---

[24] "De L'aphasie En Général Et De L'agraphie En Particulier, D'après L'enseignement De M. Le Professeur Charcot," *Progrès Médical* 7, no. 5, deuxième serie (1888). Cited in Victor W. Henderson, "Alexia and Agraphia," *Neurology* 70, no. 5 (2008). Charcot's diagram in turn had been based (among other research) on the lesion research and brain mapping of Sigmund Exner: *Untersuchungen Über Die Localisation Der Functionen in Der Grosshirnrinde Des Menschen* (Wien: W. Braumuller, 1881).

Blots and lines, these came from the two important nineteenth-century traditions of temporal-oriented psychophysical research and spatial-oriented cerebral lesion research that intersected, and were superimposed, in this photomontage of a transparent female brain body. And yet, by the early 1900s, both these methods had failed to produce the desired outcomes: though the field of psychophysiology aimed to find the exact time measurements for psychological processes, laboratories worldwide only reported disparate results by the 1890s.[25] And by 1906, Pierre Marie had started to doubt his work on localization and drove a "pickaxe into the edifice of aphasia," as one researcher later described it.[26]

In the last decades of the nineteenth and at the turn of the twentieth century, brains were thought to matter for explaining human behavior and human minds, but *how* (in what relation to the living human) and *to what extent* they mattered changed in zigzagging movements. Historians of brain science such as Anne Harrington, Susan Leigh Star, and Katja Guenther trace these differing conceptions, predominantly in the late nineteenth century.[27] Their important studies have deepened our understanding of the correlative practices of brain research, revealing the rise of more complex models of the interrelations between living bodies and cerebral lesions (dynamic and connectionist models, for example). The end of the nineteenth century represented a shift, as temporal (psychophysical) and spatial (anatomical localization) approaches were offering inconclusive results. The correlative project continued–correlating (through different emphases and systems) the living body's behaviors, utterances, and jitters with physiological data and observations of dead brains and bodies—but was approached with particular uncertainty.

Paying attention to the late nineteenth-century history and historiography of brain research helps to understand the beginnings of a configuration of knowledge that underlies the shift to the "live brain." I trace this shift using Michel Foucault's work on the history of medicine and psychiatry as it developed through his writings: from his mention of the skull breaker in 1820 in the preface of *The Birth of the Clinic* (1963), through his description

---

[25] Ruth Benschop and Douwe Draaisma, "In Pursuit of Precision: The Calibration of Minds and Machines in Late Nineteenth-Century Psychology," *Annals of Science* 57, no. 1 (2000): 24.

[26] A. Souques, "Quelques Cas D'anarthrie De Pierre Marie. Aperçu Historique Sur La Localisation Du Langage," *Revue Neurologique* 2 (1928): 362. Cited in Harrington, *Medicine, Mind, and the Double Brain*, 261.

[27] Harrington, *Medicine, Mind, and the Double Brain*; Susan Leigh Star, *Regions of the Mind: Brain Research and the Quest for Scientific Certainty* (Stanford, CA: Stanford University Press, 1989); Guenther, Localization *and Its Discontents*.

of the emergence of a new "neurological body" in the 1860s, leading up to the disappointments of trying to understand the postmortem brain and localizing cerebral lesions toward the end of the nineteenth century, described in Foucault's lectures (1973-4) and published as *Psychiatric Power*. I follow the development of his historical concepts from his archeologically oriented work to his later genealogically oriented writings, ending with his references to the final decades of the nineteenth century. In the process, I will make an argument for a media-sensitive understanding of the "apparatus of neurological capture," allowing me to return, via Foucault, to the writer at her desk.

Foucault's "Apparatus of Neurological Capture"

First, the skull breaker. This "*casse-crâne*" figures as the emblematic example in the preface of *The Birth of the Clinic*, a study in which Foucault describes the emergence of a new "anatomoclinical gaze" in the nineteenth century.[28] This gaze, a new way of seeing and saying enabled by a new configuration of practices and discourses, allowed for the correlation of visible clinical symptoms with hidden views of the human interior, of the body that had been concealed but could become accessible, postmortem, through autopsy.[29] Foucault finds exemplars of this new gaze in Francois Lallemand's act of hammering open the skull in 1820 and in Antoine Bayle's 1825 attempt to understand the pathology of hysteria through careful description of brain lesions.[30] The skull breaker and his cerebral descriptions form the start of the fundamental spatialization and verbalization of the pathological. Within this new epistemological structure, truth was revealed by probing invisible structures, by spotlighting that which tried to evade the eye.[31] Yet, while here the skull breaker seemed to offer a new understanding of pathology by opening up the patient's brain to anatomoclinical perception, in his later

[28] Michel Foucault, *The Birth of the Clinic: An Archeology of Medical Perception* (London: Routledge, 1976), 146.
[29] Ibid., 159, 49. Anatomoclinical perception aimed for the inaccessible, "enclosed world of bodies" via the "point of view of death" (169).
[30] François Lallemand, *Recherches Anatomico-Pathologiques Sur L'encéphale Et Ces Dépendances* (Paris: Imprimerie de Baudouin Frères, 1820), vii, n; A. L. J. Bayle, *Nouvelle Doctrine Des Maladies Mentales* (Paris: Gabon et Compagnie Libraires, 1825), 23–4.
[31] Foucault, *The Birth of the Clinic*, 166. Foucault refers to this new "structure" or "figure" of the medical gaze as "invisible visibility" or as the "visible invisible," a gaze that is directed by the potential of rendering visible the body's hidden content, both through revelations by the scalpel and by the activity of meticulous medical description, that is, by the "work of language in pursuit of perception" (169).

writings Foucault would signal the eventual letdown of cerebral visibility, of trying to understand patients' strange behavior by examining their brain postmortem. Dead brains could hardly be mapped onto living patients. The relation between the visible and the invisible, between surface and depth, seemed to unravel when it came to understanding the brains of the living.

In *Psychiatric Power*, Foucault describes the emergence of a new gaze around the 1860s, a "neurological gaze" instigated by new patient–doctor encounters in the neuropsychiatric clinic, with Jean-Martin Charcot and the Paris Hôpital de la Salpetrière in the 1880s as his main example. This "new clinical apparatus of neurological capture" thus offered the "discovery," in a Foucauldian sense, of a "neurological body."[32] The practice of cerebral localization, he describes, now becomes subsumed within a broader neuropsychiatric apparatus that systematized the recordings of both the "surfaces" of patients (their attitudes, their movements, their utterances) as well as postmortem studies of their insides: body and brain. He thus describes the "correlative constitution" of a neurological body before the neurological clinic's gaze and apparatus of capture.[33]

Foucault notes various ways in which anatomical localization and neurological observation were correlated or "matched up" (*ajuster*) in the nineteenth century.[34] Even when "organicist" (brain-based) explanations of pathology intensified at certain moments, he shows that such projects always interconnected with broader systems of symptomatology or nosography. He cites Jean-Martin Charcot, who celebrated, in 1879, "the spirit of localization" that would show anatomical structure of the nervous centers "in their true light."[35] Yet, such high hopes did not always mean that localizing function within the space of the brain was science's ultimate conceptual horizon, or at least that it did not need to be in practice. Adding to Foucault's account, I note here briefly that the example of Charcot aptly shows this complexity. Charcot used the term "dynamic lesion" to describe pathological signs in the living body that he conceptualized as being on the threshold of visibility, a marginal space of potential future visualization.[36] Sometimes he would argue

[32] "The emergence of the neurological body, or rather, of the system constituted by neurology's clinical apparatus of capture and the correlative neurological body." *Psychiatric Power: Lectures at the Collège De France, 1973–1974* (New York: St. Martins' Press, 2008), 307.

[33] Ibid., 299.

[34] Ibid., 297.

[35] Jean-Martin Charcot, "Faculté De Médecine De Paris: Anatomo-Pathologie Du Système Nerveux," *Progrès Médical*, no. 14 (1879): 161. Cited in Foucault, *Psychiatric Power*, 324.

[36] J. W. Marshall, *Performing Neurology: The Dramaturgy of Dr Jean-Martin Charcot* (New York: Springer, 2016), 58. T. Kaitaro, "Biological and Epistemological Models of Localization in the Nineteenth Century: From Gall to Charcot," *Journal of the History of the Neurosciences* 10, no. 3 (2001): 262–76, 273.

that these lesions did not leave visible postmortem traces in the brain, but could only be inferred by studying the performing, living body. Other times, he suggested that they escaped present means of anatomical investigation but would become materially registerable in the future, eventually becoming visible. Foucault's research, in tandem with more recent histories of the brain, makes the important point that cerebral lesion research must always be viewed as part of broader conceptions of symptoms or conditions, as part of a broader discourse on the potential somatization (and visibility) of mental pathology.

In his analysis of the apparatus of neurological capture in *Psychiatric Power*, Foucault is most invested in describing the power relations between neurologists and patients, "a whole battery" of "arrangements" through which patients were instructed and forced to respond to neurologists.[37] He recounts that by recording those responses and bringing them together with insights from different fields (including cerebral lesion research), researchers around 1860 worked toward a "system of signs" that distinguished one type of jittering or mumbling in a patient from another and thus allowed for the creation of a systematic and detailed symptomatology of the neurological body (differential diagnosis).[38] In his observations on Charcot's clinic, Foucault notes the "arrangements" of relations of force that were designed for differential diagnosis, including the spaces in which the examinations of patients were performed in front of an audience, yet he pays only partial attention to the technological devices and uses thereof that were integral to reaching a diagnosis. While he notes the importance of charts, descriptions, and classifications as part of the apparatus of neurological capture, Foucault barely traces the way Charcot, Duchenne, and other Salpetrière researchers developed many technologies of demonstrative truth—photography, chronophotography, microphotography, sketches, and wax figures—that were also part of the correlative apparatus. After Foucault, numerous historians have pointed to Charcot's photographic methods as representing the pinnacle of his differential diagnosis project; the felt immediacy of photographic records guaranteed the truthfulness of the psychiatric practice.[39]

---

[37] Foucault, *Psychiatric Power*, 300; also ibid., 298.
[38] Ibid., 301. Through neurological analysis, gestures, movements, and moods could be ascribed to (differing degrees of) voluntary or involuntary behavior, automatic responses, or intentional attitudes. Hence, this differentiation could finally allow the neurologist to separate genuine illness from simulation.
[39] Sander L. Gilman, *Seeing the Insane* (Lincoln: University of Nebraska Press, 1996), 195; Ulrich Baer, "Photography and Hysteria: Toward a Poetics of the Flash," *Yale Journal of Criticism* 7, no. 1 (1994); Georges Didi-Huberman, *Invention De L'hystérie: Charcot Et L'iconographie Photographique De La Salpêtrière* (Paris: Macula, 1982), 50; Sander L.

These types of media-assisted diagnosis also needed to be *performed*—through specific repertoires and in specific spaces—to be brought in relation to one another and in correlation with the patient. Salpetrière's infamous demonstrations of hysteric and other nervous patients were part of lectures in which Charcot expounded on the symptomatology of cases, drawing together the different elements of his diagnosis.[40] These lectures were attended by medical scientists, as well as a changing audience of affluent Parisians from intellectual and artist circles (such as actress Sarah Bernhardt) and took place in a room that could easily be darkened or illuminated and was equipped with a slide projector. Immediately after a patient left the room, Charcot could project a picture or a sketch of a microscopic brain lesion image that corresponded to the patient's symptoms, a connecting gesture that allowed for a correlation between the living body and a documented brain.[41]

It is in this arranged situation of juxtaposition, this superimposition of living bodies and imaged brains, that I propose we see the beginnings of a new configuration. By giving more attention to the emergence of media technologies and the way such technologies needed to be performed in particular assemblages, we can see the *correlative constitution* not only of a neurological body but also of an emerging conception of seeing an active, living brain.

Before I expound on this, I want to draw special attention to this correlation "performance" through technological media by means of a diversion into Foucault's life. Foucault did not pay particular attention to (media-)technical apparatuses of capture, even though he himself was immersed in a range of such technologies when he helped administer psychological tests to French prisoners in the early 1950s (as part of an EEG research team run by Georges and Jacqueline Verdeaux at Paris' St. Anne Hospital).[42] Cornelius Borck, reflecting on this biographical detail, remarks that Foucault may have viewed EEG as just one example of a broader practice of building new classifications

---

Gilman, "The Image of the Hysteric," in *Hysteria Beyond Freud*, ed. Sander L. Gilman (Berkeley: University of California Press, 1993).

[40] In Charcot's words, one had to continuously gather "the laboratory work and the autopsy of the amphitheatre." Jean-Martin Charcot, *Tome V. Maladies Des Poumons Et Du Système Vasculaire*, ed. Désiré-Magloire Bourneville, Oeuvres Complètes (Progrès médical, 1888), 6. Cited in Jonathan W. Marshall, *Performing Neurology: The Dramaturgy of Dr Jean-Martin Charcot* (New York: Springer, 2016), 52.

[41] Marshall, *Performing Neurology*, 86. "The years 1894–1900 also saw the production of the first radiographs, electrocardiographs, and even so-called 'phonocardiographs' at the Salpêtrière" (243).

[42] Didier Eribon, *Michel Foucault* (London: Faber & Faber, 1993), 87.

(separating the insane, epileptics, women, or geniuses, for example), part of his tracing a broader "operational mode of scientific discourse" based on exclusion and monitoring.[43]

Notwithstanding Borck's observation, a closer look at the research in which Foucault was involved shows a special preoccupation with creating a system of differential diagnosis preeminently based on (the combination of) new recording technologies as well the performance of these recordings. In 1955, the Verdeauxs proposed a "*technique de polygraphie*," a procedure that asked subjects to respond to audio and visual stimuli (Foucault's task was to operate the tachistoscope that flashed simple or complex figures) and that would record their responses on a sheet that registered EEG measurements as well as skin conduction, ECG, and respiratory rhythms.[44] This multirecord would then be projected for the psychiatrists by means of an epidiascope as sound tapes of the corresponding subject played simultaneously. Ultimately, this polygraphy would result in, as the authors put it, "reproducing the experience over time, and reliving it, in a way, in the subject's absence."[45] Such a "relived" demonstration would allow psychiatrists to conduct their interrogations more precisely, by means of "objective indications."[46]

This polygraphy proposed both an assembled technology and a specific way of performing diagnosis by means of projection through the epidiascope (the same technology used in 1934; see the introduction's prelude). As such, it signaled a correlative constitution of the neurological body that emphasized the possibility for newly assembled (and performed) relations between a living patient and their physiological record. Through the simultaneous superimposition of various technologies, the living body and active brain could be "relived." Reading Foucault's 1955 work now should incite us to pay more attention to media technologies and to be especially sensitive to the powers media have in shaping the discourse on categorizing brains and psychiatric patients as well as to how such discourses shaped new media and assemblages of media and bodies through performances of knowledge.

---

[43] Borck, *Brainwaves*, 183–4. Borck warns about a too "big-picture description" in historicizing the quantifying procedures in the life sciences, such as EEG research, sweepingly, as a "topos of biopolitics," which would risk a certain presentism.
[44] G. Verdeaux and J. Verdeaux, "Description D'une Technique De Polygraphie," *Electroencephalography and Clinical Neurophysiology* 7, no. 4 (1955): 647.
[45] Ibid.
[46] Ibid.

Examining and Imagining the Living Brain

Imaginaries of invading the skulls of living people by means of new technological media were prevalent at the same time that the project of the neurological apparatus of capture was plagued by disappointment. At the end of *Psychiatric Power*, Foucault's narrative of the neurological gaze ends in failure: "The neurological body, like the body of pathological anatomy, will elude the psychiatrist"; this was the "disappearance of the great neurological hope."[47] At the end of the nineteenth century, many neurologists felt it would be impossible to arrive at a differential diagnosis, and they particularly saw the project of cerebral localization as unsatisfactory.[48] Even the well-known neurologist Carl Wernicke, famous for his pioneering diagram of aphasia in the brain, moved away from the localization project.[49] Where scientists had been preoccupied with indicators of the functioning brain and nervous system that were at the threshold of visibility—almost invisible, but perhaps legible in minute gestures, microlesions, or milliseconds—now it seemed that the functional mechanisms of living organisms were beyond neurologists' ability to register.[50] Perhaps it was not the careful description of jittering and mumbling bodies, nor cerebral lesions traced after death that would yield final conclusions about the functioning of the active brain.

At the turn of the century, some researchers turned away from questions of visibility. Sigmund Freud did so with psychoanalysis, for example. In 1896, Henri Bergson downplayed the role of the physical brain and criticized localizationist theories by saying that "the brain is nothing but a kind of central telephone switchboard"; "it adds nothing to what it receives."[51] Yet, whatever minor or major role was attributed to the brain, clearly the organ had become part of a sensorium and vocabulary that was fully media-saturated. Modernity's shocking "media ubiquity," as it is now commonplace for historians to say, had by then also firmly established the idea that new media assemblages could or would eventually produce any imaginable

---

[47] Foucault, *Psychiatric Power*, 288.
[48] Anne Harrington, "The Brain and the Behavioral Sciences," in *The Cambridge History of Science: Volume 6, The Modern Biological and Earth Sciences*, ed. Peter J. Bowler and John V. Pickstone (Cambridge: Cambridge University Press, 2009), 519.
[49] Wernicke developed new clinical demonstrations (*Krankenvorstellungen*) to interpret the gestures and utterances of neurological patients in relation to established pathological anatomical knowledge. See Guenther, *Localization and Its Discontents*.
[50] They were, in the words of a neurologist cited in 1905, "far below par." Cited in Cartwright, *Screening the Body*, 64.
[51] On Bergson's involvement and influence in the aphasia debate, see Arthur Benton, "Bergson and Freud on Aphasia: A Comparison," in *Bergson and Modern Thought*, ed. Pete A. Y. Gunter and Andrew C. Papanicolaou (London: Routledge, 2016).

record of the surface of the living body (through photography, graphic inscriptions, or cinematography, for example) and even its invisible insides (by microphotography, microcinematography, X-rays, for instance).[52] The living brain had long been imagined as the final frontier, the ultimate realm for new media to conquer.

Nineteenth-century scientists and fiction authors had fantasized about illuminating the skull or measuring nerves through "cerebroscopes," "brain mirrors," and "neurometers."[53] Take, for example, the "encephaloscope" dreamt up by a French surgeon in 1884, which would make the skull transparent and project brain processes for everyone to see.[54] Or Thomas Edison, who in 1896 proclaimed that he would soon be able to produce an X-ray of the working brain inside the skull.[55] Or Richard Slee and Cornelia Atwood Pratt's fictional neurobiologist Dr. Berkeley, whose invention could turn brain slices into film strips that could be projected on a large screen.[56]

Interacting with these experimental and imagined (media-)technical assemblages and enabled by new (and imagined) technologies of visualization and graphic inscription, the 1900s and 1910s saw the emergence of a plethora of experiments attempting to find new ways to correlate organic matters and the behavior of living humans. If Charcot imagined his patients' dynamic lesions as being on the threshold of visibility—in the space of the potentially visible—now this threshold had not been crossed but widened: new technologies promised new ways of arriving at a correlative constitution of the active, living brain.

At the end of the nineteenth and the beginning of the twentieth century, a new conceptualization of accessing the living, active brain emerged through a

---

[52] Colette Colligan and Margaret Linley, "Introduction: The Nineteenth-Century Invention of Media," in *Media, Technology, and Literature in the Nineteenth Century: Image, Sound, Touch*, ed. Colette Colligan and Margaret Linley (Farnham: Ashgate, 2011), 5; Hagner, *Der Geist Bei Der Arbeit*, 270; Robert Michael Brain, *The Pulse of Modernism: Physiological Aesthetics in Fin-De-Siècle Europe* (Washington, DC: University of Washington Press, 2015), xxii. On surface readings of the neurological body, see Cartwright, *Screening the Body*, 73.

[53] Hagner, *Der Geist Bei Der Arbeit*, 223–35. A neurometer was first listed by the *OED* in 1850. "Neurometer, N." (Oxford University Press).

[54] Eduard Albert's encephaloscope, mentioned in Michael Hagner, "Mind Reading, Brain Mirror, Neuroimaging: Insight into the Brain or the Mind," in *Psychology's Territories: Historical and Contemporary Perspectives from Different Disciplines*, ed. Mitchell Ash and Thomas Sturm (London: Routledge, 2007), 288.

[55] As mentioned in Anon., "News and Notes," *British Journal of Photography* (1896): 105.

[56] Richard Slee and Cornelia Atwood Pratt, *Dr. Berkeley's Discovery* (New York: G. P. Putnam's Sons, 1899). See Flora Lysen, "The Brain Observatory and the Imaginary Media of Memory Research," in *Memory in the Twenty-First Century: New Critical Perspectives from the Arts, Humanities, and Sciences*, ed. Sebastian Groes (London: Palgrave Macmillan, 2016).

back and forth between brain experiments and fictional and scientific media imaginaries. In the domain of psychophysiology (also called experimental psychology), the project of correlating living bodies and mental processes remained a guiding premise. Numerous graphic curves, sometimes referred to as "braincurves," were registered to study the potential correlates of mental processes with, for example, breathing rates, electrical skin conductivity, muscular tension, pupil dilation, heartbeats, cerebral temperature, blood pressure, arterial pressure, or the metabolism.[57]

At the same time, the domain of nerve science saw, as historian Roger Smith puts it, an "advance of neurophysiology up the spinal cord and into the brain."[58] Especially striking were Otfrid Foerster's experiments in Germany, in the aftermath of the First World War, that intervened into the bodies of living patients who had suffered headwounds. Through electrical probing, he set out to map the functional body onto the living brain and nervous system of these patients.[59] Foerster's project aptly shows how a new conception of the active brain emerged at the intersection and superimposition of a variety of technologies: anatomical drawings of the brain, photographic atlases of brain sections, film footage of neurological patients, filmed operations with new electrical techniques, and photographs of patients' bodies covered with penciled marks and arrows denoting specific correspondences to the spinal cord.

Bodies and brains—and their interrelation—could be studied with, and imagined through, this array of newly superimposed technologies. Parallel to Foerster's assembled active nervous system of living patients, numerous other projects contributed to new imaginations of accessing active brains. Think, for example, of Karl Reicher's strange animation of brain slices as a *Kinematografie der Neurologie* in 1906, or Emile Cohl's fiction film that used animated sequences to show mental processes in *Le Retapeur des Cervelles* ("The Brain Corrector") in 1911.[60] Around 1913, physiologist Ivan Pavlov imagined a transparent brain with a moving spotlight indicating

---

[57] For example, in the work of Angelo Mosso, Charles Féré, Alfred Binet and Jules Courtier, Alfred Lehmann, and later Hans Berger; see R. D. Gillespie, "The Present Status of the Concepts of Nervous and Mental Energy," *British Journal of Psychology. General Section* 15, no. 3 (1925). Antoni Jan Milkulski and Eufemjusz Józef Herman, "Die Hirnpulsation Des Mensen Auf Grund Experimenteller Untersuchungen," *Zeitschrift für Neurologie und Psychiatrie* 90 (1924). Also see Otniel E. Dror, "Techniques of the Brain and the Paradox of Emotions, 1880–1930," *Science in Context* 14, no. 4 (2001).

[58] Roger Smith, "Representations of Mind: C. S. Sherrington and Scientific Opinion, c.1930–1950," *Science in Context* 14, no. 4 (2001): 530.

[59] Probing nerves and brains by electrical currents to test their functionality, subsequently cutting tissues that caused epilepsy and motor impairments. Guenther, *Localization and Its Discontents*.

[60] On Reicher's *"Kinematografie in Der Neurologie,"* *Verhandlungen der Gesellschaft Deutscher Naturforscher und Ärtze* (1908), see Flora Lysen, "Grey Matter and Colored

consciousness.[61] In the early decades of the twentieth century, it was not merely the ubiquity and variety of (media) technologies but especially their new intersections and juxtapositions—recalibrating back and forth between findings—that allowed for the correlative constitution of the active brain, the idea that the living brain could be perceived in action. By employing Foucault's term "correlative constitution," I emphasize that it is never technology alone that brings about such correlations, but always technologies that are actively brought together and *performed* in specific instances, configurations, and assemblages.

Conclusions

We return to the writer at her desk in 1920. Looking into her head, we see a brain that emerged through different nineteenth-century research pathways of correlating living bodies and brains. Relating the photomontage to the text in Hilton's book, we also recognize a discourse on working brains in their modern surroundings as part of a human sensorium that was constituted by telephones and wireless receivers in a media-saturated environment. This was a brain populated with active, "alive" elements, a working brain that could be related, as Hilton's volumes on business psychology did, to "live wires"—a term that denoted not only wires with electrical potential but also prodigious businessmen.[62] The image is but one example of a broader conception of the possibility of accessing the living brain that intensified, as I have recounted, with a range of twentieth-century (media) technologies to visualize cells and organs; to electrically probe the nerves and the cortex; to correlate mental activity with the active, living body; and to imagine the neurological human as always already mediated. It is this historical superimposition and

Wax," in *Textures of the Anthropocene: Grain, Vapor, Ray*, ed. Katrin Klingan, Ashkan Sepahvand, and Bernd M. Scherer (Cambridge, MA: MIT Press, 2015), 75.

[61] Ivan Petrovich Pavlov, *Die Höchste Nerventätigkeit (Das Verhalten) Von Tieren: Eine Zwanzigjährige Prüfung Der Objektiven Forschung; Bedingte Reflexe. Sammlung Von Artikeln, Berichten, Vorlesungen Und Reden* (München: Verlag J. F. Bergmann, 1926), 203. Cited in Hans Berger, *Über Die Lokalisation Im Großhirn: Rede Gehalten Bei Der Akademischen Preisverteilung Zu Jena Am 18. Juni 1927; Mit Einer Chronik Der Universität Für Das Jahr 1926/27* (Jena: G. Fischer, 1927), 36.

[62] "Not only is the living human body as a whole alive, but 'every part of it as large as a pin-point is alive, with a separate and independent life all its own; every part of the brain, lungs, heart, muscles, fat and skin.'" Hilton, *Applied Psychology: Psychology and Achievement*, 81. "Live wire," in *Applied Psychology: Initiative Psychic Energy. Being the Sixth of a Series of Twelve Volumes on the Applications of Psychology to the Problems of Personal and Business Efficiency* (New York: The Literary Digest for the Society of Applied Psychology, 1919), volume 6, 57–8.

intensification that allows me to call this new way of seeing and saying, around 1920, the emergence of the "live brain."

By pointing to the emergence of a live brain, I do not wish to argue that practices from disciplines, including nerve research, psychophysiology, chemistry, neurology, neurosurgery, psychiatry, psychotechnics, and educational psychology, were all converging toward a single point. Rather, I aim to describe the new way living bodies and brains were "matched up," to use Foucault's term, as part of a new discourse on modern subjects and with a new confidence in being able to see the brain in action. In the early decades of the twentieth century, this happened even when no well-defined correlates of mental activity had been experimentally discovered. As we will see in Chapter 3, the development of EEG (starting with Hans Berger's experiments in the 1920s and culminating in his first paper on EEG in 1929) is an important moment in the history of brainmedia, since it is here that long-established imaginaries of electrical brains met a new scientific discovery on nerves' electrical activity. Yet, the "live brain" is not the result of the invention of EEG; rather, the development of EEG must itself be understood in interaction with these new cultural configurations.[63] A new way of seeing and saying arose, an assemblage of practices and discourses that allowed for the birth of the live brain.

Hence, in the 1920s, new discourses that located the modern subject amidst a media environment of ubiquitous recording and transmission served the conceptualization of an active brain as "live." This new live brain was an amalgam of older imaginaries of nerve telegraph wires, vibratory particles, and novel media technologies. It was live in the sense that its activity was understood both to be captured and characterized by a new immediacy and modern sensibility. More than ever, it was a brain that was understood to function as a (media) technology—immersed in a back and forth between sensorium and environment. It was also *made perceptible* by media technologies. This live brain did not emerge from one new technology or discovery about the brain but must be viewed as the accumulated imaginary of a modern environment populated by media that changed the lived spatio-tempo-realities of daily life. It was at this moment that one could substantively start to imagine a working brain that acted in tandem with a working body as part of a working life—a live brain.

---

[63] See Borck, *Brainwaves*, 35–52.

2

Displaying Dynamic Brains: Illuminated Brain Models and the Enchanted Loom, 1928–38

The brain is waking and with it the mind is returning. It is as if the Milky Way entered upon some cosmic dance. Swiftly the head-mass becomes an enchanted loom where millions of flashing shuttles weave a dissolving pattern, always a meaningful pattern though never an abiding one; a shifting harmony of subpatterns.[1]

The "enchanted loom" might be the most cited and used metaphor for the nervous system in the past century. It pictures the active brain as a magical tapestry, a sparkling, spectacular cerebral display. Coined by eminent neurophysiologist Charles Sherrington in 1938, the metaphor has become a rhetorical commonplace in twentieth- and twenty-first-century neuroscience publications ever since. From discussions of computational networks (*Enchanted Looms: Conscious Networks in Brains and Computers*) to popularly oriented volumes on consciousness (*The Enchanted Loom: Mind in the Universe*), the glittering loom functions as a flexible icon that can be appropriated in different imaginaries of cerebral space.[2] Yet, with the passing of time, the historical context from which it emerged has been largely forgotten. In fact, Sherrington's loom is part of a broader history of scientists and science communicators who attempted to conceptualize and picture active cerebral processes and dynamically localized brain functions from the late 1920s onwards. In the 1930s, I will show, the enchanted loom, as part of this historical context, was proposed

---

[1] Charles Scott Sherrington, *Man on His Nature* (Cambridge: Cambridge University Press, [1940] 2009), 225 (printed version of the *Gifford Lectures 1937–1938*).
[2] Rodney Cotterill, *Enchanted Looms: Conscious Networks in Brains and Computers* (Cambridge: Cambridge University Press, 2000). Robert Jastrow, *Enchanted Loom: The Mind in the Universe* (New York: Simon & Schuster Trade, 1981).

as a scheme of the working brain that had a particular "engineering bent," as Sherrington put it.[3]

Sherrington's enchanted loom exemplifies the ebullient figurative language scientists employed (and still employ) to speak about the active mechanisms of the brain. I argue that the metaphor should be understood as having been particularly shaped by changing questions about the possibility of visualizing and localizing processes and dynamic functions in the brain, spurred by new model-making ideas and technologies and by new spheres for performing knowledge in the 1930s. Together, these developments allowed a new assemblage of imaginaries and practices to emerge: new brainmedia providing a novel conception of the active brain. Tracing such assemblages, I discern the rise of a "live brain" as an illuminated, engineered object that particularly befitted a modernizing urban environment.

This chapter focuses on two such assemblages that shape "live brains," which have not been studied before: the three-dimensional luminous brain model, which was an educational device developed in Vienna in 1931, and the "motograph brain," an analogy 1930s neurophysiologists used to picture a dynamic nervous system as a type of illuminated news ticker or message board. These two examples of blinking brainmedia linked illumination, technical media, broadcast media, and brain activity in new spheres to perform knowledges, that is, new spaces where science could be communicated to a broader public. My analysis of these two shows a shift in the conceptualization of nervous activity toward what I call the "logic of direct display." I argue that new display technologies allowed this new form of liveness to emerge, enabling an image and imaginary of direct yet technically mediated access to the brain's invisible, dynamic activities.

By analyzing the resonances of electrotechnology in the communication and presentation of nerve research around 1930, I contribute to scholarly analyses of reverberations between technological media and conceptions of human brains, constituting new live brains. Existing scholarship has predominantly focused on two eras and topics: nineteenth-century analogies between telegraph technology and human nerves, and the period starting in the 1940s, which saw a conceptual amalgamation between the information processing of early computers (logical calculators) and the work of the nervous system.[4] An important and cross-cutting observation

---

[3] Sherrington, *Man on His Nature*, 225. Sherrington's loom metaphor was often cited by, among other people, William Grey Walter in radio lectures and popular publications. See W. Grey Walter, *The Living Brain* (New York: W. W. Norton, [1953] 1963); "Enchanted Loom" (Great Britain: BBC Radio, 1948).

[4] Timothy Lenoir, "Helmholtz and the Materialities of Communication," *Osiris* 9 (1994); Christoph Hoffmann, "Helmholtz's Apparatuses. Telegraphy as Working Model of

from these histories is that machine–organism analogies can vary in their interpretation and signification; that is why they should always be studied up-close. The network analogy (which compared telegraphs to the nervous system and neuronal webs to telegraph networks), for example, was used to make different arguments about nerves and telegraphs depending on local discourses.[5] Taking heed of this, this chapter adds to the understanding of the much less studied interwar period, when new conceptions of cerebral function and structure were established.[6] I trace attempts to find "dynamic" visual imaginaries for the idea of a dynamically active brain and analyze how such practices and rhetoric helped negotiate new forms of technical mediation.

A number of cultural historians have described the pervasive influence of electrification around 1900. As Lauren Rabinovitz put it, it served as "a sensory synecdoche for the confluence of technology, excitement and modernity."[7] Important in this respect are Cornelius Borck's observations on that time, when he describes an interaction between electricity's material culture (through the electrification of everyday life) and its rhetorical and conceptual pervasiveness in cultural discourses, an electrical imaginary

---

Nerve Physiology," *Philosophia Scientiæ* 7, no. 1 (2003); Iwan Rhys Morus, "'The Nervous System of Britain': Space, Time and the Electric Telegraph in the Victorian Age," *British Journal for the History of Science* 33, no. 4 (2000); Rhodri Hayward, "'Our Friends Electric': Mechanical Models of Mind in Postwar Britain," in *Psychology in Britain: Historical Essays and Personal Reflections*, ed. G. C. Bunn, A. D. Lovie, and Graham Richards (Leicester: British Psychological Society, 2001); L. E. Kay, "From Logical Neurons to Poetic Embodiments of Mind: Warren S. McCulloch's Project in Neuroscience," *Science in Context* 14, no. 4 (2001). Claus Pias, "Elektronenhirn Und Verbotene Zone. Zur Kybernetische Okonomie Des Digitalen," in *Analog, Digital: Opposition Oder Kontinuum? Zur Theorie Und Geschichte Einer Unterscheidung*, ed. Jens Schröter and Alexander Böhnke (Bielefeld: Transcript, 2004).

[5] Otis, "The Metaphoric Circuit," 107. Florian Sprenger, for example, studies emerging telegraph technologies in the nineteenth century and points to a discrepancy between grand narratives of the nineteenth-century preoccupation with instantaneity and speed, and local discussions by engineers about whether the telegraph offered immediate transmission. Sprenger, *Medien Des Immediaten*, 21–5.

[6] On the lack of historical research for the period of the 1930s, see Justin Garson, "The Birth of Information in the Brain: Edgar Adrian and the Vacuum Tube," *Science in Context* 28, no. 1 (2015).

[7] Cornelius Borck, "Media, Technology and the Electric Unconsciousness in the 20th Century," in *L'ère Électrique: The Electric Age*, ed. Olivier Asselin, Silvestra Mariniello, and Andrea Oberhuber (Ottawa: University of Ottawa Press, 2011), 37; Lauren Rabinovitz, *Electric Dreamland: Amusement Parks, Movies, and American Modernity* (New York: Columbia University Press, 2012), 133. See also Killen, *Berlin Electropolis*; Killen, *Berlin Electropolis*; Christoph Asendorf, *Batteries of Life: On the History of Things and Their Perception in Modernity* (Berkeley: University of California Press, 1993); Anson Rabinbach, *The Human Motor: Energy, Fatigue, and the Origins of Modernity* (New York: Basic Books, 1990).

that "opened up a new space for imagined and explored electroorganic and electropsychic interactions."[8] Electrotechnology could thus advance, in Borck's words, as "a medium of ambivalences between the body and mind, the soul and society."[9]

In this chapter, I build on these broader observations, adding a new element of media-analytical specificity: active brains were conceptualized and represented through new ideas about, and devices for, electronic mediation developed in the interplay between scientists and science communicators. In describing such electrotechnological reverberations, the brain indeed emerges, around 1930, "as a piece of electric technology" (as Borck puts it), but it did so, I argue, through new brainmedia assemblages that involved the development of new electric displays.[10] Thus, the active brain could become conceptualized as a particular type of live brain: a brain as display.

A Feverish Image of the Brain Gone Mad: Electro-Brains and a Crisis of Representation

> Broadway, with its exceptional illumination at night, looks like the delirium of brains gone mad. Flashing signs, powerful light signals, wandering letters, moving figures on the edge of the roof, tubes of color, floodlights, airplanes with flares augmented by voices from loudspeakers, the megaphone, knocking sounds from the window panes: this supplants and complements our perception of the present.[11]

Writing in 1928, the German psychotechnical researcher Fritz Giese compared the illuminated cityscapes of his day with the delirious imagination (*Fieberbild*) of a raving brain. An assault on the senses, a "screaming in color," he wrote, "this is the mentality of this world!" In the 1920s, modern cities sparked both delirious imaginations by the brain and feverish images of the

---

[8] Cornelius Borck, "Electrifying the Brain in the 1920s: Electrical Technology as a Mediator in Brain Research," in *Electric Bodies: Episodes in the History of Medical Electricity*, ed. Paola Bertucci and Giuliano Pancaldi (Bologna: Università di Bologna, 2001), 263.
[9] Borck, *Brainwaves*, 77. Jeffrey Sconce offers the notion of a "logic of transmutable flow" to denote the intricate discursive and material assemblages of electricity, information, and consciousness that emerged in the early decades of the twentieth century. Jeffrey Sconce, *Haunted Media: Electronic Presence from Telegraphy to Television* (Durham, NC: Duke University Press, 2000), 8.
[10] Borck, "Electrifying the Brain in the 1920s," 263.
[11] Fritz Giese, *Psychotechnik* (Breslau: Ferdinand Hirt, 1928), 105. I thank Max Stadler for pointing me to this passage. Translations from non-English languages are my own, unless otherwise noted.

brain. As Cornelius Borck has expounded, images of brains and illuminated cities intertwined; the electrification of everyday life spurred analogies of the body and the brain as "bioelectrical-media-technical hybrids" through which ideas of mental energies and electric currents were newly amalgamated.[12] Today, the best known examples of 1920s hybrid brainmedia are perhaps the educational illustrations of Fritz Kahn, which integrated the functions of the brain in modern industrial worlds populated with telephone and radio networks, screening rooms, and railway stations, creating images of particularly "urban brains," as Borck has argued, within the space of the hectic, modern city (Figure 2.1).[13]

Kahn's work is exemplary of the way the mediated, electrified metropolis shaped a particular interpretative potential for thinking about the mechanisms of the brain and the way that urban space was itself presented as a nervous system that served the transmission of electricity—completing the metaphorical circle.[14] Yet, looking at Kahn's images, this potential must be understood mainly on the level of visual hyperbole. His metropolitan brain analogies were attractive because they integrated the brain with attractive, everchanging technologies.[15] The underlying idea was still a simple stimulus–response circuit, however, not a whole brain network of innervation but a focus on elementary, electrotechnical systems of excitation—a view that had predominated the focus of nerve researchers until the mid-1920s.[16]

Kahn's transparent brains were part of an international public sphere of science education in the United States and Europe, communicated through new popular science magazines, newspaper sections, radio broadcasts, and health exhibitions, supported by a new professional sphere of science educational expertise and new forms of mediating science.[17] Cross-sections

---

[12] Cornelius Borck, "Urbane Gehirne: Zum Bildüberschuss Medientechnischer Hirnwelten Der 1920er Jahre," *Archiv fur Mediengeschichte* 2 (2002): 272.
[13] Ibid.
[14] Ibid., 261.
[15] This hyperbole is what Borck describes as the "pictorial excess" (*Bildüberschuss*) of media-technical brain worlds. Ibid., 264. On Kahn's approach to illustration, see Michael Sappol, *Body Modern: Fritz Kahn, Scientific Illustration, and the Homuncular Subject* (Minneapolis: University of Minnesota Press, 2017).
[16] About the predominance of the 1920s stimulus–response model, see Katharina Schmidt-Brücken, *Hirnzirkel: Kreisende Prozesse in Computer Und Gehirn: Zur Neurokybernetischen Vorgeschichte Der Informatik* (Bieleveld: Transcript Verlag, 2014), 137.
[17] Arne Schirrmacher, "Introduction: Communicating Science: National Approaches in Twentieth-Century Europe," *Science in Context* 26, no. 3 (2013). Sybilla Nikolow, ed., *Erkenne Dich Selbst! Strategien Der Sichtbarmachung Des Körpers Im 20. Jahrhundert* (Köln: Böhlau, 2015). Bernadette Bensaude-Vincent, "In the Name of Science," in *Science in the Twentieth Century*, ed. John Krige and Dominique Pestre (Amsterdam: Harwood Academic, 1997).

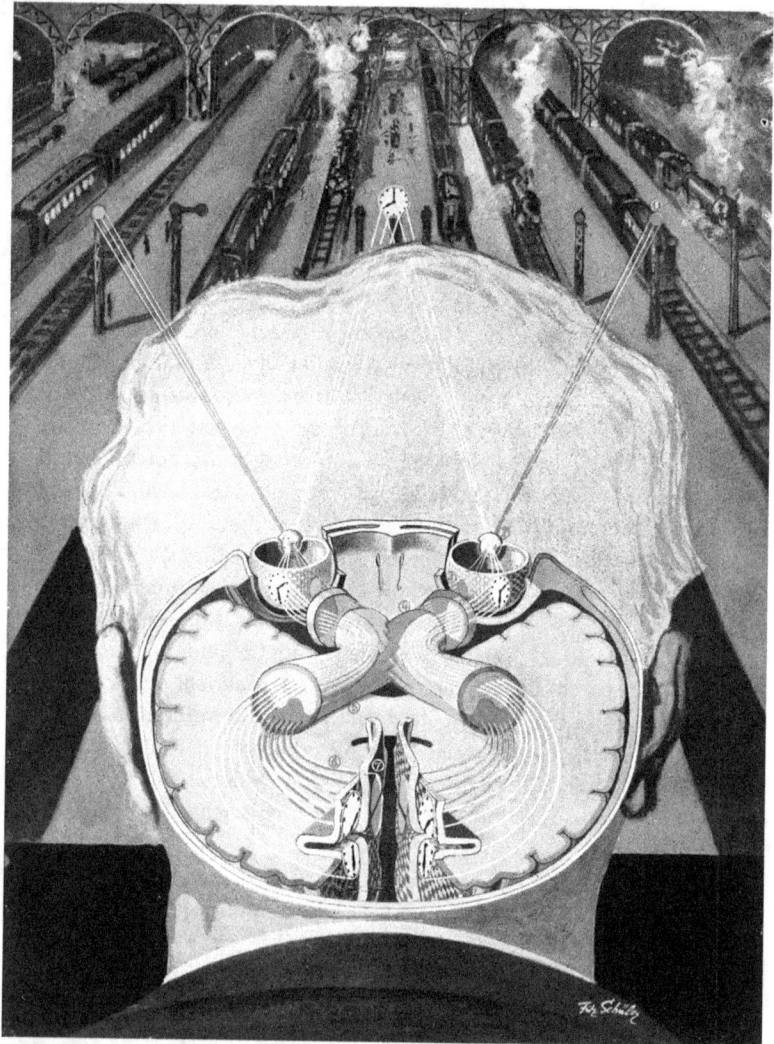

Figure 2.1 Fritz Kahn, *Die Lichtwarhnehmung*, 1929 (illustration).

of heads and transparent brains featured in popular articles and books about popular psychology, psychotechnics, and new scientific research into the brain and often did so with a particular promissory ring. Writing about contemporary research into nerve cells for the popular *UHU* magazine in 1928, for example, Kahn predicted that it was only a "question of time and technological progress" before one could use an X-ray microscope to see through the skulls of living people and read their knowledge, memories, and experiences "directly and un-deceptively."[18]

Images of transparent brains connected to a rising materialism in 1920s popular science that imagined human nature and character as chiefly determined by hormones and the structure of the nervous system.[19] Yet, any rift between popular and professional science should not be overstated; there were no confidently materialist utterances of popular educators, on the one hand, and hesitant statements of brain scientists regarding brain–mind or brain–behavior relations, on the other: the situation was more ambiguous, with both promissory and ambivalent tones being struck across various spheres of knowledge regarding the potential cerebral basis of behavior and the mind.[20] As Roger Smith has expounded, the unfinished and "incoherent" state of knowledge about the nervous system was in itself a topic of particular consideration in the early decades of the twentieth century.[21]

This "incoherence" of new knowledge on the nervous system and how it could be interpreted and mediated became pressing in the 1920s. Neurologists from various countries noted that the scientific basis of functional localization (particular locations in the brain corresponding with particular human behaviors) was weak. Different brains showed functions at different positions, and the brain's functional localization could change over time. While some of these inconsistencies had already been voiced in the late nineteenth century, in the early decades of the twentieth century these problems became more pressing.[22] By then it had become commonplace to dismiss nineteenth-century schemes of localization based on the correlative research of cerebral lesions in patients with aphasia, for example. In 1920, British neurologist Henry Head looked back at the late nineteenth-century desire to localize and map brain function, describing how "the rage for

---

[18] Fritz Kahn, "Wie Arbeitet Das Gehirn?" *UHU* 11 (1928).
[19] Peter J. Bowler, *Science for All: The Popularization of Science in Early Twentieth-Century Britain* (Chicago: University of Chicago Press, 2009), 50.
[20] Roger Smith, "Physiology and Psychology, or Brain and Mind, in the Age of C. S. Sherrington," in *Psychology in Britain: Historical Essays and Personal Reflections*, ed. G. C. Bunn, A. D. Lovie, and G. D. Richards (Leicester: The British Psychological Society, 2001).
[21] Ibid., 237.
[22] John T. MacCurdy, "The General Nature of Association Processes within the Central Nervous System 1," *British Journal of Psychology. General Section* 22, no. 2 (1931); L. S. Jacyna, *Lost Words: Narratives of Language and the Brain, 1825–1926* (Princeton, NJ: Princeton University Press, 2009), 103–7. Harrington, *Medicine, Mind, and the Double Brain*, 260–8.

diagrams became a veritable mania. Each author twisted the clinical facts to suit the lesions he had deduced from his pet schema."[23]

Instead, Head, but also scientists such as Kurt Goldstein in Germany and Karl Lashley in the United States, suggested there were "dynamical aspects" to recuperating function after brain damage ("vicarious function"); the brain seemed able to adapt and reorganize in ways that could not be explained by rigid localization.[24] In this situation, the term "dynamic" had a particularly flexible interpretative ring: it suggested interacting components and changing spatiotemporal relations of function. Already in the nineteenth century (as mentioned in Chapter 1), Charcot and associated researchers had used the adjective *dynamique* to suggest the existence of "dynamic lesions" in the brains of hysterical subjects that may or may not be registerable in the body's posture and gestures. By the 1920s, scientists in a number of European countries and in the United States, including Goldstein, Lashley, and Head, but also neuropathologist Constantin von Monakow and neurologist Pierre Marie (Charcot's former pupil and coauthor of the female writer image mentioned in Chapter 1), were influenced by a "holistic" approach to science in which terms like "dynamic"—but also "plasticity" and "regeneration"— were part of a vocabulary that emphasized wholeness and positioned itself opposite the increasing influence of a mechanistic, fragmented modern life.[25]

This dynamic conception of the active brain suggested that the brain's structure and function were variably related—through what Lashley called "patterns" or a "field theory" of brain activity—yet the word dynamic did not precisely explain how this worked in practice.[26] While previous theories had proposed the close correspondence of structural and functional units and the specialization of brain areas or cells, now scientists argue that notions

---

[23] Henry Head, "Aphasia: An Historical Review: The Hughlings Jackson Lecture for 1920," *Proceedings of the Royal Society of Medicine: Section of Neurology* 14 (1921): 396.

[24] "Dynamical aspects," in Karl Spencer Lashley, "Basic Neural Mechanisms in Behavior," *Psychological Review* 37, no. 1 (1930): 12. "Vicarious function," in K. S. Lashley, "Studies of Cerebral Function in Learning. IV. Vicarious Function in Destruction of the Visual Areas," *American Journal of Psychology* 59, no. 1 (1922).

[25] About holism in relation to neurology in the interwar period, see Anne Harrington, "Metaphoric Connections: Holistic Science in the Shadow of the Third Reich," *Social Research* 62, no. 2 (1995), and *Reenchanted Science: Holism in German Culture from Wilhelm II to Hitler* (Princeton, NJ: Princeton University Press, 1999). L. S. Jacyna, "Questions of Identity: Science, Aesthetics, and Henry's Head," in *Greater Than the Parts: Holism in Biomedicine, 1920–1950*, ed. George Weisz and Christopher Lawrence (Oxford: Oxford University Press, 1998).

[26] Nadine Weidman notes that Lashley was influenced by gestalt theorists such as Kurt Koffka and Wolfgang Kohler, but that his holism was of a particular kind, as he refrained from speaking of an "organism as a whole" or of vitalist concepts. Weidman, *Constructing Scientific Psychology*, 44.

of cerebral organization and brain activity are in fact hardly understood. Accordingly, various scholars described the state of neurology research as being in severe crisis. Brain research, as one researcher put it, was like an unfinished building; "we have placed numerous blocks alongside each other, but the synthesis toward a complete, successful building is still missing."[27]

A Glow-in-the-Dark Brain from Vienna

The example of the luminous brain model (*das Leuchtende Gehirn*), built in Vienna in 1931, serves as a case study to examine how scientists and science educators tried to navigate their ambiguous situation: they both wanted to draw public attention to brain science and respond to an uncertain "dynamic" conception of how function was structured in the brain.[28] In terms of attention, the luminous brain was certainly successful: from the United States to Australia, newspapers described it as a "record achievement," "a monster globe" eight times the size of a real human skull, radiating "blue, green, crimson, purple, pink, and yellow lights, an orgy of sparkling colour."[29]

Designed by a team of three professionals—an engineer who worked for Vienna's Technical Museum, a neurologist, and a psychiatrist—it was patented as an educational device, "an improved type of anatomical model" with a switchboard that could illuminate various fluorescent tubes representing anatomical elements inside the brain (Figure 2.2).[30] The luminous brain functioned in a context of popular science education that was unique to Vienna, where science had become a prevalent topic for newspapers and radio, and where the effects and aims of science education were a topic of

---

[27] A. Jakob, "Die Lokalisation Im Grosshirn," *Klinische Wochenschrift* 10, no. 44 (1931): 2025. In the case of Weimar Germany, Anne Harrington describes this crisis atmosphere in neurology as part of a more general preoccupation with crisis in various disciplines, a general movement against mechanistic or machinic explanations, and a turn to phenomenology. Anne Harrington, "Kurt Goldstein's Neurology of Healing and Wholeness: A Weimar Story," in *Greater Than the Parts: Holism in Biomedicine* (Oxford: Oxford University Press, 1998).

[28] To my knowledge, the luminous brain model is presently lost.

[29] "A Luminous Brain. Nerve Specialist's Invention," *West Australian*, February 12, 1937. "World's First Luminous Brain Model Made by Woman Doctor," *The Mail*, July 31, 1937.

[30] A line drawing in the international patent reveals its structure: iron wires formed a basic outline of the cerebrum, while various inside anatomical elements of the brain were represented by fluorescent tubes. Following color conventions in medical textbooks, the model used individual colors to illuminate eleven different elements ranging from the olfactory bulb, motor nerve, and sensory nerve to the nucleus dentatus and corpus callosum. Edith Klemperer, "Anatomical Model" (1934), http://www.google.nl/patents/US1951422.

58 *Brainmedia*

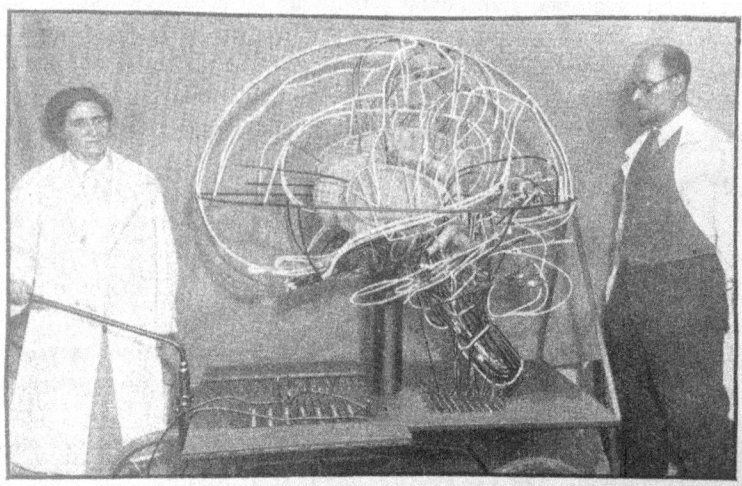

Dr. Edith Klemperer and Dr. Robert Exner, Vienna, who made glass model of brain

Figure 2.2 Dr. Edith Klemperer and Dr. Robert Exner, luminous brain model from Vienna, c. 1931 (photograph).

vivid discussion.[31] Mounted on wheels to make it transportable, the model toured various *Volkshochschulen*, voluntary associations that allowed the Viennese middle class to actively engage with science.[32] In these spaces, it served as a centerpiece for a variety of lectures on brain science by its two main creators: Edith Klemperer, physician at the Psychiatrischen und Nervenklinik of the University of Vienna, and neurologist Robert Exner. (It is beyond the scope of this study to recount the strikingly diverging histories of these two scholars. The Jewish physician Klemperer would flee Vienna in 1939 after Austria was annexed by Nazi Germany, while Exner would join the National Socialist Party already in 1933, when it was still illegal in Austria.[33])

[31] Ulrike Felt, "Science and Its Public: Popularization of Science in Vienna 1900–1938," in *Quand La Science Se Fait Culture: La Culture Scientifique Dans Le Monde. Actes I*, ed. Bernard Schiele, Michel Amyot, and Claude Benoit (Sainte-Foy: Editions MultiMondes/ Université du Québec à Montréal/Centre Jacques Cartier, 1994).

[32] Newspaper announcements prove the model was shown at various Volkshochschulen, including Gesellschaft der Ärtze, the Anatomisches Institut, the Österreichische Volkshochschule, Volksheim Ottakring, Leo-Vorein, the Österreichischen Klub, and the Technical Museum (one employee of the latter, engineer Joseph Nagler, had been codeveloper of the model). See also Klemperer's account in "Die Schöpferin Des Ersten Wiener Gläsernen Gehirnmodells," *Neues Österreich*, February 6, 1953.

[33] On Edith Klemperer, see Lazaros C. Triarhou, "Women Neuropsychiatrists on Wagner-Jauregg's Staff in Vienna at the Time of the Nobel Award: Ordeal and Fortitude," *History*

Displaying Dynamic Brains

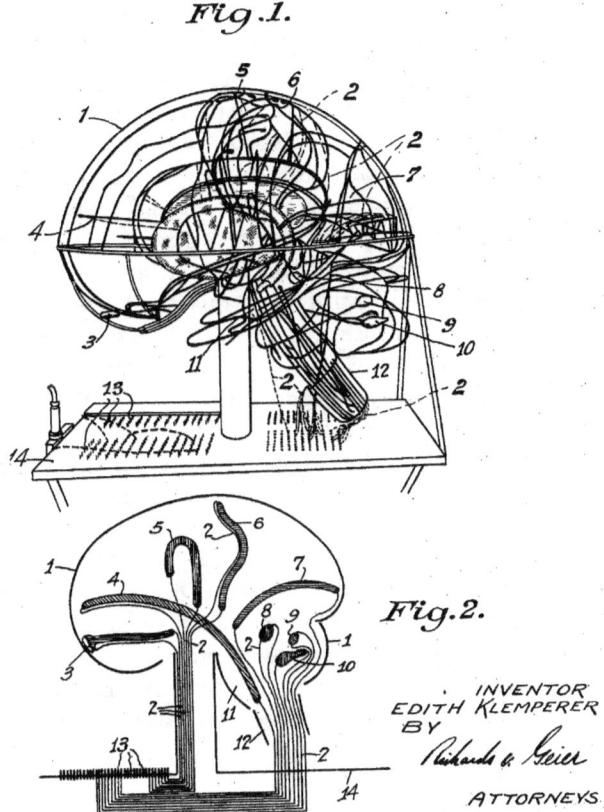

Figure 2.3 Edith Klemperer, patent of anatomical model, 1931/4 (patent drawing).

Advocating the device in neurological and medical journals, Klemperer and Exner invoked new insights from perceptual psychology in describing the model as a great aid to students' "mental image" of the makeup of the brain; its three-dimensionality and illumination would help them gain

of Psychiatry 30, no. 4 (2019). On Robert Exner, see Deborah R. Coen, *Vienna in the Age of Uncertainty: Science, Liberalism, and Private Life* (Chicago: University of Chicago Press, 2008). Fleeing Vienna, Klemperer took the luminous brain model with her to New York. In 1953, Klemperer wrote a letter to a Viennese newspaper objecting to an article about a new version of the brain model built by Robert Exner and Joseph Nagler in 1952. The latter two failed to mention Klemperer as the main inventor of the original 1931 model. See "Die Schöpferin Des Ersten Wiener Gläsernen Gehirnmodells."

clear and vivid knowledge (Figure 2.3).[34] The makers emphasized the model's ability to "impress" (it would be "unforgettably impressive," or *einprägsamer*, "more impressive"), thereby presenting it as both attractive and pedagogically effective, eye-catching, and instructive.[35] As such, it could live up to the Viennese intellectual circles' discussions on psychotechnics, perceptual psychology, and effective visual education.[36] Through illuminated transparency, it evoked the multiple dimensions of what science educational discourses had called *Anschaulichkeit*: a type of perceptibility that effected greater comprehension—"clarity"—through the vividness or liveliness of seeing a living example, a vivacity that would attract and impress viewers.[37]

By emphasizing perceptual instructiveness and the ingenuity of technical illumination, the model makers warded off potential criticism of the model's spectacular looks. In a later account, Exner would write that the "educational model was not meant be a Homunculus nor a fun fair attraction, but an image that the student can take in and build up in their own brain."[38] In evoking this nexus of popular appeal—perception-scientific validity and educational value—the luminous brain, which one newspaper dubbed the "Glass Brain," was akin to the better-known "Glass Man" model produced for the international traveling hygiene exhibitions by the Dresden Hygiene Museum in Germany, internationally hailed for its beautifully colored organs that could light up sequentially.[39] Sequenced illumination in educational displays especially also enabled animation, a type of stimulation that was conceived in relation to a newly envisioned urban spectator of scientific knowledge.

---

[34] Robert Exner, "Das Leuchtende Gehirnmodell," *Psychiatrisch-Neurologische Wochenschrift* 35 (1933). Edith Klemperer, "Demonstration: Das Gehirnmodell, Ein Plastischer Beleuchteter Unterrichtsbehelf Zur Darstellung Der Einzelnen Funktionen," *Zentralblatt fur die gesamte Neurologie und Psychiatrie* 61 (1932): 499. One of the first official appearances of the model was at the First International Neurological Congress in Bern in 1931.

[35] "Unforgettable impression" (*unvergesslich einzuprägen*), in "Kurzkurs 'Das Leuchtende Gehirnmodell,'" *Mitteilungen der Volkshochschule Wien Volksheim* 8, no. 2 (1935). "*einpragsamer*," in Exner, "Das Leuchtende Gehirnmodell."

[36] Exner was one of many Viennese scientists who combined his interests in psychology, physiology, and perception studies (predominantly color perception) with interests in effective science teaching and demonstration. On the development of Exner's ideas, see Coen, *Vienna in the Age of Uncertainty*. Janet Ward mentions the presence of psychotechnics in Vienna through the advertising office started by Edward Bernays. Janet Ward, *Weimar Surfaces: Urban Visual Culture in 1920s Germany* (Berkeley: University of California Press, 2001), 101, n. 42.

[37] On *Anschaulichkeit*, see Christian Stifter, "'Anschaulichkeit' Als Paradigma. Visuelle Erziehung in Der Fruhen Volksbildung, 1900–1938," *Spurensuche* 14, nos. 1–4 (2003).

[38] Robert Exner, "Das Gehirnmodell," *Annalen des Naturhistorischen Museums in Wien* 59 (1953).

[39] "A Glass 'Brain' Aids Medical Students," *Maitland Daily Mercury*, March 29, 1932.

Animated Brains for Modern Citizens

Around 1930, sequenced illumination was advocated as a way to combine attraction and scientific education, and especially to successfully integrate scientific visualizations within a specifically modern urban environment. Writing in a professional journal for artificial lighting, *Das Licht*, scholar Werner Lincke considered this new light educational method (*Lichtlehrmethode*) not only a wonderful method to "enliven" (*beleben*) "dead pictures" or "dead statistics" (just like animation films did, he notes) but also a necessary technology in a new economy of public attention.[40] Inhabitants of the modern metropolis were constantly bombarded by illuminated advertisements; they were "estranged from the quiet tranquility necessary to appreciate numbers and words" and had learned "to seclude themselves from the obtrusive impressions from outside."[41] Because this modern city dweller no longer reacted to simple representations and words, exhibition makers had to employ moving, illuminated, and changing images not only to leave a more lasting impression but also to exert a "greater incentive to take a look, even if the object in question has no initial interest for the exhibition visitor."[42] Vienna's glass brain belonged to this modern culture of attractive illumination. One newspaper suggested that the model's creators must have been influenced by a stroll on Vienna's Kärtnerstrasse, with its glorious advertising displays.[43]

Illuminated models were part of a new material culture of illuminated displays that included urban shop windows and facades, but also novel approaches to exhibition making and science pedagogy that advocated "dynamic displays," a type of animation that could excite the overstimulated urban citizen.[44] Things needed to move, twinkle, swirl, and make sound; a lively clarity was considered vital to attract and retain visitors' attention. Reasoning in this vein, exhibition makers argued that illumination controlled by buttons would be particularly attractive. When visitors themselves could choose to highlight parts of models or display cases, this would stimulate

---

[40] Werner Lincke, "Lehrmeister Licht," *Das Licht. Zeitschrift für praktische Leucht- und Beleuchtungs- Aufgaben* (1930): 127. I thank Max Stadler for reference to this source.
[41] Ibid.
[42] Ibid.
[43] See "World's First Luminous Brain Model Made by Woman Doctor."
[44] On the development of international exhibition languages of "dynamic displays," see Karen A. Rader and Victoria E. M. Cain, *Life on Display: Revolutionizing U.S. Museums of Science and Natural History in the Twentieth Century* (Chicago: University of Chicago Press, 2014), 100; Erin McLeary and Elizabeth Toon, "Here Man Learns about Himself," *American Journal of Public Health* 102, no. 7 (2012).

their natural "urge to play" (*Spieltrieb*).⁴⁵ At the celebrated 1930 Dresden Hygiene exhibition, push-button illumination served to enliven a diagram of the brain's functional nerve centers (Figure 2.4). Pressing a button with a photograph of a mouth would light up the corresponding brain area for speaking on a transparent brain diagram (Figure 2.4).⁴⁶ In another traveling exhibition, a giant three-dimensional brain model speckled with buttons allowed corresponding photographs to light up on an adjacent board.⁴⁷ However modern this light educational method seemed to be, when it came to conceptualizing the active brain, these strategies actually only enlivened a long-established idea of functional localization in the brain through its transparent diagrams and visual juxtapositions. In fact, the push-button diagram exhibited in Dresden was directly based on the 1888 image of the writing brain (produced by Marie and Charcot, as mentioned in Chapter 1). Above the button indicating the "writing" center is a photo of a hand holding a quill pen.

In contrast, the luminous brain model was not meant to serve a theory of rigid specialization. The model's 1931 patent description shows that the colored light tubes inside the metal skull indicated only a few anatomical structures such as the pons, the olfactory bulb, and the motor and sensory areas. Yet, an enthusiastic Austrian *Reichspost* review of the model's *Demonstrationsvortrag* suggested that through illumination the model had indicated *processes* arising from light and sound sensations in a "lively (*lebendig*) and understandable (*übersichtlicher*)" manner.⁴⁸ The short review indicates the interpretative potential of (sequenced) illumination: intuitively, it was thought to reveal temporal processes in the brain. Sequenced illumination had started to be understood as a strategy to represent active, living processes—a way of seeing the brain in action. The perceived liveliness of the model's colored, glowing activity was thus understood in multiple ways, not only as more attractive and educational but also as having more access to the living, dynamic, temporally active brain.⁴⁹ Indeed, the model makers projected, an improved version of the illuminated model with adjustable lighting would be able to present an entirely new spatiotemporal image of the living, active brain.

---

[45] Lincke, "Lehrmeister Licht," 125.
[46] Ibid.
[47] "Tentoonstelling 'De Mensch,'" *De Telegraaf*, August 31, 1935.
[48] "Das Leuchtende Gehirn," *Reichspost*, June 23, 1932. It was this capacity for lively attraction (and not the comprehensiveness of the anatomical information) that spurred the conclusion that the model would be "outstanding" (*hervorragend*) for educating doctors.
[49] Ibid.

Displaying Dynamic Brains   63

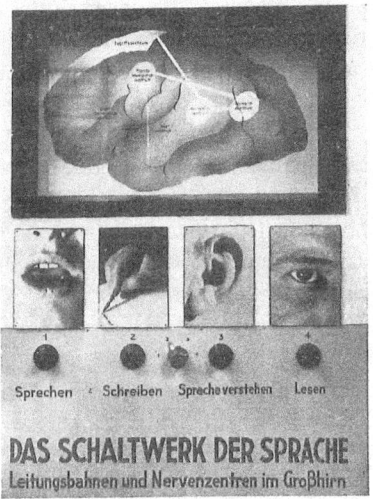

Figure 2.4 Push-button brain diagram, *c.* 1930 (photograph).

Searching for a Dynamic Image: Between *Gesamtbild* and *Gehirnwahrheiten*

The luminous brain ensured popular science educators the attention and attraction of spectators in darkened lecture rooms. In the future, the creators proposed, this first "dynamic" prototype (better than any previous "static" model) should become even more dynamic. They imagined a model in which the lecturer could select the specific degree and duration of illumination, thus enabling the demonstration of time-based information about brain "processes" in slow motion, superimposed onto the brain's morphology.[50] With their remarks on a "plastic image" and "dynamic relations," the luminous brain makers connected this modern electrotechnological device to the most up-to-date contemporary brain

---

[50] Exner, "Das Leuchtende Gehirnmodell"; Klemperer, "Demonstration: Das Gehirnmodell, Ein Plastischer Beleuchteter Unterrichtsbehelf Zur Darstellung Der Einzelnen Funktionen," 499.

research, which had started to show the uncertainties of any rigid spatial mapping of brain function onto brain anatomy. Illumination thus offered a potently ambiguous interpretability that served this uncertainty. Accounts of the model referenced the work of well-known Viennese neuropsychiatrist Otto Pötzl, who was much influenced by holistic perspectives in science and interested in the "dynamical processes" through which neurological patients recovered from brain lesions that proved the integrated or distributed nature of brain function.[51]

If we situate the luminous brain project and its imagined future model in relation to the Viennese context of neurological research around 1930, what emerges are contested opinions about making maps of the functioning, active brain. This was a negotiation between a more rigid anatomical schematism, on the one hand, and doubts about how to represent the adaptability of cerebral regions, on the other. Two Viennese brain researchers who both supported the building of the luminous brain characterize the different sides: Otto Pötzl and nerve cell researcher Constantin von Economo. Following the influence that these two different professors had on the luminous brain reveals a turn in the practice of making diagrams and mapping the brain.

A proponent of a distributed and dynamic conception of brain function, Pötzl had been a supporter of the project early on and even promised to finance the expensive model.[52] During the model's construction, however, he withdrew his backing, as he thought the model did not live up to scientific standards and would (as one of the model makers recounted) "move further away from the construction of a real brain."[53] In his own 1928 textbook on aphasia (which included patient observations and cerebral lesion research), Pötzl wrote he had "explicitly avoided" putting his findings on lesions into a drawn topography or a more "general formula."[54] Influenced by a holistic approach to the functioning of the nervous system, Pötzl (an early collaborator of gestalt psychologist Max Wertheimer) argued that readers should draw out the more general formula themselves, and that this process—the movement

---

[51] Otto Pötzl, "Über Die Rückbildung Einer Reinen Wortblindheit," *Zeitschrift für die gesamte Neurologie und Psychiatrie* 52, no. 1 (1919): 265.

[52] Exner recounts Pötzl's involvement (a promise of 2,000 Schilling) in Exner, "Das Gehirnmodell."

[53] To my knowledge, Pötzl's critique of the model has not been archived, yet from the reactions of the luminous brain makers, we can deduct that the model could not live up to his vision of dynamic relations in the central nervous system. See ibid.

[54] Otto Pötzl, *Die Aphasielehre Vom Standpunkte Der Klinischen Psychiatrie* (Leipzig: F. Deuticke, 1928), 231. Pötzl's research offered detailed descriptions of individual patient cases and lesion studies, yet nowhere in the book did he provide an averaged image or diagram summing up these individual cases.

from complexity toward an *Gesamtbild* (overall picture) of the brain—was in fact cerebropathology's central conundrum at the time.[55]

As Otto Pötzl wanted to avoid a static *Gesamtbild* of the brain, instead emphasizing variability and plasticity, such aspects were clearly not well served in the luminous brain's construction of fixed iron wires and glass tubes. Instead, the model was more akin to the colorful anatomical plasters by von Economo, another well-known Viennese professor whose maps and models were an inspiration for the luminous brain (Figure 2.5).[56] Produced in the mid-1920s and made from detachable parts, his models were based on his research on the cortex's different cell types (cytoarchitectonics), which he had meticulously mapped out in 170 different areas on the cortex surface. Though these three-dimensional plaster models were focused on representing cell anatomical information, von Economo also used them to make claims about possible correlations between areas with specific cell types and particular mental functions. For lectures at international venues like the New York Academy of Medicine in 1929, von Economo used his "brain casts" (along with his encephalometric research) to lecture on the brains of exceptional people, conjecturing about localizing musical talent, for example, or a particular visual ability.[57]

Von Economo's lectures are characteristic of the persistence of the functional localization paradigm in the 1920s. In various research communities, localization remained an important conceptual background, that is, a final end toward which different types of research, such as histology (the study of cell types) and electrocortical mapping (stimulating the cortex of living patients with electrodes), were compared with and superimposed on each other.[58] As historian Michael Hagner has argued, the early twentieth-century "calibrated" cartographies (such as Wilder Penfield's famous maps of sensory motor functions) were extremely attractive, not only as more

---

[55] Citing Goethe, he described the problem, "simply a part of nature, ... it is easier than we understand, and at the same time more complicated than we can comprehend" (ibid., 232). It is beyond the scope of this chapter to delve deeper into the holistic approach to mapping, but clearly Pötzl had been influenced by a particular interpretation of wholeness and Goethian science. Harrington, *Reenchanted Science*.

[56] Exner, "Das Gehirnmodell."

[57] Constantin Von Economo, "Some New Methods for Studying Brains of Exceptional People (Encephalometry and Braincasts) (Presentation with Models and Demonstration at the New York Academy of Medicine, Section on Neurology, December 3, 1929)," *Journal of Nervous and Mental disease* 72, no. 2 (1930).

[58] Katja Guenther gives the example of the exchange between Otfrid Foerster's new epilepsy maps and Oskar Vogt's cytocortical research. Guenther, *Localization and Its Discontents*, 110. Michael Hagner argues that the conceptual course of brain research was firmly entrenched in mapping function onto standard anatomies. Michael Hagner, "Lokalisation, Function, Cytoarchitektonik. Wege Zur Modellierung Des Gehirns," in

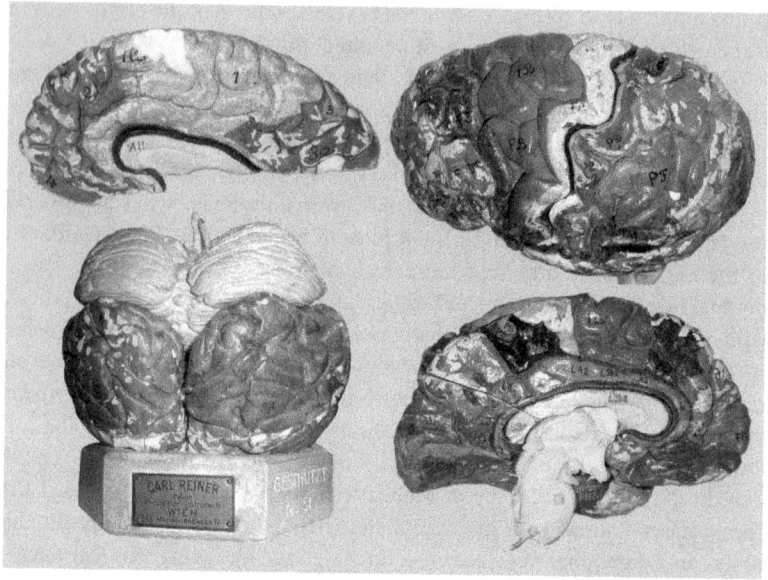

Figure 2.5 Constantin Von Economo's plaster models of cytoarchitecture, c. 1927 (photograph).

precise aids for brain surgery (for epilepsy patients, for example) but also as representations of the "psychic apparatus" itself.[59] Functional maps, because of their authoritative epistemic status, could embody and enact what Hagner calls "*Gehirnwahrheiten*" (brain truths) about the localization of physiological and mental functions.[60]

The example of von Economo, lecturing with his colored plaster brains on stage, illustrates the performative potential of striking material models. It is in using a model, in the action of its display, that its epistemic powers are most forcefully exercised. It is in their interaction between model, scientists, and audience that new brainmedia most successfully establish "brain truths." The luminous brain similarly utilized the performative strength of models during lectures, augmenting these visual-rhetorical powers even more by

---

Objekte, Differenzen Und Konjunkturen: Experimentalsysteme Im Historischen Kontext, ed. Hans-Jörg Rheinberger, Bettina Wahrig-Schmidt, and Michael Hagner (Akademie Verlag, 1994), 147.

[59] Hagner, *Der Geist Bei Der Arbeit*, 270.
[60] Ibid.

Displaying Dynamic Brains 67

the attraction and "impressiveness" of colored illumination. Yet, this model's objective was not to claim one persuasive brain truth; the makers emphasized its flexibility—more parts and different types of lighting could be added to adapt to the needs of changing research insights. Indeed, during lecture demonstrations in various *Volkshochschulen*, Exner and Klemperer used the illuminated model as a flexible backdrop for a variety of lectures (serving their different research interests and backgrounds) about the "achievements of the brain" and about human psychology.[61]

Ultimately, the luminous model evinces the multiple ambiguities that needed to be navigated in the late 1920s and 1930s. There were the scientific dissonances between spatial localization versus a dynamic and variable structure–function relation, and there was the question of the model's scientific truthfulness versus its objective of impressing audiences. In both cases, the makers invoked the technology of sequenced illumination to suture these debates.

> The current light-technical means allow the truthful representation of every neurological research finding, because a model does not have to be a copy of the real central nervous system, so a type of "Homunculus." [Instead,] its character is to represent the image that we make as a result of our research, and of this often only a selection for specific teaching and learning purposes.[62]

Sequenced illumination was both scientifically flexible (able to associatively and promissorily gesture to "dynamism" and "plasticity") and pedagogically sound (impressive but emphasized as never being mere spectacle). This rhetorical potency of new illumination technologies tied in with the importance of (experimenting with) novel ways and spheres of demonstrating science, allowing expertise to be drawn from illuminated advertising in shopping streets as well as traveling science exhibitions.

In the next section, I analyze another 1930s instance when new illuminated technology—in this case a modern illuminated news ticker, a so-called motograph—served the imagination of mechanisms in the living brain. In the 1930s, this vehicle for imagining "dynamic" activity was strengthened by a new conception of direct, "televisual" access to the nerves' electrical pulses.

---

[61] Announcements suggest the luminous brain was used as a visual centerpiece for a variety of lectures. In 1935–6, for example, Edith Klemperer gave a three-part lecture series about "Das leuchtende Gehirnmodell. Die Leistungen des menschlichen Gehirns" at Volksheim Ottakring. "Kurzkurs 'Das Leuchtende Gehirnmodell.'"
[62] Exner, "Das Leuchtende Gehirnmodell."

This, I argue, served one particular interpretation of the illuminated display analogies that were employed, that of immediate mediation, or direct display.

Streaming Headlines and the Motograph Brain

The rhetoric of crisis and uncertainty regarding functional localization research that I sketched in the first part of this chapter is arguably most visible in the work of the American physiologist Karl Lashley. Time and again, he emphasized that both experimental evidence of animal lesion studies and observations of recuperating neurological patients demonstrated that it was impossible to draw any conclusions regarding rigid locations in the brain dedicated to particular human functions, sensations, or behaviors.[63] Lashley and other scientists thus attempted to find concepts and imaginaries that fitted their plastic and dynamic view of brain function. To address the "fundamental problem of neural integration" (the way structure and function interact), he resorted to a rather obscure vocabulary, explaining integration as functioning through a "localization of fields" within which were "schemata" that were conditioned by "dynamic patterns of organization."[64] Other nerve researchers similarly struggled with dynamic conceptions of structure and function. Cambridge psychologist John MacCurdy thought up "anatomical designs" that were four-dimensional (including time) and determined by "immaterial patterns."[65] MacCurdy's little graphic diagrams depicted single dots building up larger patterns through the rapidly forming and deforming actions of wandering electrical nerve impulses. How this occurred exactly remained "the fundamental mystery," as he put it.[66]

Summarizing the state of research on neural mechanisms in behavior in a 1930 overview article for *Psychological Review*, Lashley turned to a new analogy to clarify his dynamic outlook. The old model of the point-to-point telephone connection was rather outdated, he felt, as it could only illustrate

---

[63] Even in very obviously specialized areas such as the motor and visual areas, Lashley claimed there simply "was no narrowly localized specialization of intercellular connections." Karl Spencer Lashley, "Integrative Functions of the Cerebral Cortex," *Physiological Reviews* 13, no. 1 (1933).

[64] "Fundamental problem of neural integration," in Lashley, "Basic Neural Mechanisms in Behavior," 10. "Localization of fields" and "schemata," in Lashley, "Integrative Functions of the Cerebral Cortex," 34. Similarly, historian Nadine Weidman notes the inconclusiveness of the pattern vocabulary: "Lashley's solution to the mind-body problem was rather vague." Weidman, *Constructing Scientific Psychology*, 45.

[65] John T. MacCurdy, *Common Principles in Psychology and Physiology* (Cambridge: Cambridge University Press, 1928), 183.

[66] Ibid.

simple reflexes or habits, and he doubted whether neurons had any kind of individual functional specialization that supported it. Instead, he sought to conceive of functional *organization* (the relation between incoming sensations and outgoing motor responses) in a way that was not "expressed in terms of definite anatomical connections."[67] To do so, he used the image of a modern electric advertising sign or news ticker bulletin board: "The same cells may not be twice called upon to perform the same function. They may be in a fixed anatomical relation to the retina, but the functional organization plays over them just as the pattern of letters plays over the bank of lamps in an electric sign."[68] This comparison allowed Lashley to imagine neuronal integration in the cortex as "a variable pattern shifting over a fixed anatomical substratum."[69]

Seven years later, neurophysiologists Ralph Gerard possibly borrowed from his University of Chicago colleague Lashley when he compared the brain to a "modern sign" that allowed patterns to "move in time and transcend the spatial structure; and so words move across the sign as individual bulbs flash on and off, while the whole pattern remains intact."[70] With this analogy, Gerard aimed to shape the new conception of nerve activity that had emerged from a decade of recording the electrical impulses of the nerves.[71] Research had revealed that the living brain was characterized by nerve cells' continuous spontaneous impulses and that local masses of cells exhibited intricate patterns of synchronous activity by yet unknown distant causes. Clearly, older images of the functioning brain had to be discarded; just like Lashley had dismissed the telephone, Gerard dismissed the static idea of a "nervous system with a set structural pattern waiting peacefully for nerve impulses to travel through it, like a switch yard set for freight trains."[72] Such Fritz Kahn-like images of simple railway or telephone circuits had to be dismissed in favor of a "dynamic concept." Even though the mechanisms behind nervous discharges were not yet understood, he claimed such "patterns are in time as well as space, dynamic not static."[73]

---

[67] Lashley, "Basic Neural Mechanisms in Behavior," 9. In 1929, Lashley spoke about "non-specialized dynamic function of the tissue as a whole." *Brain Mechanisms and Intelligence: A Quantitative Study of Injuries to the Brain* (Chicago: University of Chicago Press, [1929] 1989), 176.
[68] Lashley, "Basic Neural Mechanisms in Behavior," 9.
[69] Ibid.
[70] Ralph W. Gerard, "Brain Waves," *Scientific Monthly* 44, no. 1 (1937): 56.
[71] Gerard mentions Edgar Adrian's recordings of single sensory nerve potentials (1926) as an important moment.
[72] Gerard, "Brain Waves," 51, 56.
[73] Ibid., 56.

Situating the electric display brain structures imagined by Lashley and Gerard in their metropolitan working environment of Chicago in the 1930s, we can see that the electric sign analogy they proposed for the changing patterns of nerve activity actually referred to a novel electrical display board, the so-called motograph.[74] This moving, illuminated news ticker (also called a zipper or revolving blackboard) had become more widespread in 1930s urban spaces, especially after one was mounted in New York's Times Square in 1928.[75] The motograph could show advertising slogans or streaming headlines—like HERBERT HOOVER DEFEATS AL SMITH—by allowing electric bulbs on a fixed grid to alternately flash on and off.

Returning to Charles Sherrington's famous enchanted loom metaphor mentioned at the opening of this chapter, we can now understand it as part of a lineage of electric display analogies for the active brain. Citing Gerard and Lashley in his 1938 Gifford lecture, Sherrington employed the twinkling tapestry in a section titled "The Brain and Its Work."[76] Quoting Lashley, Sherrington emphasized that a scheme in which behavior was localized in centers was "over-simplified, and to be abandoned."[77] Instead, he proposed imagining the active brain as "a scheme of lines and nodal points. Imagine activity in this shown by little points of light ... The lines and nodes where the lights are, do not remain, taken together, the same even a single moment."[78] Sherrington's enchanted loom, like the motograph brain, used an engineering analogy to conceptualize dynamic spatiotemporal patterns of nerve activity. In Sherrington's version, the model was not only a conceptual aid to strengthen the argument of dynamic function but also a way to imagine the brain itself as a site of "remarkable display."[79]

---

[74] A giant motograph sign (80 meters tall) had attracted ample press attention in 1934 after it had been installed in Chicago, the city where Gerard and Lashley worked. "Giant Electrical Sign Has Its Own Elevator," *Popular Mechanics* (1934): 16.

[75] Dale L. Cressman, "News in Lights: The Times Square Zipper and Newspaper Signs in an Age of Technological Enthusiasm," *Journalism History* 43, no. 4 (2018).

[76] Sherrington's 1938 lecture was first published in 1940. I cite from the adapted lectures in Sherrington, *Man on His Nature*.

[77] Ibid., 223, citing Karl Spencer Lashley, "Functional Determinants of Cerebral Localization," *Archives of Neurology & Psychiatry* 38, no. 2 (1937): 386.

[78] Sherrington, *Man on His Nature*, 223–4. In 1938, Sherrington also signaled that such models might already be outdated because neurophysiology had started to rely more and more on mathematics. Around 1938, Grass and Gibbs introduced frequency analyzers for the statistical analysis of large amounts of recorded nerve impulses.

[79] I cite from the adapted lectures in *Man on His Nature*, 223. In an earlier work, Sherrington had already attempted to describe the whole nervous system as consisting of "patterned networks of threads" with junctional points or intersecting lines where signals may coalesce. *The Brain and Its Mechanism* (Cambridge: Cambridge University Press, 1934), 12.

By using the motograph in their descriptions of the nervous system, these scientists created an electrocerebral analogy that conveyed the brain's capacity to express any action pattern at any possible spot without a fixed relation between anatomy and activity. At the same time, the motograph analogy also associated the brain with a new emblem of fascination and immediacy. Popular 1930s reports of the arrival of motographs recounted, as historian David Nye has described, how modern citizens were transfixed by their continuous activity and spellbound by their capacity for direct transmission: current news events could immediately be transferred into attention-grabbing motograph messages, which in turn instantly impacted the moods of urban spectators.[80] Situating the motograph brain within the modern environment that offered new electrical display experiences thus shows that this analogy suggested more than just dynamic nerve activity patterns. The motograph's appearance in nerve researchers' accounts from the 1930s signals the way researchers sought to negotiate the position of *technical mediation* in the field of nerve physiology prompted by the rapid rise of electrical pulse measurements visible on new types of display screens (or audible as crackly noise through sonification).

Since the 1920s, new records of electric pulses of nerves had been pouring into laboratories, made possible by new vacuum tube amplifiers and oscillographs. Yet, throughout the 1930s, it was hard to produce clear nerve recordings, as continuous adjustments were necessary to improve vacuum tube amplification and create faster, brighter signals on screen.[81] While active nerves—let alone nerve activity patterns—were hard to decipher in practice, the motograph analogy suggested a bright future for the active brain: associating nerve research with both immediacy and meaningfulness, the motograph implied the clear and instantaneous illumination of single active nerves and the visibility of patterns that literally formed messages on screen.

---

[80] David E. Nye, *American Technological Sublime* (Cambridge, MA: MIT Press, 1996), 191. Nye (p. 191, n. 57) describes an exemplary three-part cartoon published in 1931 that mocked the zipper's capacity for instantaneous transmission: (1) Three men look up, transfixed by the Times Square zipper, (2) Distracted, they are hit by a taxi, (3) As they are hit, the display reads: "Three being hit by Taxi in Times Square." Cartoon mentioned in Leonard Falkner, "The Sky Is His Blackboard," *American Magazine*, March 1931.

[81] Illumination from the cathode ray oscillograph was extremely dim, and a nerve impulse had to be captured through many reiterations to be visible on a photograph. This made the cathode ray tube ineffective for the whole-brain recordings Edgar Adrian would later pursue. He used a capillary electrometer instead. Edgar Douglas Adrian, "The Impulses Produced by Sensory Nerve Endings," *Journal of Physiology* 61, no. 1 (1926); Joseph Erlanger and Herbert Spencer Gasser, *Electrical Signs of Nervous Activity* (Philadelphia: University of Pennsylvania Press, 1937); Robert G. Frank, "Instruments, Nerve Action, and the All-or-None Principle," *Osiris* 9 (1994): 233.

Tele-visual and Televisual Neurophysiology

It is important to zoom in on the discrepancy outlined in the previous section for a moment: on the one hand, the instantaneity suggested by the motograph image—the nervous system as an illuminated news bulletin—and, on the other, the *practical* struggle of making nerve recordings in the 1930s. Ralph Gerard, who used the motograph analogy in a 1937 article, argued that the arrival of new measuring tools had changed experimental practice: while previous nerve activity studies relied on inference and "distant observation," being able to record potential changes offered "direct evidence."[82] He described the older situation as a type of physiology by "television or teleaudition," like determining the source of music heard over the radio while never having heard an orchestra; "it would be extremely difficult to deduce, only from the loud speaker outpourings, anything of its essential character."[83]

For Gerard, the words television and teleaudition signaled *distant observation*, while new experimental electroneurophysiology practices—based on combined setups of (for example) animal nerve preparations, amplifying tubes, graphic and audio-recording mechanisms, oscillographs, photo cameras, and fluorescent screens—instead strengthened his idea that active nerve processes could soon be fully heard and viewed in action. Reflecting on this development, historian Max Stadler notes that in the 1920s and 1930s recording a nerve was never conceptualized as "simply using an instrument": recording practices were cobbled together in recording systems ("relatively loose and local assemblages") made up of multiple components.[84] Stadler argues that technical nerve recording procedures were thus not rapidly "black-boxed" in the same way that, for example, electron microscopes had been, as this field "resisted objectification."[85] Yet, notwithstanding this persistent effort necessary to produce successful measurements through component systems, what clearly emerged in the 1930s was a new focus on image engineering, tinkering with experimental systems to arrive at better images. As historian Robert Frank summarizes the state of the field of neurophysiology around 1930, "Logic yielded to images created by instruments."[86]

---

[82] Gerard, "Brain Waves," 48.
[83] Ibid.
[84] Max Stadler, "Assembling Life. Models, the Cell, and the Reformations of Biological Science, 1920–1960" (PhD dissertation, University of London, 2010), 160.
[85] Ibid.
[86] Frank, "Instruments, Nerve Action, and the All-or-None Principle," 233. It is beyond my scope to trace the new image and display character of neurophysiology in the 1920s and 1930s with cathode ray tubes (whose results were photographed) and three-valve

This movement toward image making constituted an epistemic turn in neurophysiology. And yet the extent to which this new direction was also understood as yielding *direct* access to single nerve potentials, *instantaneous* visibility of activity patterns, or *immediate* accessibility of the visible active brain was under both practical investigation and rhetorical negotiation. When neurophysiologists used motograph analogies in the 1930s, they were discursively negotiating the level of directness of their measuring circuits. Within this negotiation, they drew on the symbolic leverage of things like motographs, that is, on established technologies, models, emblems, or instruments that gave a new sensibility to the technical mediation of neurophysiology.

While Gerard used the term "television" to denote the status of past nerve measurements as *inferential* and *indirect*, in 1934 American neurophysiologist George Bishop used it to frame the *directness* of new nerve-registering systems by referring to them as a "physiological television apparatus."[87] Here is that rhetorical ambiguity of media-technological analogies appears again: for Gerard, physiology offered a more direct perception than television, while Bishop explained that when a measurement setup used the right type of oscillograph, nerve impulses could "be visually observed while they are actually occurring, as if by a physiological television apparatus one could watch the cells concerned in one of their essential activities."[88] The television reference helped to fortify the status of the electrical pulse as the direct concomitant of nerve activity, as, like Bishop put it, "the identification of the electrical response of nervous tissue with the characteristic physiological activity of that tissue is so close that neurophysiology and the electrophysiology of nervous tissue have come to be practically synonymous."[89] With these new recording technologies, he proposed, neurophysiologists could "look directly for changes in the functioning brain itself."[90]

What "television" meant in neurophysiology, both in terms of an experimental practice and the field's imagined future, thus became subject

---

amplifiers plus capillary electrometer and (film) camera; for an overview, see Frank's article.

[87] George H. Bishop, "Electrophysiology of the Brain," in *The Problem of Mental Disorder: A Study Undertaken by the Committee on Psychiatric Investigations, National Research Council* (New York: McGraw Hill, 1934), 127; Borck, *Brainwaves*, 164.

[88] Bishop, "Electrophysiology of the Brain," 127. Cited in Borck, *Hirnströme*. Bishop does not yet mention the new technology of EEG whole-brain recordings, which became popular in the United States after Edgar Adrian's US lecture tour in 1934.

[89] Bishop, "Electrophysiology of the Brain," 127.

[90] Ibid., 126.

to rhetorical negotiation. On the one hand, Bishop's forward-looking vision referred to "tele-vision" as it had been invoked in longstanding nineteenth-century imaginaries (like Robida's *téléphonoscope* of 1878) of seeing things unfold simultaneously and immediately, even if they were far away (or inside the brain). On the other hand, any reference to television in the 1930s also referred to an actual new system of technologies, and this link made particular sense since nerve recording circuits were actually made of the very same technical elements for teleauditory and televisual mediation: electronic tube amplification developed in radio technology and oscillograph screens used in television.

In the 1930s, these technological means of transmission and reception were in development, which meant that radio and television were not stable referents in terms of media.[91] Their very status—how attractive, clear, direct, and effective they were understood to be—was in development. As media historians have pointed out, in the 1930s, televisual technologies were predominantly hybrid systems (just like neurophysiological circuits): varying setups combined elements of radio, photography, and cinema (Rudolf Arnheim, for example, spoke of *Funkkino*, radio film).[92] These systems could be employed both to transmit (preexisting) moving pictures or for *immediate* transmissions (called *eigentliches* or *reines Fernsehen*, real or pure television, in the German context).[93] Hence, the immediacy of the television apparatus was itself a compound and hybrid notion that fluctuated according to different setups. Just like the direct neurophysiological visual records that Bishop mentioned, the transmission of "direct" television images required laborious tuning and tweaking, as a (physiological) television apparatus was a rather coarse device. Nevertheless, Bishop believed that visual-technological advances had converged in such a way that the future for his research field looked bright. The time was now ripe for neurophysiologists not just to study single nerves but perhaps even

---

[91] On television as an "experimental system" in development in the 1920s and 1930s, see Lorenz Engell, "Fernsehen Mit Unbekannten. Uberlegungen Zur Experimentellen Television," in *Fernsehexperimente: Stationen Eines Mediums*, ed. Michael Grisko and Lorenz Engell (Berlin: Kulturverlag Kadmos, 2009).

[92] Doron Galili, "Television from Afar: Arnheim's Understanding of Media," in *Arnheim for Film and Media Studies*, ed. Scott Higgins and Doron Galili (New York: Routledge, 2011).

[93] Fritz Banneitz, "Der heutige Stand des elektrischen Fernsehens," in *Funkalmanach 1929: Grosse Deutsche Funkausstellung 1929* (Berlin: Rothgiesser und Diesing, 1929), 53, quoted in Anne-Katrin Weber, "Recording on Film, Transmitting by Signals: The Intermediate Film System and Television's Hybridity in the Interwar Period," *Grey Room* 56 (2014).

to venture, he hesitantly suggested, connecting them to mental functioning and researching "brain-function as a whole."⁹⁴

In Bishop's forecast, the television analogy resonated with both the longstanding imaginary of tele-vision's potential immediacy and simultaneity, and contemporaneous experiments with television systems that were understood as a visualization promise for the future. The 1930s imagined "physiological television apparatus" may thus be interpreted as instigating a rhetorical shift by which a still-unstable experimental neurophysiological system could become envisioned as a visualization instrument or even a visual medium that would be able to show the active brain. Even though the significance of newly acquired data on nerve activity was still unclear, the physiological experimental system was imagined as a televisual medium to show cerebral activity in direct new ways.⁹⁵

This example of electroneurophysiology recordings as a possible new instrument or even imaging *medium* to view the brain in action is significant. A closer look at the history of neurophysiology demonstrates that it is necessary to historically situate what we mean when we say that image making is presented as an immediate practice. In the 1930s, neurophysiologists definitely emphasized what happened on the displays of the oscillograph. Yet, rather than suggesting this meant a wholesale epistemic turn toward imaging, I want to argue that technical mediation itself was negotiated through analogies such as the motograph and television. To understand this discussion, it is imperative to situate the emerging screens of neurophysiology within a broader context of the 1930s rise of display devices. This broader history, I suggest, asks us to historicize what "immediately" seeing visualized activities on new types of screens and displays meant.

Brains at Work in Office and Factory: Living Diagrams and a Logic of Direct Display

Around 1930, at the same time that Edith Klemperer filed a patent for the luminous brain model built in Vienna, two Austrian engineers patented what

---

94 Bishop, "Electrophysiology of the Brain," 131. While some researchers started to equate these new nerve measurements with mental activity, others were more hesitant to hypothesize the material nature of the mental domain. Borck, *Hirnströme*, 192. On the complex position of neurophysiologists toward the "mind," see Smith, "Physiology and Psychology, or Brain and Mind, in the Age of C. S. Sherrington."
95 Here, I borrow the vocabulary of Hans-Jorg Rheinberger, who describes "experimental systems" as offering a "space of representation" for material traces. Hans-Georg Rheinberger, *Toward a History of Epistemic Things: Synthesizing Proteins in the Test Tube* (Stanford, CA: Stanford University Press, 1997), 105.

they called a "living diagram": an illuminated circuit diagram with colored light indicators (also called a *Leuchtschaltbild*) that enabled instantaneous monitoring of power plants.[96] Turning briefly to the history of these new types of displays allows me to situate motograph brains and the luminous brain as part of an electrotechnical world of display media that enabled direct monitoring of ongoing temporal processes, enacting new conceptions and sensations of immediacy and mediation. Since the 1920s, different illuminated display media—moving bulletin boards, signaling systems, indicators, control desks, educational models, luminous commercial displays, illuminated circuit diagrams—had invaded urban streets, offices, and factories, instituting the changing, illuminated visibility of ongoing (invisible) temporal processes, such as the activity of power plants, telephone networks, or even seat occupation in theaters.[97]

Such living diagrams allowed for immediate feedback about work processes and production numbers, and some types of display even allowed for direct manipulation of the ongoing processes of work.[98] Writing in the late 1920s, Siegfried Kracauer observed seeing one such living diagram during a visit to the director's room of a German factory, where its colored light bulbs showed an instant overview of all the operations.[99] Kracauer's observations reveal a new physical presence of the *direct display* of work processes as an integral part of the modern workplace, supporting a new culture of rationalization. He saw this indicator diagram as "the principal ornament of the real office," even more impressive, he remarked, than the

---

[96] Johann Latzko and Otto Plechl, "Living Diagram. Patent US2042667 A," filed March 15, 1930, and issued June 2, 1936, http://www.google.com/patents/US2042667. Previously, they had patented a similar structure ("Leuchtschaltbild. Patent AT118723B," filed August 28, 1928, and issued August 11, 1930). Few academic sources mention the history of the *Leuchtshaltbild* or illuminated wiring diagram in the early twentieth century. For a contemporaneous source on the "sinnbildliche Darstellung der Vorgange und Zusammenhange," see Guido Wünsch and Hans Rühle, *Messgeräte Im Industriebetrieb* (Berlin: Julius Springer, 1936), 79; "The Centralisation of Control of Power Networks," *Electrical Communication* 13, no. 3 (1935); Jakob Tanner, "The Visual Culture of Rationalization in the Modern 20th Century Enterprise," *Entreprises et Histoire*, no. 44 (2009).

[97] Anon. "Signal Lights Indicate Vacant Theater Seats," *Popular Science*, August (1922).

[98] Mark Seltzer mentions "living diagrams" and "working models" (such as a miniature and operative 1890s model of a coal mine elevator) in his analysis of machine culture around 1900, though he does not indicate contemporaneous uses of these terms. He describes the nineteenth-century fascination with these maps and models as allowing for the "superimposition of the visible and the calculable, representation and quantification, physical bodies and abstract models." Mark Seltzer, *Bodies and Machines* (New York: Routledge, 1992), 114.

[99] Siegfried Kracauer, *The Salaried Masses: Duty and Distraction in Weimar Germany* (London: Verso, [1930] 1998), 41.

Displaying Dynamic Brains 77

Figure 2.6 Fritz Lang, *Die Spione*, 1928 (film still).

fictional version integrated in the desk of Fritz Lang's genius villain in his 1928 film *Die Spione* (Figure 2.6).[100]

Kracauer's remarks connect this blinking display of temporal processes—like the luminous brain model—to the mesmerizing, flashy surfaces of Weimar Germany's illuminated facades—such as Fritz Giese's cityscape as a "delirium of brains gone mad," cited at the start of this chapter. Keeping with Kracauer's reading, the luminous brain and motograph brain can similarly be viewed as part of a "surface culture" of distracting spectacle, the sensational side of the contemporary electrotechnical environment.[101] Yet, simply linking enchanted looms and blinking brains to a new visual culture of spectacular surfaces does not suffice: blinking, living diagrams mediated temporal processes; their surfaces signaled a particular relation to an invisible but underlying patterned presence. This becomes even more apparent when we compare Kracauer's living diagram to one in the office of another of Fritz Lang's notorious villains. In Thea von Harbou's script for *Metropolis*, this evil

[100] Ibid.
[101] For a succinct analysis of Kracauer's ambivalent position vis-á-vis Weimar "surfaces" and "the mass ornament," see Ward, *Weimar Surfaces*.

factory owner is described as the "Brain of Metropolis" who lived in a "brain-pan," equipped with an abundance of contemporaneous display media and "switchboards on all sides."[102] Indicators signaled incoming messages from different cities, a tickertape machine ran continuously, and screens showed images from inside the factory. Information about factory activities and production statistics poured into these headquarters nonstop. Immediately, the villain could respond and give new orders by touching a "blue metal plate" that was undoubtedly modeled on actual indicator diagrams employed in industry.[103] Within the office, the workers themselves had become invisible; as the script mentions, they were only "phantoms" in a brain that was all lights and numbers.[104]

A superficial look at *Metropolis*'s Brain connects it to the hyperbolic analogies of Fritz Kahn's early drawings, where we see switchboard operators frantically working inside heads. Yet, a deeper reading in relation to a broader culture of display technologies expands our understanding of this clichéd "brain as switchboard" analogy. *Metropolis* shows the ongoing complexity of immediate transmission and sounds an important warning about the "phantoms" left unseen. Culminating in this fictional brain is the modern problem of a newly enacted relationship between the illumination of ongoing temporal processes (invisible yet visible) and the potential of forgetting the phantoms beneath the blinking lights. Electric display media heralded a new experience of immediate mediation based on the perception and anticipation of a virtually direct feed from active yet invisible real world or, in the case described here, real brain processes: this is what I call an emerging new form of liveness, a logic of *direct display*.

More historical research needs to be done on the conceptions and effects of this emerging world of ubiquitous displays. Erkki Huhtamo notes that while there is abundant cultural analysis of early twentieth-century peepshows, cinemas, and popular entertainment, "visual attractors" in public space have remained understudied and should be studied as "media interfaces that constitute their own peculiar modes of spectatorship."[105] By sketching this

---

[102] Thea von Harbou, *Metropolis* (Frankfurt: Verlag Ullstein GmbH, [1926] 1984), 23.
[103] Ibid., 24. A 1935 article on indicator diagrams describes the control desk of the Tummel Power Station in Scotland, noting how it is "filled in with blue enamel, resulting in a well-balanced and pleasing diagram." "The Centralisation of Control of Power Networks," 273.
[104] von Harbou, *Metropolis*.
[105] Erkki Huhtamo, "Monumental Attractions: Toward an Archaeology of Public Media Interfaces," in *Interface Critism: Aesthetics Beyond Buttons*, ed. Christian Ulrik Andersen and Soren Bro Pold (Aarhus: Aarhus University Press, 2011), 23. For a more recent contribution to this field, see Rachel Plotnick, "Force, Flatness and Touch without

new logic of direct display, this chapter's living diagrams—luminous brain, motograph brain, and enchanted loom—can all be understood as brainmedia assemblages that allowed for a new form of liveness that influenced how the active brain could be thought. Through the logic of direct display, the organ could be spatiotemporally thought according to the abstraction of electrical transmission as having an on/off light indicator structure, as fitting new types of surface displays. This instant display brainmedia promised a new *form of liveness* for the active brain, instantaneous yet mediated access to the brain in action as it blinked activity patterns. The brain itself could now be imagined as a type of display device.

Conclusions

This chapter analyzed 1930s brainmedia assemblages that were part of a contested field that traversed science educational and scientific discourses: the attempt to create a new conception of and a novel visual language for the dynamic brain, a brain that was populated by spatiotemporal activity patterns and whose structure and function were dynamically connected. In the luminous brain, sequenced illumination technology was imagined as a future possibility for the spatiotemporal animation necessary to depict cerebral processes. Here, the *Lichtlehrmethode* of dynamic displays was both likened to and seen as a lively antidote for the distracting illuminated surfaces omnipresent in the modern city. The visual language of illumination could serve as a flexible rhetorical tool that offered modern citizens sensations of liveliness and promised a transparent view of the brain in action. Science educational models of the brain banked on the cultural position of illuminated surfaces and people's more general association of electrification with active processes. Dynamic illumination also served the challenge of conceptualizing and representing the spatiotemporal brain in other ways. Various scientists also used the analogy of the illuminated electric news ticker for the nervous system, what I have called the motograph brain, to conceptualize a fixed (anatomical) substructure with changing electrical patterns. This instant visibility of temporal processes through display structures helped imagine the brain as part of a new cultural sphere of material surfaces that mediated temporal processes directly. These two blinking brainmedia thus reveal

Feeling: Thinking Historically about Haptics and Buttons," *New Media & Society* 19, no. 10 (2017); "Touch of a Button: Long-Distance Transmission, Communication, and Control at World's Fairs," *Critical Studies in Media Communication* 30, no. 1 (2013).

material discursive circuits through which electrotechnology and cerebral biology were mutually articulated.

Importantly, this chapter shows how the analogy between the active nervous system and particular media devices connected to changing attitudes of nerve researchers who used new display devices to study the electrical activity of the brain. In the 1930s, researchers needed to negotiate processes of technological mediation: producing images on display screens was central, but could not be conceived as a simply instantaneous endeavor. Analogies with the motograph brain, as well as with the emerging medium of television, allowed researchers to negotiate dimensions of immediacy and mediation in such technical processes. Electrical indicators directly showed underlying functional activities, yet their mediating function—as displays, indicators, living diagrams—remained present within such practices. By pointing to this logic of direct display, I nuance and adjust accounts in media philosophy about the "constitutive invisibility" of media or "media marginalism," that is, the idea that media "erase" themselves in their endless search for immediacy or transparency.[106] The logic of direct display introduced both new visibilities (spatiotemporal complexity) *and* invisibilities (the phantoms of active processes below). The ubiquitous signs and living diagrams in both scientific and popular spheres around 1930 shaped a display culture that drew new relations between moving illumination on visible surfaces and previously invisible temporal processes in power plants, factories, and brains.

The conjunction of the invisible and the visible, through these emergent display media, makes an important amendment to existing histories of brain and nerve research in the mid-twentieth century. On first glance, the fascinating and immediate display brains of this chapter neatly lay the groundwork for existing cultural histories of the brain that have pointed to a brain research shift between the mid-1920s and 1940s that Michael Hagner characterized as a move from the brain's anatomy—an "organicist view" focusing on the structure, size, and number of brain structures—toward its functionality—a "technicist view" focusing on the way nervous activity is wired.[107] This technicist interpretation allowed researchers to conceive of the brain as an isolated entity, not a fleshy brain in a body but a brain on the screen isolated from "distracting physical processes."[108] Toward the 1940s,

---

[106] Krämer, "Was Hat 'Performativität' Und 'Medialität' Miteinander Zu Tun?" 22. Timo Kaerlein, "Presence in a Pocket: Phantasms of Immediacy in Japanese Mobile Telepresence Robotics," *communication+1* 1, no. 1 (2012): 15.

[107] Michael Hagner, "Das Kybernetische Gehirn," in *Geniale Gehirne: Zur Geschichte Der Elitegehirnforschung* (Göttingen: Wallstein Verlag, 2004), 289.

[108] Ibid., 293.

neurophysiology increasingly focused on interpreting the fast-growing mass of electrophysiological measurements through mathematical correlations and statistics, looking for nerve cell activity patterns while moving further away from the material properties of neurons and cerebral tissues.[109] In turn, the 1940s new computational and cybernetic era developed its own visual language of schematic wiring diagrams of bodily processes and nerve circuits, an iconography that Hagner calls "anti-physiognomic," hardly interested in correlating function with the physical brain or body.[110]

Yet, the histories of the luminous brain model and the motograph brain are more complex. This means that we should not, as Douwe Draaisma has argued, present Sherrington's enchanted loom as a simple antecedent to the concept of brains as information-processing systems (the rule-based manipulation of symbols).[111] Organicists' and technicists' lines of research and thinking intersected in the 1930s; the search for dynamic models of electrical activity did not immediately lead down the path of an informational conception of the nervous system.[112] While an image of the nervous system as an illuminated news ticker may seem to anticipate a conception of electric messages or "code" inside the active brain, I argue that this analogy is part of a more complex history of negotiating technological mediation.

With this chapter I have revealed how, in the 1930s, science educational and scientific spheres were populated by alternative, hybrid imaginaries and models, part of a more ambiguous logic of display that shaped research into the active brain. Telling, in this respect, is the reprint of a photograph of the luminous brain in a 1934 American newspaper article titled "Science's Futile Attempt to Build a Perfect Mechanical Brain" (Figure 2.7). Here,

---

[109] Garson, "The Birth of Information in the Brain," 47. In tandem, Garson notes that nerve researchers such as Edgar Adrian also increasingly used informational vocabulary (especially in popular publications) to speak of nerve elements in terms of messages, signals, and codes.

[110] Michael Hagner, "Bilder Der Kybernetik: Diagramm Und Anthropologie, Schaltung Und Nervensystem," in *Konstruierte Sichtbarkeiten: Wissenschafts- Und Technikbilder Seit Der Frühen Neuzeit*, ed. Martina Hessler (Munchen: Wilhelm Fink Verlag, 2006), 394.

[111] Douwe Draaisma, for example, mentions the enchanted loom as a connecting thread between Charles Babbage and Ada Lovelace's nineteenth-century analytical engine and twentieth-century AI theories. Douwe Draaisma, "An Enchanted Loom," in *Metaphors of Memory: A History of Ideas about the Mind* (Cambridge: Cambridge University Press, 2000).

[112] Max Stadler similarly warns about a sweeping grand narrative of "information discourse" that slips into the history of nerve research and takes centerstage after the Second World War. While computation processes and informational vocabularies definitely impacted nerve and brain research, Stadler argues for a "deliberately unspectacular perspective," pointing to the tedious tinkering that continued in the laboratory with hundreds of measurements of actual, physical nerve cells and tissues. Stadler, *Assembling Life*, 320.

Figure 2.7 Science's futile attempt to build a perfect mechanical brain, 1934 (newspaper article).

the model literally featured as a connecting image between fleshy and "mechanical" brains, a techno-anatomical brain model that supplied readers, in a dismissive and ironic way, with the ingredients of a new imaginary, some future project of "building a brain." In the 1930s, rather than a wholesale turn from biology to technology, from tissues to instruments and screens, we see hybrid amalgamations that blended with the material culture of electric display media.

This important ambiguous hybridity of the fleshy and the electrotechnical brain is poignantly illustrated if we follow the work of Ralph Gerard from his 1937 motograph analogy for the nervous system to his work after the Second World War in relation to a growing cybernetic, information-oriented direction in nerve research. When a loosely affiliated group of scientists in cybernetics in the late 1940s proposed that the nervous system could be studied by focusing on the "discrete" nature of the nerve impulse (zero or one, hence a fundamentally digital approach to the nervous system), Gerard vehemently objected.[113] Decades of experience in researching the minute activities of nerve cells and tissue had showed him the many ways in which nervous conduction, synchrony, and delay characteristics were influenced by the nerves' material properties. Underneath the mesmerizing motograph display, he knew, lay complex material worlds.

[113] Ralph W. Gerard, "Some of the Problems Concerning Digital Notions in the Central Nervous System," in *Cybernetics—Kybernetik. The Macy-Conferences 1946–1953. 2: Dokumente Und Reflexionen*, ed. Claus Pias (Zurich: Diaphanes Verlag, 2004).

3

Demonstrating Brainwaves beyond the Laboratory: EEG as White Magic and Dark Media, 1934–41

> Having an electroencephalogram made of your mental processes is about as simple and painless as having your picture taken. At a recent demonstration to scientists in Detroit, a number of men volunteered to be "human guinea pigs" and, apart from some embarrassment at being part of the show, seemed to enjoy the experience.[1]

In the introduction, I described the 1934 demonstration in a Cambridge University lecture hall in which physiologists Edgar Adrian and Brian Matthews demonstrated, on stage and directly for an audience, a new machine that would soon be called electroencephalography (EEG), recording the living brain's electrical pulses from the scalp. The *Nottingham Evening Post* reported about the discoveries of the two British physiologists as an "uncanny achievement" that would "open up immense possibilities and will, it is hoped, enable scientists to 'see' the brain working."[2] After 1934, the development of EEG snowballed in research laboratories in the United States and Europe, especially after Adrian toured several countries with lectures about the new technology.[3] Soon EEG also started to capture the attention of the public, as scientists actively sought to demonstrate the new technology beyond the laboratory.

In this chapter, I study this new technology's emergence from Adrian's first demonstration in 1934 to the publication of the first *Atlas of Encephalography* in 1941 and the first Hollywood portrayal of a brainwave reader, particularly examining the public manifestation of EEG as a technology in the making. I analyze different ways in which scientists and science communicators

---

[1] Jane Stafford, "Science Takes a Look at Your Brainstorms," *Arizona Republic*, June 30, 1935.
[2] "'Seeing' the Brain at Work. Specialists' New Aid to Diagnosis. Important Discovery," *Nottingham Evening Post*, December 1, 1934, 8.
[3] In 1934, Adrian visited not only the conference of Electro-Radio-Biology in Venice, for example, but also a number of American universities.

demonstrated and articulated this emerging research technique and the new phenomenon of seeing "the brain at work," as the *Nottingham Evening Post* put it.[4] This helps contribute to a better understanding of an underexamined part of the history of the EEG's establishment as a research technique: measuring living brains beyond scientific laboratories and lecture halls. I particularly focus on three kinds of EEG performances: demonstrations of the technology recounted in newspaper and journal articles in the 1930s, the first exhibition of EEG to a broad audience at the *Paris International Exposition* in 1937, and the first appearance of a brainwave-reading device in a mainstream Hollywood film in 1941. I trace EEG through these different spheres to show how visions of an active brain—and the hybrid setups of oscilloscopes, electrometers, zigzag lines, screens, and subjects that produced them—were newly framed through such changing medial contexts.

The brainmedia I describe in this chapter are the special assemblages of fluctuating, experimental EEG setups and the particular media formats through which knowledge of the active brain is performed. It is in and through these brainmedia that EEG's conception was shaped between 1934 and 1941 in the interplay between scientists and science communicators. For my analysis, I am particularly interested in tracing the ambiguities and negotiations between scientific validity and popular fascination that shaped these EEG performances. On the one hand, EEG demonstrations such as those of Adrian and Matthews were clearly meant to establish authority and scientific legitimacy for the proposition that regular patterns of electrical activity could be measured from the human scalp and for EEG as a new technology to do so. On the other hand, studying the articulation of EEG in articles, exhibitions, and films also reveals presentations of recorded brainwaves as strange or wonderful entities (brainwaves as unexpected visible presences of what is usually invisible) and as prompts for speculation (about the future possibilities of thought recording or mental profiling, for example).

In this chapter, I trace both this boundary work between science and spectacle and the (permissible) forms of fascination that were invoked by EEG performances. How were audiences envisioned to make sense of the newly visible "invisible" phenomenon of brainwaves? How were they asked to wonder about, or question, this uncertain technology? The different forms of fascination were tied, I argue, to the evocation of liveness in the EEG performances. I discern an oscillation between (what Tom Gunning has called) the "optical uncanny," associated with older imaginaries of brainwave media, and a new understanding of EEG as a particularly modern magic since the

---

[4] "'Seeing' the Brain at Work," 8.

1930s. Practices of performing the brain at work invoked, in the examples I show, ambiguous forms of liveness, interpretations that could waver between the "liveliness" of scientific demonstrations (to be life-like, direct, and vivid) but also the strange "aliveness" of the scientific phenomena on display.

In probing these issues, I draw on historical scholarship concerning public scientific demonstrations and international exhibitions that showed modern novelties such as X-rays, cinematography, and electrotechnological displays, a body of work (mostly focused on the era from the 1880s to the 1920s) that has been interested in the dynamics of scientific performance and the enactment of scientific spectatorship. Historian Tom Gunning, in this context, offers important analyses of the modes of reception of modern technological inventions at world fairs around 1900.[5] As Gunning notes, at international expositions, "not only the products of modernity were displayed but the protocols of modern spectating were rehearsed within the context of a new consumer culture."[6] He describes the way visitors could be interpellated by a "discourse of wonder," that is, by the social repertoires, media communications, and exhibition designs that prompted a particular type of visitor experience—dazzlement, bewilderment, and astonishment, for example.[7]

Yet, such discourses of wonder, and the experiences they call for, do vary. Ian Morus, studying Victorian science spectacles, has argued for a nuanced typology of different genres of performance for different "genres of science, carrying different epistemological and cultural messages."[8] In the late nineteenth century, new forms and sites for scientific performance mixed traditions from "natural magic," academic science presentations, and illusionist theater.[9] In doing so, these performances did not generate a

---

[5] Tom Gunning, "Re-Newing Old Technologies: Astonishment, Second Nature, and the Uncanny in Technology from the Previous Turn-of-the-Century," in *Rethinking Media Change: The Aesthetics of Transition*, ed. David Thorburn and Henry Jenkins (Cambridge, MA: MIT Press, 2003), 40.
[6] "The World as Object Lesson: Cinema Audiences, Visual Culture and the St. Louis World's Fair, 1904," *Film History* 6, no. 4 (1994): 424.
[7] Gunning, "Re-Newing Old Technologies," 45.
[8] Morus, "Placing Performance," 776; "Worlds of Wonder: Sensation and the Victorian Scientific Performance," *Isis* 101, no. 4 (2010): 810.
[9] Martin Willis, "On Wonder: Situating the Spectacle in Spiritualism and Performance Magic," in *Popular Exhibitions, Science and Showmanship, 1840–1910*, ed. Jill A. Sullivan (New York: Routledge, 2015); Tiffany Watt Smith, "Of Hats and Scientific Laughter," in *Staging Science*, ed. Martin Willis, Palgrave Studies in Literature, Science and Medicine (London: Palgrave Macmillan, 2016); Iwan Morus, "'More the Aspect of Magic Than Anything Natural': The Philosophy of Demonstration," in *Science in the Marketplace: Nineteenth-Century Sites and Experiences*, ed. Aileen Fyfe and Bernard V. Lightman (Chicago: University of Chicago Press, 2007); Morus, "Worlds of Wonder"; Diarmid A. Finnegan, "Lectures," in *A Companion to the History of Science*, ed. Bernard Lightman (Chichester: John Wiley, 2016).

uniform spectatorship of naïve, stupefied onlookers, but asked spectators to negotiate different sensibilities: knowing that part of performance was an act or illusion, for example; or appreciating a new technology's industrial ingenuity; or exhibiting what historian James Cook described as "a new, media-driven form of curiosity—perpetually excited, never fully satisfied."[10]

When new media technologies such as the X-ray and cinematography were on display or discussed in popular literature, commentators often actively invoked relations to the supernatural and the occult, presenting them as decidedly magical tools.[11] In tandem, the audience's encounter with these new technologies was itself framed as prompting an experience of "modern magic," as Rachel Moore notes in the context of discourses of early cinema spectators.[12] These observations on mixed affects and sensibilities on the part of the public are part of a body of historical scholarship on the *fin de siècle* period that nuances a dominant discourse of modern "disenchantment," arguing instead that disenchantment, as Michael Saler and Simon During have emphasized, is always paired with striking new forms of "modern enchantment."[13]

While such genres of performance and elements of "modern magic" or "modern enchantment" have been well studied in relation to scientific and technological displays in the late modern period of the 1880s to the 1920s, there is considerably less scholarship analyzing how audiences in the United States and Europe were prompted to engage with new scientific displays of the body and the brain in the 1930s.[14] By that time, the landscape of science mediation had considerably changed and diversified. As mentioned in Chapter 2, new publication formats, magazines, radio

---

[10] James W. Cook, *The Arts of Deception: Playing with Fraud in the Age of Barnum* (Cambridge, MA: Harvard University Press, 2001), 68. Cited in Michael Saler, "Modernity and Enchantment: A Historiographic Review," *American Historical Review* 111, no. 3 (2006): 711.

[11] Pamela Thurschwell, *Literature, Technology and Magical Thinking, 1880–1920* (Cambridge: Cambridge University Press, 2001); Roger Luckhurst, *The Invention of Telepathy, 1870–1901* (Oxford: Oxford University Press, 2002); Maria Warner, *Phantasmagoria* (Oxford: Oxford University Press, 2019). Solveig Jülich, "Media as Modern Magic: Early X-Ray Imaging and Cinematography in Sweden," *Early Popular Visual Culture* 6, no. 1 (2008).

[12] Rachel O. Moore, *Savage Theory: Cinema as Modern Magic* (Durham, NC: Duke University Press, 2000), 2.

[13] Simon During, *Modern Enchantments* (Cambridge, MA: Harvard University Press, 2009); Saler, "Modernity and Enchantment"; Jason A. Josephson-Storm, *The Myth of Disenchantment: Magic, Modernity, and the Birth of the Human Sciences* (Chicago: University of Chicago Press, 2017).

[14] Saler, "Modernity and Enchantment," 711. For an example that extends further into the twentieth century, see Cartwright, *Screening the Body*.

shows, and (traveling) museum exhibits reached audiences of various levels of scientific understanding and interest. As various historians have recently argued, wonder takes on different forms in different circumstances, so it is vital to historicize the particular discourses of wonder.[15] The newly differentiated science communication of the 1930s thus requires historical studies to examine both the negotiations between science and its publicness and the varying modes of reception across different sites and genres.[16]

To do so, this chapter in divided into four sections. The first tackles the era prior to the emergence of EEG, around 1900, when the conception of an all-pervasive sphere of undulating waves allowed for interacting imaginations of thought particles and technological media to capture them. When communicating about the new technology of EEG, after 1929, scientists and science communicators needed to negotiate these long-established brainwave imaginaries. By sketching a new landscape of science communication, I showcase the 1930s more heterogeneous field of performing knowledges that harbored ambiguous framings of EEG. I also show how drawing a rigid divide between performative practices of "serious" scientists and "popular" science communicators is inadequate and instead highlight the tropes and rhetoric the two shared. In the second section, I zoom in on American newspaper reports on EEG after its US introduction in 1934 and describe the characteristic reporting on EEG demonstrations, events that presented a vivid narrative to communicate an uncertain scientific phenomenon. I also reveal newspaper articles' ambivalence on EEG's "photographic truth": on the one hand, EEG graphs were rhetorically compared to photographs and movies to strengthen their technological objectivity; on the other hand, readers were also asked to be critical and distinguish between modern records of mental activity and old fantasies of thought recording. The subsequent section studies EEG demonstrations at an international science exhibition in 1937. Central here is a discourse on *mystique populaire* or "white magic"—a spectacular presentation of science that was framed within a new emphasis on specialized and expert science. In the last section, I discuss the return of the uncanny in the portrayal

---

[15] This call for the historicity of "wonder" has been opened up by historians who have called attention to varying inflections of "cognitive passions" attached to objects and performances of wonder in scientific spheres. "Cognitive passions," in Lorraine Daston and Katharine Park, *Wonders and the Order of Nature, 1150–1750* (New York: Zone Books, 1998), 15.

[16] Arne Schirrmacher, "Nach Der Popularisierung. Zur Relation Von Wissenschaft Und Öffentlichkeit Im 20. Jahrhundert," *Geschichte und Gesellschaft* 34, no. 1 (2008): 83.

of a fictional brainwave reader in the 1941 Hollywood film *The Devil Commands*. I suggest that this hyperbolic presentation asked its viewers to skeptically examine and tentatively critique the scientific certainty and optical objectivity of brainwave science.

Ultimately, this chapter shows how particular forms of liveness in science demonstrations present ambiguities that crisscross between different genres of circulating knowledges and performances of science, between the work of nerve scientists and a new sphere of professional science communication. EEG performances in this period are characterized by oscillating positions between establishing scientificity and invoking wonder, between debunking myths and conjuring speculative futures.

Brainwave Imaginaries and Popularizing Science

When Hans Berger first published his experimental results in 1929 on what he called a "brain mirror," the technology measuring whole-brain nerve cell activity and the idea of the electrically patterned brain were not readily accepted in scientific circles.[17] In established EEG histories, it is the 1934 demonstration by eminent and Nobel Prize–winning physiologist Edgar Adrian that offered the necessary stamp of approval for this technological setup and allowed the concept of patterned whole-brain activity to become a scientific fact.[18] Yet, while members of the scientific community were initially doubtful, Berger's publications hit popular newspapers like a bomb. As Cornelius Borck recounts, in the summer of 1930, numerous articles reported on Berger's research with bold headlines such as "The Electric Script of the Brain," "The Zig-Zag Curve of the Human Mind," "Making Thoughts Visible," "Electricity Brings All to the Light of Day—Finally We Learn What's Going on Inside Our Brains," "The Electrical Record of Thoughts," "Electrical Brain Script," "The Machine That Reads Thoughts," "The Electrical Thought-Reader."[19] Newspapers eagerly printed the graphic records of Berger's original article without much further explanation; the images functioned as

---

[17] "Hirnspiegel," in Berger, Hans (November 16, 1924, IV, p. 164), cited in Borck, *Hirnströme*, 58. Cornelius Borck and David Millett have traced the complex coming into being of EEG as a technology and offer a number of reasons for the initial rejection of Berger's discovery, including EEG's inconsistency with reigning theories of nerve mechanisms and Berger's affiliation with the terrain of psychophysiology (rather than neurophysiology). See ibid.; Millett, "Wiring the Brain," 326. See also Adrian's own account of the initial obscurity of Berger's discovery: Adrian, "The Discovery of Berger."
[18] Millett, "Wiring the Brain," 310.
[19] Borck, *Brainwaves*, 112.

conclusive visual proof of a new, electrotechnically produced visibility—not only of the brain but also of the human psyche.[20]

Popular reporting on EEG followed what Borck calls a "seductive story pattern," which presented it as a type of thought recorder noting down the brain's natural script.[21] This story pattern tied into a long-imagined existence of cerebral radiations and the understanding of the brain as a space permeable to, and populated by, some type of thought-related energetic particles that were potentially perceptible.[22] Long before Berger published on brain rhythms in 1929, literary imaginations of, and material experiments with, thought-reading helmets, cerebroscopes, and telepathic radiations shaped ideas about the potential visibility of human thoughts. Around 1900, numerous physiologists, (para)psychologists, and engineers had claimed the existence of mental radiations emanating from the brain, while at the same time, in fiction, the concept of neurophotography or mind script became a literary motif.[23] Within this brainwave universe, the development of technological media, imaginaries of thought-reading phenomena, and conceptions of the nervous system were reciprocal and coemergent. Historians of science have studied cases of "vibratory photography," "psychic television," "psychic radio," and "spiritual wireless" to show that (new) media were shaping the imaginary of capturing the human psyche but were also significantly shaped *by* it; as Stefan Andriopoulos puts it, "One rendered the other imaginable."[24]

Various modern imaginaries of brainwave media in the *fin de siècle* focused on what authors have described as technology's uncanny character:

---

[20] Ibid.
[21] "Electricity as a Medium of Psychic Life: Electrotechnological Adventures into Psychodiagnosis in Weimar Germany," *Science in Context* 14, no. 4 (2001): 584.
[22] On resistance to these imaginaries, see Sabine Haupt, "'Traumkino.' Die Visualisierung Von Gedanken: Zur Intermedialität Von Neurologie, Optischen Medien Und Literatur," in *Das Unsichtbare Sehen*, ed. Sabine Haupt and Ulrich Stadler, Edition Voldemeer (Vienna: Springer, 2006), 117, n. 144.
[23] John Durham Peters, "Broadcasting and Schizophrenia," *Media, Culture & Society* 32, no. 1 (2010). As Anthony Enns and Shelley Trower point out in their cultural history of "vibratory modernism," the interpretation and scientific understanding of vibrations at the turn of the century was frequently conceived of as a form of energy that was underlying a type of unconscious, pre- or nonlinguistic communication. A. Enns and S. Trower, eds., *Vibratory Modernism* (London: Palgrave Macmillan, 2013); Haupt, "Traumkino"; Littlefield, *The Lying Brain*.
[24] Stefan Andriopoulos, "Psychic Television," *Critical Inquiry* 31, no. 3 (2005): 636; Anthony Enns, "Vibratory Photography," in *Vibratory Modernism*, ed. A. Enns and S. Trower (London: Palgrave Macmillan, 2013); Richard Noakes, "Thoughts and Spirits by Wireless: Imagining and Building Psychic Telegraphs in America and Britain, circa 1900–1930," *History and Technology* 32, no. 2 (2016); Anthony Enns, "Psychic Radio: Sound Technologies, Ether Bodies and Spiritual Vibrations," *Senses and Society* 3, no. 2 (2008).

that is, its boundary-crossing ability to traverse time and space and animate the dead.[25] Brainwave media offered a particularly special combination of, on the one hand, the magical and eerie capabilities attributed to modern recording and transmission media, and, on the other, the elusive realm of thoughts and the impenetrable space of the skull. Hence, fantastic brainwave media were doubly uncanny, combining the obscurity of the brain with the disorienting and supernatural powers attributed to modern media. At the turn of the century, this technological uncanniness was also an important resource in performing scientific and technological knowledge. Tom Gunning, in his analysis of world fair displays of new technologies around 1900, argues that it is ultimately the uncanny aspect that sustained visitors' astonishment and attraction to spectacles of technical innovation, even when such technologies had become fairly familiar.[26]

Tracing performances of nerve recordings in the 1920s and EEG in the 1930s reveals that the discursive shape of technology's uncanny—of the sensed potential for magical or unnatural powers—had gradually started to shift. As I will show below, scientists and science communicators persistently invoked a language and atmosphere of mystery and wonder while at the same time actively questioning or even debunking this association with the supernatural (with telepathy, talking to the dead, or materializing spirits). This did not mean that evocative connections between nerve research and phenomena, such as hypnosis, clairvoyance, and extrasensory perception, vanished altogether (there were a number of conspicuous psychotechnical technical machines circulating in the 1920s that connected to parapsychological discourses), but that new understandings of electrotechnical magic had become available.[27]

Brainwave technologies could now be marvelous but not supernatural; they could be magical in a new, modern, technologically and scientifically supported way. In the 1920s, science reporters negotiated this modern type of magic with the older uncanny and did so in new spheres and genres of popular science communication. A telling example in this respect is a full-page article in a 1926 American newspaper with the headline "Why Radio

---

[25] On the technological uncanny, see Laura Mulvey, *Death 24x a Second: Stillness and the Moving Image* (London: Reaktion Books, 2006), 27; Gunning, "Re-Newing Old Technologies."

[26] Gunning, "Re-Newing Old Technologies."

[27] In *Brainwaves*, Borck describes the popular attention and early popular descriptions of EEG as fitting with the already established public presence of remarkable psychotechnical machines in the 1920s—such as diagnoscopy, electrotherapy, or thought radio—that promised to reenergize bodies, read a person's character, find a suitable vocation, or broadcast thoughts. Borck, *Brainwaves*, 76–121.

May Have Uncovered a Sixth Sense! Science Now Investigating Cases of Broadcast Programs Being Picked Up, Unaided, by the Human Nervous System" (Figure 3.1).[28]

It recounts how American radio engineers had started to research the possibility that modern humans (living in contemporary, media-saturated atmospheres) would develop a "sixth sense" or "radio sense" that would allow the nervous system to directly pick up radio broadcasts. This was not a supernatural development, as the article makes clear, but a case of natural evolution, a type of neurotechnogenesis. The article strengthened the veracity of this fantastic yet entirely scientific development by noting the fact that in Europe various scientists "had managed to pick up 'brain waves' being broadcast by the nervous system through the medium of a modified radio site."[29]

The article is typical of a new type of popular science reporting in newspapers at the time, created by the growing newspaper syndicates that circulated their features in multiple newspapers.[30] These articles typically presented their readers with an amalgamation of discoveries. In this case, the article connected, with a grand sweep, the elusive work of an unknown American radio engineer with Adrian's nerve recordings, as well as the work of Italian parapsychological researcher Fernando Cazzamalli (whose research into telepathic brainwaves with supersensitive radio had been a recurrent favorite topic for popular science features, though received with much suspicion by various scientists).[31] Popular science reporting in the 1920s thus rhetorically united these disparate spheres in a common universe of electrotechnological circuits and waves. Such brain–radio connections were intuitively and speculatively produced in the text and by a photomontage technique that evocatively superimposed and combined

---

[28] "Why Radio May Have Uncovered a Sixth Sense! Science Now Investigating Cases of Broadcast Programs Being Picked up, Unaided by the Human Nervous System," *Indianapolis Star*, July 4, 1926.

[29] Ibid. The article referred to the work of Ferdinando Cazzamalli, *Archivio generale di neurologia, psichiatria e psicoanalisi* 4 (1925); P. Lasareff, "The Theory of Nervous Activity," *Science* 59, no. 1530 (1924); Adrian, "The Impulses Produced by Sensory Nerve Endings."

[30] Patrick Scott Belk, "King Features Syndicate," in *Comics through Time: A History of Icons, Idols, and Ideas*, ed. M. Keith Booker (Santa Barbara, CA: ABC-CLIO, 2014), 218. The brain radio piece was by the well-known *Newspaper Feature Service*. See Cynthia Denise Bennet, "Science Service and the Origins of Science Journalism, 1919–1950" (PhD dissertation, Iowa State University, 2013).

[31] For another example of reporting on Cazzamalli, see "Radio to Solve the Secrets of Telepathy. How Science Seeks to Show That 'Brain Waves' Transmit Thoughts and Find a Way to Talk with the Dead," *Daily Press* (1925). On Cazzamalli's popular yet contested research in the 1920s and 1930s, see Borck, *Brainwaves*, 100.

Figure 3.1 "Why radio may have uncovered a sixth sense! Science Now investigating cases of broadcast programs being picked up, unaided, by the human nervous system," 1926 (newspaper article).

illustrations of brains, zigzagging radiations, and radio equipment, thus visually materializing this brain radio on paper.

It was this heterogeneous and ebullient popular science reporting, with its amalgamations of supernatural and electrotechnoscientific magic, that set the tone for the seductive story pattern of Hans Berger's discoveries of a "brain script" and "thought reader" in 1929. Berger himself was uncomfortable with what he saw as journalists' overly speculative reporting on his work. As Borck recounts, Berger refused further inquiries for information on his discovery and the corresponding graphs, even to the *New York Times*.[32] Similarly, Adrian remarked that his 1934 brainwave publications and demonstrations—described as uncanny by at least one reporter—had been spellbinding to such an extent that many newspapers and journals covered the story "with varying degrees of accuracy," as it appeared "the temptation to write about brain waves was too great to be resisted."[33]

This picture of the reluctant academic researcher versus the overimaginative reporter is a well-known one befitting the self-fashioning of the serious scientist in the early decades of the twentieth century. Yet, this image deserves further scrutiny and nuance. By the 1930s, science popularization had become an institutionalized practice in many countries: to be involved in representing science and participate in some form of promoting science had become an integral part of many researchers' scientific work.[34] Nerve researchers often actively participated in the interwar science communication that framed life as newly animated by electrotechnology (a world I have described in Chapter 2 in relation to Fritz Kahn and the ensuing luminous and motograph brains). Neurophysiologists in the 1920s made use of a range of equipment—telephones, loudspeakers, phonograph records—to translate nerve records to students and broader audiences.

With these technological setups, they were much invested in developing an attractive demonstration aesthetic that would convey experiments vividly. One reporter, for example, described what he saw at a 1926 neurophysiological gathering (a meeting that included Adrian's demonstration of the first sensory nerve recording by light projections of graphic records and sounds played from a phonograph) as "some of the most beautiful experiments of the past years."[35] These vivid demonstrations were very amenable to press coverage: in 1928, US newspapers announced a radio broadcast by the

---

[32] Borck, *Brainwaves*, 112.
[33] Adrian, "The Discovery of Berger."
[34] Bensaude-Vincent, "In the Name of Science," 324.
[35] "Mödosam Onsdag for Fysiologer Och Försöksdjur. Efter Dagens Hundra Föredrag Ger Stockholms Stad Bankett I Gyllene Salen," *Dagens Nyheter*, August 5, 1926. "Listening-in to the Nerves: Dr. Adrian's Experiments," *The Observer*, March 28, 1926.

Iowa psychiatrist Lee Travis, who studied the effect of fear, excitement, and alcohol on the nerve centers and would allow listeners to hear the "rat-a-tat sound" of the human brain at work.[36] Similarly, in the UK, Adrian's colleague Brian Matthews spoke about nerve research in a series of broadcast talks in 1930 (later published in a popularly oriented book titled *Electricity in Our Bodies*) in which he described the human body as permeated by invisible currents, not unlike wireless radio waves, that could be made similarly perceptible. And Adrian himself also engaged in public scientific events: one year after his demonstration of brainwave measurements, for example, he would again lecture on EEG as part of an annual popular exhibition by the Royal Society at the Burlington House, an event "for entertainment and instruction," as the *London Times* reported, that included, among other things, a driving simulator, X-ray cinematographic films, and examples of new electrocardiographs.[37]

These examples all signal the necessity of a more nuanced image of (nerve) scientists as partaking in a growing practice of science communication that was an integral part of scientific work. A particular emphasis on the liveliness or vividness of demonstrations was shared between different spheres in which science was performed. What becomes visible is a more heterogeneous field of scientific performances and communications by scientists and science communicators invested in attracting attention to scientific research and developing vivid forms of engagement. In the next three sections, I typify the characteristic and novel forms that new EEG research took in this emerging sphere of science communication. Brainwave imaginaries persisted, as did the boundary work between the supernatural and modern magic, but these dynamics took on different forms in varying medial contexts.

Framing EEG in Print: Vivid Demonstrations and "the Stuff That Dreams Are Made On"

After Berger's and Adrian's demonstrations of brainwaves in 1934, EEG research developed into different directions in local research communities (predominantly in the United States, France, and Germany), ranging

---

[36] "Pattern of the Brain at Work to Be Broadcast by Wsui," *Wilmington Morning News* (1928).
[37] "Driving Tests Indoors," *The Times*, June 4, 1935.

from EEG as diagnostic tool for epilepsy research to investigations into the underlying physiology of the brain's electrical activity.[38] The rapid discovery of EEG's applicability in diagnosing epilepsy, one year after Adrian's demonstration, strengthened the sense that EEG was a tool of great promise. Yet, throughout the second half of the 1930s, EEG research was characterized by a "peculiar epistemological status," as historian David Millett puts it. There was no solid grasp of the underlying causes of brainwave patterns, and there was only general and provisional agreement on what constituted a "normal" measurement—"EEG remained a phenomenon in search of its significance."[39]

Though the underlying physiology of the brain's activity patterns was hardly understood, what emerged was a bustling practice of correlating EEG measurements to find the characteristic curves of people with schizophrenia, criminal delinquents, misbehaving children, and intoxicated subjects, as well as a search for the EEG markers of different personality types, gender differences, and intelligence levels.[40] American researchers had particularly rapidly taken up EEG research after 1934, a fact that Borck and Millett also attribute to the United States' greater availability of, and expertise with, electronic amplification technology that could be drawn on, as well as its generous funding for neurology research.[41]

American newspaper reports about EEG from the mid-1930s are evidence of the enthusiasm for this new technology, yet also of the uncertainty about EEG's meaning that science communicators needed to navigate. At the same time, the mid-1930s saw the emergence of the first full-time science news reporters catering to American newspapers and magazines, who also formed new professional associations.[42] These science writing associations not only claimed to benefit the better education of humankind but also emphasized their assistance to scientists in painting a better image for science, one that would help them secure more government funding.

Within this new landscape of mediating science, new exchange structures between scientists and science reporters emerged: reporters

---

[38] Cornelius Borck, "Between Local Cultures and National Styles: Units of Analysis in the History of Electroencephalography," *Comptes Rendus Biologies* 329, nos. 5–6 (2006).
[39] Millett, "Wiring the Brain," 406.
[40] Borck notes that it was the success of EEG in diagnosing epilepsy in the 1930s that sparked the high hopes for applying EEG as an objective tool to classify other pathological conditions, as well as the brainwave characteristics of normal subjects.
[41] Borck, "Between Local Cultures and National Styles"; Millett, "Wiring the Brain," 408–11.
[42] Bennet, "Science Service and the Origins of Science Journalism, 1919–1950"; David Dietz, "Science and the American Press," *Science* 85, no. 2196 (1937).

were actively invited to scientific meetings; lectures were mailed to reporters in advance; and universities started to hold press conferences.[43] Increasingly, professional scientific meetings—which often included exhibitions and demonstrations—became suitable events for reports about scientific developments. In American nerve research (a field characterized by a particularly competitive funding environment), EEG demonstrations were vehicles for laboratory teams to show their scientific expertise and technological prowess. At the scientific exhibit of the 1936 meeting of the American Medical Association in Kansas City, for example, a team of Harvard researchers used an actual epilepsy patient to demonstrate the capability of an EEG machine to reveal small epilepsy fits in the brainwave record.[44] While the fits were unnoticeable to the subject, they would become visible for the exhibition visitors. Science reporters jumped on such mediagenic demonstrations, which presented newsworthy scientific occasions and provided newspaper readers an attractive entry into a new and equivocal phenomenon.

In 1935, at the meeting of the Federation of American Societies for Experimental Biology, a team of Harvard researchers demonstrated the EEG at Detroit's Henry Ford Hospital, inviting members of the press to observe and participate. The *New York Times* featured a front-page article on this demonstration with the headline "Electricity in the Brain Records a Picture of Action of Thoughts."[45] In the article, the *Times*'s science reporter recounts how he offered himself as a guinea pig, taking part in the mental tasks the scientists assigned him. He describes the "tense silence of the room," while his mental processes generated tiny electrical currents, "such stuff as dreams are made on," that could now be recorded on a long roll of white paper. Unfolding the course of the experiment to his readers—sitting in a chair, applying the electrodes, hearing the commands of the researchers (multiply thirty-two by twenty-one), and seeing the changing wave patterns on the recording paper—the author invokes an atmosphere of suspense and anticipation while at the same time framing

---

[43] The Harvard Tercentenary celebration in 1936 is a characteristic example in this respect. David Dietz's comments (as president of the National Association for Science Writers) also show the opening up of a new science–journalism relationship, where the professional science writer could counsel the scientists: "The best advice that I can give you is that you play that you are a newspaperman when you write the abstract of your paper. Make it a condensed statement of your paper with all the important facts in it." Dietz, "Science and the American Press," 111.

[44] William Laurence, "Brain Records for Diagnosis," *New York Times*, May 15, 1936.

[45] "Electricity in the Brain Records a Picture of Action of Thought," *New York Times*, April 14, 1935.

this excitement within the arc of a scientific experiment that the reporter witnessed directly.⁴⁶

Casting a story about EEG in the narrative of a personal encounter, and thus emphasizing the author's own participation, evoked an intimate and more direct engagement between the reader and the scientific setup. The reporter's personal account added experiential substance, drawing the reader into what was described as a pioneering and uncertain situation. With a similarly direct and engaging style, *Popular Science Monthly* ran a field report on EEG ("these mysterious currents") prompted by a meeting of sixty "brainwave' experts" at the Loomis EEG laboratory in New York, as well the reporter's personal visit to Louis Max's EEG laboratory at NYU.⁴⁷ The piece turns to a direct address when it asks the reader to "follow Dr. Max into the dimly lighted instrument room" and "imagine yourself behind the scenes in his basement brain-wave laboratory." Reading the article, the reader would traverse the magazine's pages, encountering images of researchers dressed in white lab coats and of test subjects in intricate technological setups in which the readers themselves might one day be encased.

By mixing eyewitness accounts of experimental situations with adjectives full of wonder—"stuff of dreams," "magical," "mysterious"—these articles are expressive of the particularly modern magical attitude of these mediations of science, presenting images of wonderfully strange technology and science performances that always also emphasized scientific experimentation and direct observation. In science magazines and the newspapers' new science sections, accounts of EEG were characterized by this dual aspect, combining delight at the mystery of these new "telltale" visualizations and the suggestion of future visibilities of the mind with a simultaneous focus on the ingenuity of a technological setup revealing a previously invisible phenomenon. Throughout this discourse, the supernatural connotations of brainwaves were usually actively debunked—reporters assured readers there was nothing otherworldly or mystical to see. EEG's magical allure, reports suggested, was strictly due to its marvelous technological capacity of making visible the usually invisible.

---

⁴⁶ Ultimately, the story ends in a true revelation by the machine, since the device disclosed something that most spectators could not have gauged: a short "thought wave" is displayed after the author has given his response, explained by the scientist as a quick recalculation of the problem, an expression of the author's doubt at his own, initial answer. Together, the scientist and his new device had exposed the writer's hesitance, making visible to the spectators an otherwise private fact. The story thus endowed the mechanism of the electroencephalogram with particularly agency and foretelling power while simultaneously elevating the status of the scientist who had the ability to read the machine.

⁴⁷ Edwin Teale, "Amazing Electrical Tests Show What Happens When You Think," *Popular Science Monthly* 128, no. 5 (1936).

The scientific status of technological magic becomes poignantly visible in the returning reference to the strikingly photographic-seeming measurements of sleeping subjects made by Louis Max. In May 1935, a *Science Service* reporter described the first presentation of these electromyographs (electrical activity of muscles) at an evening event of the New York Academy of Sciences.[48] The article assures the reader that these are "not occult, not 'spirit' pictures" but that "the dream tracings were made with a practical hook-up of familiar electric apparatus, string galvanometer and amplifier." Printed next to the newspaper report was a "photograph made of a dream of Coney Island" that showed a flat line of peaceful sleep that turned into wildly spiking peaks when its subject had reportedly dreamed of a New York amusement park. A significant element of the graph's attraction was the striking juxtaposition it offered between seeing the famous fun fair—a cultural symbol of sensory excitement and excess—within the familiar and simple zigzagging line of a scientific graph.[49] Yet, additionally, with the emphasis on this invention as a type of "dream photography," reports also invoked long-established imaginaries of "thought photography" even while they actively debunked occult and spiritual associations.

Max's "photographic" myographs of dreams exemplified the type of objective, photographic truths EEG graphs were hoped to offer in the future.[50] In the mid-1930s, public demonstrations of physiological nerve activity graphs were reported as "photographs" or "films" of the active brain. And even when such announcements used inverted commas when they spoke of a "movie of a brain" or the "first sound and motion 'pictures' ever made of the brain in action," these phrases still reinforced the idea that single thoughts could be objectively (photographically) captured by EEG, which strengthened the imaginary of direct access to mental events.[51]

---

[48] "Scientists See 'Dream Walking' with Chart Aid," *Berkeley Daily Gazette*, May 21, 1935; "Photographic Record Made of a Sleeper's Dream," *Science News-Letter* 27, no. 737 (1935). The Coney Island dream photograph is also mentioned in "Amazing Electrical Tests Show What Happens When You Think"; and Stafford, "Science Takes a Look at Your Brainstorms."

[49] Adding to this imagination of potential dream photography, another *Science Service* article by Jane Stafford, "Science Takes a Look at Your Brainstorms," reported on this by including an illustration of this sleep study: a "thought cloud" above the head showed the subjects' dream of a merry-go-round, while right next and above the illustration we see an EEG graph. See ibid.

[50] Even when Louis Max's "photographs" were in fact based on measurements of action currents in the muscles (not brain cells), reporters intuitively connected his "photograph of a dream" to the new practice of EEG measurement.

[51] "'Movie' of a Brain Is Exhibited Here: Sound Film Shown at Museum of Natural History Records Reactions to Impulses," *New York Times*, April 18, 1937.

When reporting about EEG (demonstrations), newspaper and magazine reports often hinted at thought reading, personality testing, and criminal brain fingerprinting, such hints used a playful, ambivalent rhetoric: thought photography could be speculatively evoked while debunking occult interpretations of science at the same time. A *Popular Science Monthly* article, for example, playfully speaks of a "mind-reading needle" and "thought recorder," yet ultimately remarks that it was "needless to state" the "electrical mind reader" did not decode actual individual thoughts.[52] The article's mode of address interpellates the reader not only as an awestruck enthusiast of a new technology but also as a critical reviewer of science, a reader who is capable of separating a legitimate scientific demonstration from the imaginaries of thought reading. Similarly, a week after the front-page *Times* article on the Detroit EEG demonstration, the newspaper's main science editor Waldemar Kaempffert wondered: "But is this really a thought recorder? Do we really see our thoughts on the tape?" Reiterating a sense of marvel at a new technology, he also casts the reader in the role of a critical reviewer who wants to be informed, not misled in thinking that an actual thought recorder exists.[53] Similarly, articles in *Literary Digest* and several other newspapers attempted to set the record straight after the *Times*'s spectacular account of the Detroit demonstration: readers should not be deluded and think this a "mind-reading apparatus."[54]

What becomes visible after analyzing popular accounts and remarks by researchers is that demonstrations of EEG and electropsychophysiology for a wider audience were characterized by a two-sidedness or ambiguity: giving the technology scientific legitimacy while also framing it within a speculative realm (gestures to an elusive future) that evoked attraction and amazement. By describing a scientific demonstration, laboratory environment, or test setup, reporters aimed to relay a sense of the scientific reliability of a direct presentation within their textual account. Eyewitness reports using direct address allowed readers to vividly engage with the situation through a particular configuration of liveness that propagated this dual effect of scientificity and marvel. At the same time, these articles show another emerging discursive pattern in the portrayal of EEG: a kind of boundary work between legitimate interpretations and occultist beliefs. Science journalists

---

[52] The author distinguishes between reading individual thoughts and reading states of mind. Teale, "Amazing Electrical Tests Show What Happens When You Think."
[53] Waldemar Kaempffert, "The Week in Science: The New 'Electrical Thinking'; Activity of the Brain Recorded on a Tape by the Delicate Electroencephalogram," *New York Times*, April 21, 1935.
[54] "Recording Rhythmic 'Brain-Waves,'" *Literary Digest* (1935); Stafford, "Science Takes a Look at Your Brainstorms."

invoked a popular interpretation of EEG (an association with an ambiguous, supernatural, and uncanny realm of thought reading and parapsychology) to subsequently distance their accounts from this popular position, instead proposing a more nuanced and scientific understanding of the demonstrated phenomena. This debunking was paired with an emphasis on liveliness that aimed to engage readers and spectators with science. This boundary work returns more strongly in the next section, which examines the way scientific EEG demonstrations were reframed in a science exhibition.

The "White Magic" of Science: EEG at the 1937 Paris International Exhibition

> Go to the Palais de la Découverte, where an oscillograph will show you, on a photographic film, the shape of the oscillating currents of your own brain.[55]

After its pioneering demonstration in 1934, the public demonstration of EEG found a new highpoint in 1937, which saw the first direct public demonstration for a mass audience at an international science exhibition in the newly founded Palais de la Découverte as part of the *Paris International Exhibition*.[56] Decades after the event, one witness would recount his experience as follows:

> In a cage, of which you could see the interior, there was a man with numerous electrodes on his head ... A scholar explained to us what happened, while at the same time giving orders to the man in the cage—"close your eyes, open your eyes, don't think of anything, re-open your eyes, look at me"—and every time he gave an order, we observed, on a fluorescent tube, a curve that changed position ... I thought it was absolutely amazing.[57]

---

[55] Jean Labadié, "Au Congrès Des 'Ondes Courtes,'" *L'Illustration*, no. 4930 (1937).

[56] After 1934, other public demonstrations were part of scientific demonstrations, especially in the United States, where EEG research surged after Edgar Adrian's lecture tour in 1934. See, for example, an account of an EEG recording for a larger audience at the Kansas City meeting of the American Medical Association in 1936. J. L. Stone and J. R. Hughes, "I. The Gibbs' Boston Years: Early Developments in Epilepsy Research and Electroencephalography at Harvard," *Clinical EEG and Neuroscience* 21, no. 4 (1990). Millett, "Wiring the Brain," 370.

[57] Charles Penel, deputy director of the *Palais*, recounting his experience (at age ten) of the 1937 demonstration in an audio-taped interview, *1937: The Inauguration of the Palais De La Découverte*, http://videotheque.cnrs.fr/doc=1114?langue=EN2003.

The context of this 1937 *International Exhibition* enables us to examine the new scientific, modern magic we encountered in the American reports in the previous section in a new sphere. As we saw, this scientific enchantment was most vividly expressed in public demonstrations that offered readers and viewers a particular form of EEG's "liveliness." The demonstration setup at the 1937 Paris exhibition unfolded it as part of a discourse of wonder that structured particular modes of reception. While there are not many accounts of their critical reception, we can gain some sense of how the EEG demonstrations were framed by situating this example within its larger science exhibition layout and examining remarks in exhibition guides and reviews.

As historian Bernadette Bensaude-Vincent noted, the Palais exhibit should be understood as part of an international landscape in which science communication had a dual aim: scientists promoted science to a wider public and, in doing so, also aimed to find appreciation and funding for newly developing, specialized branches of science. As Bensaude-Vincent puts it: "Paradoxically, the popularization of science played a key role in the professionalization of science" in the 1930s.[58] As part of this promotion of specialization, the Palais exhibition served as an international meeting place for researchers; various scientific disciplines organized their annual scientific meetings around the 1937 event and stimulated participants to visit the exhibits. In the summer of 1937, for example, the EEG exhibit was visited by participants of the International Congress of Psychology (including Hans Berger).[59]

In the 1930s, popularizing (in French: *vulgariser*) science was fervently discussed and scrutinized: scientists pursued active boundary work to distinguish serious popularization efforts from too spectacular or speculative communications. Within this context, the makers of the Palais exhibition emphasized the "pure science" of "scholars, not that of commercial popularizers."[60] At the same time, the makers used popularization innovations, combining museum display strategies with experiment demonstrations

---

[58] Bensaude-Vincent, "In the Name of Science."
[59] The congress included a special session on EEG chaired by Hans Berger. Henri Piéron and Ignace Meyerson, eds., *Onzième Congrès International De Psychologie: Paris, 25–31 Juillet 1937: Rapports Et Comptes Rendus* (Nendeln: Kraus Reprint, 1974), 149–57, 231–4; "Psychologisch Congres Te Parijs. Zeshonderd Psychologen Uit Alle Landen Bijeen. Een Hartelijke Samenwerking Van Alle Zijden," *Algemeen Handelsblad*, August 4, 1937; J. B., "Impressions of the Eleventh International Congress of Psychology," *Journal of Consulting Psychology* 2, no. 3 (1938).
[60] Jacqueline Eidelman and Odile Welfele, "Enseignement Supérieur Et Universités; Palais De La Découverte (1900–1978)" (Archives Nationales, 1990), 4.

by scientists dressed in white lab coats, rooms with prerecorded voiceover explanations, and, in some cases, the possibility to touch, push, and use the objects in the show.[61] At the exhibition, as one popular science magazine claimed, spectators were asked to verify science "with their own eyes" and see "the experimental 'fact,' in its most demonstrative form."[62]

Significant elements of the Palais's discourse of wonder are the rhythm and scale of the exhibition experience. While navigating the spaces, visitors would encounter one impressive thing after another—thus evoking a sense of infinitely new supplies of amazement. Walking through the biology rooms, for example, a visitor could see the changing electrical potential of plant cells projected on a screen, a film with bioluminescent fish, and tubes filled with luminescent bacteria shaping the words "*Lumière Végétale*" (herbal light) when fed with oxygen.[63] A review of the exhibition by the renowned cultural critic Siegfried Kracauer typifies this sense of pedagogically oriented astonishment.[64] To Kracauer, the displays offered an educational spectacle to the crowd, "without succumbing to cheap popularity," a mode he dubbed "white magic":

> One presses a button and obscure models start to function; one walks through mysterious darkness, in which something suddenly crisps and flashes. Nothing more exciting than this white magic. The theory of probability has gained an improbable seduction, and the room with fluorescent phenomena looks like a magician's cabinet ... Mendelian laws of heredity, atomic theory, X-rays—it is impossible to expound everything in detail.[65]

Kracauer's reference to white magic to describe the atmosphere and approach of the displays characterizes the discourse of wonder summoned

---

[61] Ibid. The fact that the didactic aims of such exhibitions might not always have been successful is evident from at least one contemporary commentator, who remarked in a piece about the "advantages and disadvantages of vulgarization" that "we have seen lectures by great scholars that are just as incomprehensible as most of the exhibitions at the Palais de la Decouverte at the exposition of 1937 (J. Noir, "Concours Médical," 1938).
[62] Jean Labadié, "Que Savons-Nous Des Ondes Électriques Émises Par Le Cerveau?" *Science et Vie* (1937): 218.
[63] R. Bonnardel, *La Biologie: Exposition Internationale De Paris 1937* (Paris: Palais de la Découverte, 1937).
[64] Siegfried Kracauer, "Kosmos Der Wissenschaften-Konglomerat Der Kunste," *Das Werk; Schweizer Monatschrift für Architektur, Freie Kunst, Angewandte Kunst. Offizielles Organ des Bundes Schweizer Architekten BSA und des Schweizerischen Werkbundes SWB* 25, no. 1 (1938).
[65] Ibid., 21–2.

by the exhibition, similarly described as *"mystique populaire"* by its principal designer Jean Perrin.[66] Within this discourse, the visitor is cast not as someone interested in comprehensively understanding the displays but as a spectator suspended in a continuing state of inquisitive attraction. This mystical or white magical aspect could be deemed as appropriate popularization, especially because the overarching emphasis of the exhibition's paratexts was on serious, specialized science presented by authoritative scientists.

This white magic discourse also underpinned the exhibition's EEG demonstration setup, which was part of the biology exhibition's section on bioelectrical waves organized by the French neurophysiologist Alfred Fessard.[67] Combining expertise in neurophysiology and psychophysiology (he had trained with French physiologists and psychologists, as well as with Matthews and Adrian in Cambridge), Fessard had copublished his first paper on EEG in 1935, examining the relation between EEG and mental activity by studying changes related to varying visual and auditory stimuli.[68] The paper evinces the uncertain and exploratory approach to EEG research, but also the enthusiastic tone of the researchers who described the revelatory experience of seeing the brain at work. It was "banal but striking," they explained, to abruptly see a motionless spot on the oscillograph change when a subject closed their eyes on the operator's command, now "suddenly animated into a regular vibration that can be evaluated by the eye."[69]

In 1937, it was this striking but simple visuality of EEG research that was put on public display at the Palais when Fessard placed the EEG testing setup in one of the exhibition rooms, where it would remain on view for at least a decade.[70] A photograph of the room printed in the popular science

---

[66] Jean Perin, "Préambule du Projet de Palais de la Découverte, Décembre 1935," cited in Eidelman and Welfele, "Enseignement Supérieur Et Universités; Palais De La Découverte (1900–1978)," 5.

[67] On Fessard's work, see Jean-Gaël Barbara, "The Fessard's School of Neurophysiology after the Second World War in France: Globalistion and Diversity in Neurophysiological Research (1938–1955)," *Archives Italiennes de Biologie* 149 (2011). The EEG exhibit had already been announced as coming to the international exhibition. See N.A., "Wat Denkt Gij?" *Limburger Koerier*, Maart 1937.

[68] G. Durup and A. Fessard, "I. L'électrencéphalogramme De L'homme. Observations Psycho-Physiologiques Relatives À L'action Des Stimuli Visuels Et Auditifs," *L'année psychologique* 36, no. 1 (1935). Most of the examinations on which this thirty-two-page article was based were conducted on one subject, Gustave Durup himself, because of his consistent alpha activity and experience with psychological testing.

[69] Though the underlying physiology was uncertain, Fessard and Durup wrote that "the best way to know the instrument ... is to use it," as psychological data will indirectly help to "understand the physiological mechanism itself" (3, 4).

[70] Denise Albe-Fessard recounts a visit to the setup in 1943. Denise Albe-Fessard, "Denise Albe-Fessard," in *The History of Neuroscience in Autobiography, Volume 1*, ed. Larry R. Squire (Washington, DC: Society of Neuroscience, 1996), 13.

Figure 3.2 Electroencephalography booth in the exhibition *La Biologie: Exposition Internationale de Paris*, at the Palais de la Découverte as part of the *Paris International Exposition*, 1937 (photograph).

magazine *Science et Vie* gives an overview of the layout (which included the "magic device" of the vacuum tube at its core, the magazine mentioned) (Figure 3.2).[71] Importantly, during EEG demonstrations, visitors would be able to see the subject in a booth as well as the simultaneously measured EEG oscillations on a projection screen.

In the official guide that accompanied the biology exhibit, Fessard framed EEG with future-oriented enthusiasm. Even though "we are still in a period of clumsy fumbling" when it comes to understanding brainwaves, EEG offered a valuable "direct sign" that would definitely lead to future practical applications.[72] Again, the exhibition also shows moments of boundary work when white magic was separated from previous parascientific imaginaries. Fessard took pains to dissociate his display of the body's electricity from "fantastic or premature interpretations that circulate, and too often

---

[71] Labadié, "Que Savons-Nous Des Ondes Électriques Émises Par Le Cerveau?"
[72] A. Fessard, "Les Ondes Bioélectriques. De La Décharge Du Poisson-Torpille Aux Oscillations Electriques Du Cerveau Humain," in *La Biologie: Exposition Internationale De Paris 1937*, ed. R. Bonnardel (Paris: Palais de la Découverte, 1937), 68.

reverberated in the press."[73] Perhaps to distance himself from the popular bioradioelectrical therapeutics (which had been much critiqued in scientific circles and newspapers) of George Lakhovsky conducted in the 1920s at Paris's Salpetrière Hospital, Fessard emphasized that the body did not emit or send activity like a radio, but that fluctuating electrical potentials were fundamentally part of all living processes at the cellular level, including nerve cells.[74] The bioelectrical wave, he said, had nothing to do with speculative phenomena such as animal magnetism or character dowsing. By aligning the discovery of EEG with a much longer and established history of bioelectricity, Fessard aimed to present brain activity as a regular and intelligible phenomenon, now visible due to new technological machines that could amplify them.

Yet, while Fessard tried to typify all physiological waves, including brainwaves, as "normal" (*banal*) phenomena, the EEG exhibit also framed the technology in relation to possibly spectacular future applications. In the catalog, Fessard adopted a promissory tone when he asked, "Can normal people be classified in distinct nervous types according to the shape of their electroencephalogram?" answering that this was "impossible to foresee."[75] His speculation framed EEG as a prospective tool for nervous profiling, a psychodiagnostic instrument.[76] This connection of EEG to the realm of psychotechnics and diagnostics becomes more evident if we view Fessard's exhibit within the larger Palais exhibition, where it was positioned next to the "Human Biometrics" section, an interactive exhibit that allowed visitors to compare their physiological and mental capacities (intelligence, memory, sensory acuity, physical strength) to those of other visitors and to statistical averages on display.[77] This amenability between EEG and psychometrics was made explicit in *Science et Vie*, which reported on the biometrics section and included the adjacent section on EEG ("a quasi-miraculous" method) in its description of a broader practice of psychotechnology. Hence, the 1937 exhibition's layout and paratexts placed EEG in line with an established practice of psychodiagnosis and mental profiling.

Whether EEG was also tested as a potential biotypological research method in the late 1930s and early 1940s remains a question for further

---

[73] Ibid., 58.
[74] On Lakhovsky, see Borck, *Brainwaves*, 104–11; Alfred W. Gaspart, "L'oscillation Cellulaire Possède-T-Elle Une Vertu Thérapeutique?" *L'Homme Libre*, November 4, 1931.
[75] Fessard, "Les Ondes Bioélectriques," 58.
[76] Durup and Fessard, "I. L'électrencéphalogramme De L'homme."
[77] Charles Brachet, "La Découverte Scientifique: Création Continue. Le 'Banc D'essais' De La Machine Humaine Au Palais De La Découverte," *Science et Vie* 241 (1937).

historical research (quite possibly, though it was too premature to be practically employed in the 1930s). However, in the first chapter of this study I mentioned Foucault's work with EEG as part of vocational and personality testing in the 1950s, which shows how the technology had by that time become part of a regular set of test instruments. Within Fessard's Parisian research environment, psychophysiological research (which had now started to include EEG) was clearly conducted in close association with research avenues in *biotypologie* and psychotechnics (vocational testing, intelligence testing).[78] EEG research was also marked by this wider research landscape and characterized by an interest in creating typologies. Within this paradigm, EEG testing was predominantly understood as potentially recording characteristic patterns. Historians Lorraine Daston and Peter Galison mention EEG as one of several measurement practices in the 1930s that were shaped by "physiognomic sight," a new confidence in trained researchers' technologically assisted capacity to recognize family resemblances in data, which also served categorizations of populations of subjects.[79]

Yet, while it must thus be viewed as part of a larger biopolitical research project, drawing a direct connection between EEG and the racial and eugenic ideas of the 1930s ignores the historical complexity of this field. Biotypological research was linked to a wider preoccupation with what Daston and Galison have termed the "totalistic recognition" that was foundational to Gestalt psychology and racial theories in the 1930s and 1940s. In the French context, for example, biotypologies were also part of psychotechnical research approaches by socialist and communist psychologists who aimed to provide vocational guidance, for example, for underprivileged students.[80]

Ultimately, what is important to remark in the context of the 1937 exhibition is the fact that the political potential of biometrics (with EEG in its vicinity) may not have been clearly visible to exhibition visitors. As Sybille

---

[78] Fessard's mentor Henri Pieron had been a dominant researcher in intelligence testing and a new "docimology" testing procedure in France. He was chair of the Department of Physiology of Sensations at College de France and vice-president of Laugier's Société de Biotypologie (1932) and Laboratoire de Biometrie (1938). William H. Schneider, "After Binet: French Intelligence Testing, 1900–1950," *Journal of the History of the Behavioral Sciences* 28, no. 2 (1992); Luigi Traetta, "Docimology Enters into Psychology: Dagmar Weinberg's Work in French Applied Psychology Laboratories," *International Review of Social Sciences and Humanities* 4, no. 2 (2013).

[79] Lorraine J. Daston and Peter Louis Galison, *Objectivity* (Cambridge, MA: MIT Press, 2007), 337–8.

[80] Paul-André Rosental, "Eugenics and Social Security in France before and after the Vichy Regime," *Journal of Modern European History* 10, no. 4 (2012).

Nikolow has argued in the context of German psychotechnical exhibits, "The attractiveness of scientific diagnostic practices for the politically motivated social management of modern society is probably in the (at first sight) apolitical-looking character of these practices."[81] The biometrics exhibit emphasized visitors' interactive engagement and the fun of gaining new insight in oneself. As *Science et Vie* mentioned, everyone wanted to know about the "metric factors concerning their modest person."[82] In tandem, the adjacent EEG exhibit followed a white magic strategy, offering an opportunity to marvel at the technological ingenuity of the setup as part of a future-oriented scientific project.

This context of wonder also prompted Fessard to imagine ingenious new technological mediations of brainwaves. While nerve research had already been vividly mediated with the help of neon lights and audio speakers (as he noted in the exhibition guide), he now fantasized about an even more spectacular EEG setup: in the future, the electric EEG oscillations could operate a relay triggering "any grandiose event," such as "the departure of an electric train, or opening the doors of the exhibition."[83] Fessard's vision of merging EEG activity and electrotechnology was the epitome of EEG's white magic—an evocation of amazement without comprehensive understanding spurred by the science exhibition's discourse of wonder. It shows how exhibiting spectacular technological setups did not provide conclusive answers about the meaning of brainwaves (neither to scientists nor to laymen) but allowed for imagining new electrotechnological assemblages—a future world of new brainmedia, a world in which brains could open doors.

The EEG's imagined future remained apolitical in this exhibition context, even in its proximity to a project of biotypology. Brainwaves were presented as a wonderful, newly visible but ultimately "normal" phenomenon in the technological age. In contrast, the last part of this chapter will show that EEG also spurred adverse reactions and interpretations. EEG's eerie qualities as an uncanny medium visualizing the mental realm made it possible not only to perform boundary work in the demarcation of science from pseudoscience but also to critique the aims and effects of scientific and technological developments.

---

[81] "'Erkenne Und Prüfe Dich Selbst!' in Einer Ausstellung 1938 in Berlin. Körperleistungsmessungen Als Objektbezogene Vermittlungspraxis Und Biopolitische Kontrollmaßnahme," in *Erkenne Dich Selbst! Strategien Der Sichtbarmachung Des Körpers Im 20. Jahrhundert*, ed. Sybilla Nikolow (Köln: Böhlau, 2015), 230.
[82] Brachet, "La Découverte Scientifique: Création Continue," 5–6.
[83] Fessard, "Les Ondes Bioélectriques," 68.

Dark Brain Media in Hollywood

Throughout the 1940s, EEG remained a phenomenon in search of its significance. In a 1947 report on the state of the art of psychophysiology and EEG, one reviewer noted that while EEG had been useful in detecting cerebral tumors or forms of epilepsy, it "has contributed thus far surprisingly little to psychology."[84] This scientific uncertainty meant that scientists and science communicators had to keep navigating ambiguities when articulating brainwave research. In the previous two sections, I described EEG's ambiguous forms of liveliness (particularly connected to EEG demonstrations) as they were negotiated in newspapers and science exhibitions. These both invoked and debunked older, supernatural views of brainwave technologies, an oscillation that went hand in hand with the emergent portrayal of EEG as a spectacular yet scientific phenomenon expressed through the aesthetics and discourse of white magic.

In the 1940s, this debunking of supernatural and uncanny interpretations of EEG had become a rhetorical commonplace for scientists and science communicators. In the second edition of the *Atlas of Encephalography* in the 1941, the authors acknowledged that EEG "may at first seem strange" and even "fantastic," but that viewers should not be too puzzled, as the phenomenon of brainwaves "should become quickly familiar."[85] In a section on "the electroencephalographer's place in nature," they explained EEG graphs as providing "broad-focus eyes," merely intensifying or extending our natural ability to see.[86] The first 1941 *Atlas* was the culmination of six years of searching (since 1934) for characteristic EEG records that correlated with particular mental tasks and personality profiles.[87] By the late 1930s, the mountain of recorded EEG data had started to be stored on punch cards and could be evaluated by means of a new frequency analyzer that could read and compare records faster.[88] The *Atlas*'s authors emphasized that this new technological assistance now allowed trained professionals to objectively analyze an EEG

---

[84] Chester W. Darrow, "Psychological and Psychophysiological Significance of the Electroencephalogram," *Psychological Review* 54, no. 3 (1947): 157.
[85] Frederic Gibbs and Erna Gibbs, *Atlas of Electroencephalography* (Cambridge, MA: Cummings, 1951), 18.
[86] Ibid.
[87] Borck notes the US dominance in the rapid expansion of the correlative analysis of EEG typologies, noting the influence of the competitive situation that forced research teams to "specialize on new fields of application or to secure an advantage by new data-processing techniques with which they could find and maintain an audience on the flourishing market." Borck, *Brainwaves*, 197.
[88] Albert M. Grass and Frederic A. Gibbs, "A Fourier Transform of the Electroencephalogram," *Journal of Neurophysiology* 1, no. 6 (1938).

with an experienced "seeing eye," as if "recognizing a person by his features."[89] While EEG assessment had previously been "more art than science" according to the inventors of the EEG frequency analyzer, the *Atlas*'s authors now claimed that "the problem of visual analysis of electroencephalograms resolves itself into a fairly simple task of recognition—much like learning to read a new language with an unfamiliar alphabet and different kinds of script."[90]

Throughout the *Atlas*, the tone was one of promissory enthusiasm about EEG's ultimately revelatory photography-like powers. Yet, this future-oriented tone also unavoidably made it clear that EEG did not provide all the answers yet. Because one still needed "a brain to analyze a brain," the "ultimate solutions to the problems of electroencephalography" were still ahead.[91] Hence, within this atmosphere of EEG's imminent anticipated value, the act of debunking the fantastic and strange allowed scientists to garner some necessary scientific legitimacy in an otherwise uncertain situation. The dual invocation and denunciation of EEG's uncanny must thus be understood as both a commonplace assertion of scientific authority and suggesting a necessary sensibility on the part of a science audience, a rhetorical move that diverted attention from the uncertainties at the heart of EEG research.

While the uncanny has so far appeared as a rhetorical trope to demarcate genuine science (and serious popularization) from laymen's naïve and backward interpretations, the uncanny could also be a resource to articulate a different mode of reception vis-à-vis science. The 1941 Hollywood film *The Devil Commands* is evidence of this alternative portrayal of brainwave measurement technologies. I argue that this particular popular portrayal of brainwave science subverted the symbolic power of the technology's uncanny character, employing it instead to articulate a position of distrust about both the morality and veracity of brain research. While we may conventionally think of popular media, especially Hollywood films, as exaggerated, science-fictional, and future-oriented portrayals of science and technology, the example of *The Devil Commands* shows the simultaneous potential for a skeptical position.

---

[89] "Seeing eye," in Gibbs, Frederic, and Erna Gibbs. *Atlas of Electroencephalography*. Cambridge, MA: Addison-Wesley Press, 1951, n.p. "Recognition of features," in W. G. Lennox, E. L. Gibbs, and F. A. Gibbs, "Inheritance of Cerebral Dysrhythmia and Epilepsy," *Archives of Neurology & Psychiatry* 44, no. 6 (1940): 1158. On learning to become an expert EEG record reader, see Lee Edward Travis and Abraham Gottlober, "Do Brain Waves Have Individuality?" *Science* 84, no. 2189 (1936).

[90] Gibbs, Frederic, and Erna Gibbs, *Atlas of Electroencephalography*. Cambridge, MA: Addison-Wesley Press, 1951, 112.

[91] Ibid., 68.

Seven years after the first public demonstration of brainwaves in Cambridge, *The Devil Commands* was the first mainstream Hollywood production to portray the conspicuous science of brainwaves. Following the conventions of the haunted house genre, the atmosphere of the film is set by gloomy weather, night skies lit by lightning, dark rainy streets, and power cuts. Boris Karloff stars as a respected scientist gone haywire, convinced that there is life after death and that a newly invented brainwave reader will allow him to communicate with his recently deceased wife. The film framed this novel scientific invention within a humorously hyperbolic narrative of supernatural powers and eeriness. Because of their shared liminal capacity to traverse established ontological realms—between the visible and usually invisible, between the materiality of the body and the immateriality of thought—this association of brainwave machines and scientists with the supernatural and parascientific resembled the discourse on X-rays decades earlier.

The Devil Commands followed a number of popular science-fiction stories that gave imaginative substance to the futuristic possibility of capturing brainwaves and did so with characteristic dramatic exaggeration, employing the uncanny to portray (brain) science with a mocking irony.[92] This ironic stance vis-à-vis thought visualization followed a current present in fiction since the late nineteenth century that exhibited growing skepticism about the scientific discoveries of neurologists and thought photographers.[93] Literary scholar Sabine Haupt has characterized the 1920s and 1930s as the height of this skepticism, when the literary motif of thought visualization became increasingly perfused with a critique of capitalist and totalitarian regimes intruding on private life.[94] Oftentimes, plot turns conveyed a mistrust of the veracity of the newly visible entities—a clear suspicion of new mediating technologies—and alluded to the unreliability and hubris of scientists and their scientific practices. Hence, the imaginaries of thought reading and brainwave communication these narratives proposed were heterogeneous

---

[92] Some literary examples are Willard Rich, *Brain-Waves and Death* (New York: C. Scribner's Sons, 1940), and William Sloane, *The Edge of Running Water* (London: Hachette, [1939] 2013). Not long after *The Devil Commands*, Curt Siodmak's *Donovan's Brain* (New York: Alfred A. Knopf, 1943) was adapted for cinema multiple times: *The Lady and the Monster* (1944), *Donovan's Brain* (1953), and *The Brain* (1962). Historian Melissa Littlefield analyzes various thought translator stories in American science-fiction magazines of the 1930s and notes the ubiquity of fictional machines (neurocameras, mental microscopes) that could immediately translate thoughts into sounds, words, or images. Littlefield, *The Lying Brain*, 76.

[93] Haupt, "Traumkino."

[94] Ibid., 115. Haupt points to various critiques of the corruptibility of capitalist scientific institutions, of a science that is only interested in money-making inventions and profitable patents.

and ambiguous; they could both be critical of and at the same time contribute to a growing attention to "photographic" and "objective" measurements of mental phenomena and to a materialist understanding of the mind.

In a key scene at the beginning of *The Devil Commands*, the professor reveals his new brainwave reader to a group of colleagues during a scientific demonstration. A fantastic transmutation of existing technological equipment, the brainwave-reading apparatus is said to produce a fingerprint-like "portrait of the mind." To operate it, a subject must be encased in an iron helmet connected to a complicated machine that operates a giant mechanical arm producing a zigzagging graph. The professor invites one visitor to become a test subject, while the others see a direct inscription of his mental activity on a large screen in the back of the room. While, he states, some may think this impossible ("like we used to think about radio"), one day this device will record individual thoughts and "unlock the secrets of the human mind." Shot over the shoulders of the gathered scientists, this scene transposes the viewer in the position of a similarly incredulous and amazed spectator of science (Figure 3.3). Both the professor figure and the scientific method on display are presented ambiguously, raising the question: has the professor turned mad or evil, overstepping the ethical boundaries of science,

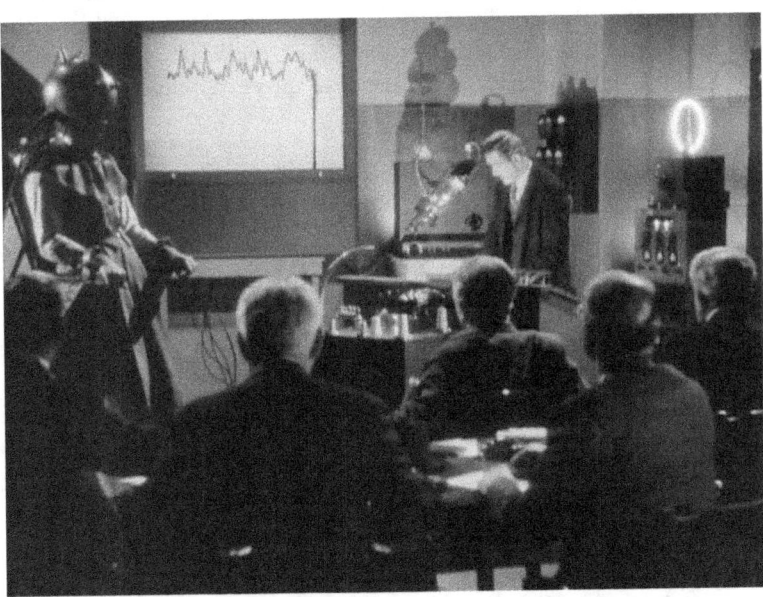

Figure 3.3 Edward Dmytryk, *The Devil Commands*, 1941 (film still).

or is brainwave reading perhaps a hallucination on the part of the scientist, or us, viewers ourselves?

In *The Devil Commands*, the professor's quest to contact his deceased wife turns him into an irrational and reckless experimenter who accidentally fries his subjects' brains. Through his unbelievable claims and aspirations, the film not only foregrounds the uncanniness of the technology but also makes the viewer wonder whether the scientist, and the scientific system itself, has become delusional—an allusion to the potential hubris of modern science and technology. As such, the film follows the typical "overreacher" plot of horror tales—the hubris of scientists going too far—with a type of uncanniness that Tom Gunning calls the "optical uncanny," that is, an uncanny sensation on the part of the viewer triggered by a doubt whether what new technologies make visible is real or imagined.[95] A variety of other stories and films portrayed brain science within this realm of the optical uncanny. They show brain scientists as searching for new objective knowledge but making our reality all the more perplexing and disturbing in the process.[96] As new or imagined technologies offered strange new modes of vision, the uncanny dimension of these stories expressed the "ambiguities of mediated vision" not only to evoke a sense of awe but also to question both the epistemological and moral trustworthiness of these new technologies.[97]

Media philosopher Eugene Thacker has coined the term "dark media" to refer to these uncanny mediating technologies.[98] Dark media are the procedures, machines, and circuits that "work too well," moving beyond the capacity of the human senses to mediate "between different ontological

---

[95] Tom Gunning, "Uncanny Reflections, Modern Illusions: Sighting the Modern Optical Uncanny," in *Uncanny Modernity; Cultural Theories, Modern Anxieties*, ed. John Jervis (London: Palgrave Macmillan, 2008).

[96] Brain science and brain reading featured extensively in fictional tales about supernatural phenomena, haunted houses, spiritualist séances, and hypnosis such as Edward Lytton's "The Haunted and the Haunters, or, the House and the Brain," *Blackwell's Magazine*, August 1859. Even when such stories culminated in a debunking of the phenomenon, the sense of a supernatural link had already been established. A familiar strategy in some fictional accounts of brain science is to combine and oscillate between tropes of revelatory science and the hallucinations of lone mad scientists, such as, for example, in nineteenth-century tales about the invention of thought-visualizing machines in Edward Mitchell's "The Soul Spectroscope," *The Sun*, December 19, 1875, and Edward Bellamy's *Doctor Heidenhoff's Process* (London: W. Reeves, 1890). In turn, however, the brain scientists in these narratives are revealed to be unreliable narrators, and the brain-revealing device itself may ultimately just be a figment of the scientist's imagination. Lysen, "The Brain Observatory and the Imaginary Media of Memory Research."

[97] Gunning, "Uncanny Reflections, Modern Illusions," 79.

[98] Eugene Thacker, "Dark Media," in *Excommunication: Three Inquiries in Media and Mediation*, ed. Alexander R. Galloway, Eugene Thacker, and Wark McKenzie (Chicago: University of Chicago Press, 2013).

domains—the natural and the supernatural, the normal and the paranormal, life and the afterlife."[99] Fictional tales that portray new technologies as dark media particularly reflect on the status of mediation, flagging the doubtful evidentiary status of visual technologies. As such, the optical uncanny of dark media conjures a double movement: on the one hand positing a suspicious and critical viewer capable of debunking fraudulent, hallucinatory science, on the other hand maintaining the association between optical media and the realm of the supernatural.

If we interpret *The Devil Commands* as playing with the optical uncanny and framing brainwave readers as dark media, the film thus also contributes to a new mode of reception, a specific positioning of the layman spectator of science. Throughout the film's comical back and forth between scientific rigor and supernatural realms, spectators of science are portrayed as navigating between amazement and critique. In one scene, the brain scientist attends a spiritualist séance and exposes it as simple optical trickery ("I hate to disillusion you, but...").[100] While the séance is exposed as false, the medium's electrical nervous powers are presented as real. After wiring the medium with a brain writing machine, the scientist assures her it "is science, nothing occult about it." An occult performance is thus dismissed to lend more credibility to a scientific demonstration while at the same time reassociating brainwave measurement with spirit mediums and supernatural powers. Spectators of science are here placed in an oscillating position between recognizing the accomplishments of a wonderful science and evaluating the boundary-transgressing hubris of hallucinatory scientists. Portraying EEG as dark media and framing its optical uncanny qualities, *The Devil Commands* reinforces an oscillatory stance that we saw glimpses of in newspaper accounts but that obtains a particularly critical dimension within the context of this fictional tale.

Conclusions

Between 1934 and 1941, scientists, science reporters, and filmmakers needed to navigate an uncertain terrain in communicating about the elusive and developing technology of EEG. In my first example, I showed

---

[99] Ibid., 102.
[100] Here, *The Devil Commands* follows a familiar motif in haunted house films: exposing the spiritualist to disprove paranormal phenomena but remaining ambiguous about the ultimate status of ghosts and the supernatural. Simone Natale, "Specters of the Mind: Ghosts, Illusion, and Exposure in Paul Leni's the Cat and the Canary," in *Cinematic Ghosts: Haunting and Spectrality from Silent Cinema to the Digital Era*, ed. Murray Leeder (London: Bloomsbury, 2015).

how scientists and science news reporters portrayed nerve research in relation to a long-established imaginary of brainwave media and thought readers that was closely tied to associations with the supernatural and often framed within the vocabulary of the uncanny, with a particular focus on technology's animating and liminal powers to render the invisible visible. Often scientists and science reporters invoked the uncanny as a newly pacified delight, a conquered remnant of bygone era of supernatural magic that was proof of a more modern, technological, and scientific sensibility. To mention and then denounce the uncanny allowed them to distance themselves from an older, magical understanding of science—a mode to which a serious reader or viewer of science should no longer adhere. Calling upon the supernatural thus invoked a multilayered mode of reception: it summoned a serious contemporary witness of science, one that could appreciate the wonder of EEG as an ingenious technology, a modern type of magic that was stripped of its false supernatural powers. In my second example, the 1937 *Paris International Exposition*, a live EEG exhibit was part of a discourse of (what Siegfried Kracauer called) white magic, presenting it as incredibly yet scientifically real. Within this mode of reception, the contours of an implicit biopolitics of EEG were only vaguely visible, eclipsed by EEG's white magic as part of a mysterious bioelectric universe in which electric brain potentials were imagined as triggering electric trains or opening exhibition doors. Finally, the portrayal of brainwave reading in the 1941 movie *The Devil Commands* presented a particular ironic uncanny, marking brainwave readers as strange and uncertain visualization technologies and opening the possibility to question this scientific invention, conjuring a potentially skeptical spectator of science.

My analysis shows that the emergence of professionalizing platforms and practices for science popularization led to a new emphasis on vivid demonstrations of the new EEG technology in action. These live demonstrations with nervous subjects, scribbling ink writers, and/ or moving lights on oscilloscope screens were intended to offer a direct engagement with a new scientific phenomenon. In the examples I examined, performing knowledge about the brain at work in newspapers, an exhibition, and a Hollywood film invoked oscillating forms of liveness, interpretations that wavered between the "liveliness" of scientific demonstrations (lifelike, direct, and vivid) and the strange "aliveness" of this new scientific phenomenon. Mapping these ambiguities shows how supernatural interpretations of brainwaves were often rhetorically invoked so as to be dismissed, thus establishing a modern, yet magical attitude toward this new science. These equivocal narratives of *mystique populaire*, "white magic," and "dark media" reverberated across different spheres of performing science. My analysis ultimately suggests that drawing a rigid divide between the performative practices of "serious" scientists and "popular" science communicators is inadequate. Instead, I highlight their shared tropes and rhetoric.

4

Broadcasting Live Brains: The Brain on Television and as Television, 1949–57

1954, we are watching "Catching a Brain Wave," a live transmission of America's first weekly science show, the *Johns Hopkins Science Review*. Inside the TV studio, a man in a lab coat is standing beside the bed of a young woman. Flanking her is a large machine speckled with buttons and featuring a horizontal sheet of white recording paper. We pan in on the woman's head while the man applies the final electrode to her scalp. Subject and machine—"a rather formidable-appearing device," the man tells us—are now connected. The next scene captures the moving recording, the machine panel, and the reclining subject. As we zoom in on the paper, the voice of the scientist explains how the zigzag lines distinguish a normal brain from an abnormal brain. Pointing to the scribbles, he says: "This recording is actually coming from the patient, you see." Using televisual effects, the broadcast superimposes the running EEG record on the girl's face. The record visibly changes when the patient is asked to open and close her eyes. Through a careful choreography of machines, subject, host, voiceover, montage, framing, and overlay effects, the live broadcast creates a sensible relation between the technological apparatus and the living brain's invisible—but now visible—activity. Through this immediate transcription of the subject's cerebral activity on the surface of our home TV set, brainwaves entered the living room (Figures 4.1.1 and 4.1.2).

In this chapter, I study brains *on* and *as* television. I discuss the portrayal of brain science in early science television broadcasts and, conversely, the way the TV apparatus was *the* medium to offer analogies for the active human brain around 1950. But the metaphorical circuits or analogical loops between brains and television cannot be neatly separated along these lines of representation and conceptualization. My analysis reveals that the meanings of brains *on* and *as* television are entwined and intersecting.

In this chapter's first part, I show how the topic of brain science took new shape during television's foundational era. I examine the way broadcasters literally connected home TV sets to broader systems of (medical) visualization, thus foregrounding new circuits of technologies and bodies.

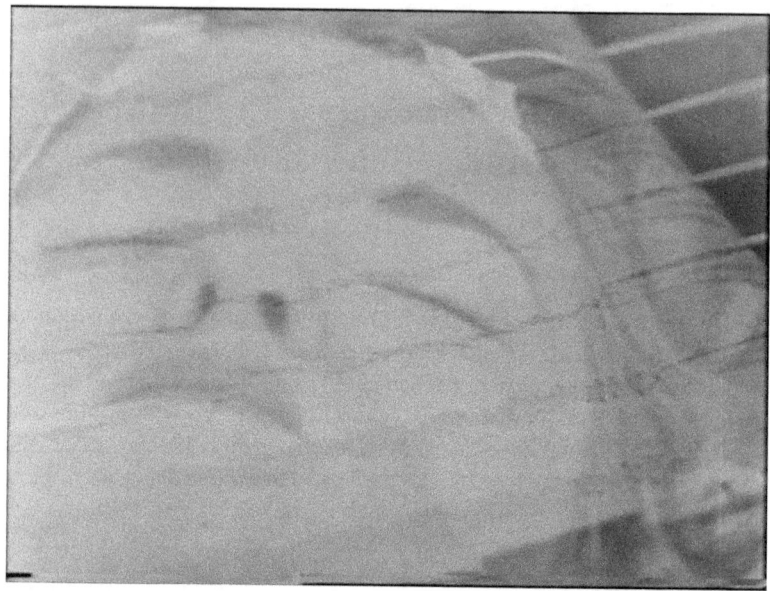

Figure 4.1.1 Catching a brain wave, 1954 (television broadcast still).

When broadcasters showed new brain research practices such as EEG research or brain X-rays, they combined television's ability to shape new forms of liveness (combining realness, directness, intimacy, nearness, and newness) with the longstanding quest to see the brain at work. By experimenting with live transmissions of visualizations of body and brain and producing new connections between transmission and visualization technologies, 1950s science broadcasts shaped what I argue was a networked intimacy between bodies and viewers.

While media historians have extensively analyzed television's 1950s golden age, there are relatively few studies about the decade's resonances between (medical) science and television. My analysis of the brain *on* television reveals how the era shaped a new, multifaceted notion of liveness that helps to understand simultaneous research developments into the active brain. By discussing previously unstudied science television experiments, I shed new light on this scholarly discourse on liveness. Though TV broadcasts about brains clearly showed off the technical ingenuity of brain science as well as television's capacity for liveness, these broadcasts were more than simply celebrations of science and technology. They shaped a new circuit between proximate TV screens and distant brain-measuring machines that

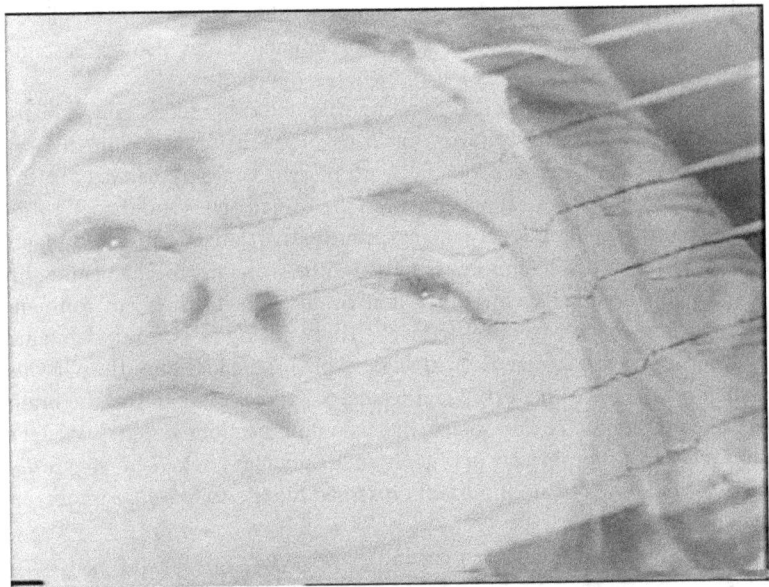

Figure 4.1.2 Catching a brain wave, 1954 (television broadcast still).

enabled a new conception of access to the living brain. These experimental 1950s broadcasts self-reflexively negotiated *what* television was—as a new medium—and *how* it brought new relations between studio and world/brain into being. Television thus functioned as both object and instrument of these investigations.

Building on my analysis of the complexity of brains *on* TV, the second part analyzes how a simultaneous cybernetics discourse on brain science led to a vision of the brain *as* TV around 1950. The resonances between brains and television are manifold. On the one hand, television—*the* medium of the time—appeared in brain research discourse as a symbol of the absorbing, media-saturated present. For example, describing the brain as a TV studio was a metaphor signaling the brain's (and modern subject's) immersion in an ever-busy (ever-broadcasting) and emotionally taxing environment. At the same time, cybernetics researchers found that television's scanning mechanism offered two interconnecting analogical lines of reasoning. The first hypothesized a scanning mechanism at work in human perception. Second, the effect of a flickering lamp on the TV receiver and screen was compared to the effect of flicker on human subjects, who reported strange sensations and visions, and whose EEG records showed altered nerve activity when stimulated with a stroboscope. While scientific texts employed television as a discursive tool, actual experimental setups also increasingly employed cathode ray tubes as displays.

Television's material and discursive roles in cybernetics research are particularly poignant in the work of British EEG researcher William Grey Walter, part of a heterogeneous network of cybernetics researchers in the late 1940s and 1950s (most prominent in the United States and the UK) who proposed new similarities between machines and organisms, such as humans, animals, and brains. In this cybernetic universe, organic mechanisms were envisioned to work like machines that could adapt to their environment through information feedback circuits. Three analogies were fundamental to cybernetics, as historian Ronald Kline explains: the idea that "the nervous system was deemed to work like a feedback control mechanism, the brain like a digital computer, and society like a communications system."[1] Walter's research (most prominently publicized in his popular book *The Living Brain* in 1953) functions as an important crossroad for television's appearances in cybernetics discourse.

My analysis of Walter's research, and that of other cybernetics researchers, adds a novel perspective on television's role and significance in 1940s and 1950s cybernetics discourses to the vast scholarship on cybernetics' "model-making heuristics."[2] By shedding light on the appearance of the human brain on and within a new broadcasting network, as well as cybernetics' conceptualization of television-like brains, I contribute to broader historical studies of the changing conceptions of the human brain and its relation to its media-saturated surroundings starting in the 1940s.[3] This chapter enlivens these discussions through lesser-known brain–television assemblages,

---

[1] Ronald R. Kline, *The Cybernetics Moment: Or Why We Call Our Age the Information Age* (Baltimore, MD: Johns Hopkins University Press, 2015), 45.
[2] Hayward, "Our Friends Electric," 298; see also ibid.
[3] A well-researched brainmedia assemblage from this period is the entanglement of brains and televisions in discussions on brainwashing. A number of historians have traced the emergence of cultural discourses on brainwashing during the budding Cold War in the mid-1940s: the rise of new anxieties over the mind's vulnerability to (mental) invasion (the term "psychological warfare" was first used in 1948, "menticide" in 1951). Television and broadcasting were often envisioned as having a particular "demonic agency," as Andreas Killen puts it. It was viewed as a pervasive infrastructure for manipulation and conditioning, part of what Stefan Andriopolous called an imaginary of a "mediatic network of mind control." Hence, histories of brainwashing (predominantly centered on US and UK developments) have demonstrated an intricate interrelation between (military-funded) psychological research into memory, attention, and perception, and contemporaneous conceptions and developments of broadcast media. Andreas Killen, "Homo Pavlovius: Cinema, Conditioning, and the Cold War Subject," *Grey Room* 45 (2011): 56; Stefan Andriopoulos, "The Sleeper Effect: Hypnotism, Mind Control, Terrorism," *Grey Room* 45 (2011): 98.

delving deeper into the tropes that reveal how the two became entangled and transposed in this period. Ultimately, I argue that the multiple and interacting material-discursive resonances between brains and television helped conceive of the brain as a type of black box, and this conception endowed the brain with both a new potential to reveal its functionality and brain research with a new deductive logic.

Broadcasting Science and a "Television of Attractions"

The aforementioned "Catching a Brain Wave" (1954) used televisual strategies to carefully choreograph objects, machines, and subjects and give the sense that viewers were looking at the electrically active, living brain. Discussing televised science in a 1950 publication, the *Johns Hopkins Science Review* producer emphasized how they aimed for a sense of reality, to "create an atmosphere of here and now, an atmosphere leading the viewer to the feeling that he himself is beside the demonstrator."[4] Beyond being informative and entertaining, science programs, like news broadcasts, were meant to foreground television's unique ability to convey the here and now, the feeling of proximity. This immediacy and intimacy were widely discussed in the 1940s and 1950s. The meaning and cultural status of this "liveness" was foregrounded by journalists, academics, and especially television producers, who were, as historian Rhona Berenstein argues, "intent upon describing a fledgling technology with an important overarching ontological mandate."[5] This process of establishing new authority and identity for the medium of television has been a central discussion in television historiography; in the 1950s, television makers and cultural critics foregrounded the idea of liveness as the essential quality of television.

Media theorists have stressed the multifaceted nature of television's liveness. Central were its technological capacity for instantaneity (seeing things as they happen, simultaneity) as well as proximity (bringing distant locations into living rooms, "seeing at a distance").[6] Yet, it was also shaped

---

[4] "Science via Television," cited in Marcel C. LaFollette, "A Survey of Science Content in U.S. Television Broadcasting, 1940s through 1950s: The Exploratory Years," *Science Communication* 24, no. 1 (2002): 44.

[5] Rhona J. Berenstein, "Acting Live TV Performance, Intimacy, and Immediacy," in *Reality Squared: Televisual Discourse on the Real*, ed. James Friedman (New Brunswick, NJ: Rutgers University Press, 2002), 25.

[6] At the end of the 1950s, television's unique capacities to transform experiences of space and time were perhaps most emblematically echoed in the texts of Marshall McLuhan, who saw television as the exemplary medium of a new electronic age, a time in which

by the intimacy of the TV in private homes, communal viewing, and risk (the possibility of unscripted things happening during a live broadcast).[7] Importantly, these heterogeneous aspects gave TV a perceived ability to show what contemporaneous interlocutors called "actuality," a "presumed access to the real."[8] The medium's conceived identity as being "live" (which has been called its "ontological identity," "ideology," or "generalized fantasy") persisted in later discourses too.[9] Even when it became possible to broadcast prerecorded programs in the 1950s, the medium was still structurally conceived around the potential of instantaneous broadcasts.[10]

Due to this perceived ontological quality, instantaneity has received the most attention in scholarly literature. Yet, as Mimi White has argued, a closer look at live broadcasts also reveals the importance of television's own emphasis on spatiality.[11] Live transmissions (still) continuously orient viewers in relation to distant locations, emphasizing their ability to reveal things viewers cannot literally see: a "spatial pyrotechnics of images that function as visual spectacle."[12] As such, they underline television's visually revealing character, what White calls a "television of attractions": "an excessive visibility and visuality" flaunting the medium's revelatory technological abilities to traverse physical boundaries.[13]

1950s science and medical broadcasts preeminently offered such a television of attractions. Instantaneity and proximity were foregrounded, for example, by connecting viewers to hospitals' closed-circuit television—opening an otherwise closed world—or by creating special demonstration labs in studios, showing scientific experiments and intricate machines working instantaneously.[14] Early US science television experimented with live

<blockquote>
audio-visual media formed new "extensions of man" (the exteriorization of the nervous system), shaping new relations, a new culture of simultaneity. Marshall McLuhan, Understanding Media: The Extensions of Man (Cambridge, MA: MIT Press, [1964] 1994), 248.
</blockquote>

[7] For summary, see Martin Barker, *Live to Your Local Cinema: The Remarkable Rise of Livecasting* (London: Palgrave Macmillan, 2013), 57–8.
[8] Berenstein, "Acting Live TV Performance, Intimacy, and Immediacy," 25.
[9] Ibid.
[10] Jane Feuer, "The Concept of Live Television: Ontology as Ideology," in *Regarding Television: Critical Approaches—An Anthology*, ed. E. Ann Kaplan, American Film Institute Monograph Series (Frederick, MD: University Publications of America, 1983). Feuer argues television's ontology of immediacy was in fact "ideology," what Mary Ann Doane calls a "generalized fantasy."
[11] Mimi White, "The Attractions of Television. Reconsidering Liveness," in *Mediaspace: Place, Scale and Culture in a Media Age*, ed. Nick Couldry and Anna McCarthy (London: Routledge, 2004), 89.
[12] Ibid.
[13] Ibid., 87.
[14] The *Johns Hopkins Science Review* staged live broadcasting itself as an important example of technical and scientific progress and dedicated two episodes to a "behind the scenes"

transmissions of scientific visualization. In 1948, for example, a Washington network broadcasted direct views from Naval Observatory telescopes to show grainy images of Mars, Saturn, and the Moon and also transmitted direct microscopic views of blood flow in a mouse.[15] The scientific demonstration format could lay claim to authenticity and veracity through the event's perceived directness (seeing it with your own eyes) and indeterminacy (the potential of the unexpected and unscripted). Appropriating this reality effect, these "direct" views underlined the medium's special ability to bring the viewer into direct contact with unusual, scientific perspectives, thus celebrating the technical abilities of both the medium of television and the scientific technologies on display.

In various broadcasts, the body and the brain served as key subjects of this "spatial pyrotechnics." Broadcasting images of the body's insides offered new ways to flaunt television's ability to transgress spatiotemporal boundaries, effecting a seemingly impossible intimacy—a directness, nearness, and aliveness. TV thus realized a new type of bodily liveness.[16] Especially potent, in this sense, were live transmissions of surgeries or a patient's remote (tele-)diagnosis via a TV feed.

In 1950s France, "live" medical broadcasts were widely discussed after the first TV broadcast of an endoscopy, *Endoscopie en direct*, by science filmmaker Jean Painlevé in 1954.[17] Film theorist André Bazin described "*le direct*" as television's medium-specific contribution to the aesthetics of scientific cinema. He believed television could be a "passionate intermediary" between the modern human and the rapidly developing domain of scientific research.[18] Its liveness and simultaneity created powerful spectacle, he stated,

of their own program. "A Visit to Our Studio," January 7, 1952, and "From Studio to Your Home," January 14, 1952. See, for instance, Lynn Poole, "A Visit to Our Studio," in *John Hopkins Science Review* (USA: DuMont Television Network, 1952).

[15] Marcel C. LaFollette, *Science on American Television: A History* (Chicago: University of Chicago Press, 2013), 10.

[16] David Serlin speaks of an "interface potential," arguing that television helped medicine by contributing to a particular "heightened visual sensibility" of postwar medical science, which enabled "a visual culture of public health that was unprecedented in the public sphere." In turn, medicine allowed television to associate itself with a postwar narrative about the rise of promising medical technologies and the scientific authority of new technologies. David Serlin, "Performing Live Surgery on Television and the Internet," in *Imagining Illness: Public Health and Visual Culture*, ed. David Serlin (Minneapolis: University of Minnesota Press, 2010), 226.

[17] Painlevé's images of the 1954 endoscopy were transmitted at the start of *En direct de* on December 14, 1956 and described as "like walking in the shafts of a mine." Roxane Hamery, *Jean Painlevé, Le Cinéma Au Cœur De La Vie* (Rennes: Presses Universitaires de Rennes, 2013).

[18] André Bazin, "La Télévision: Moyen De Culture, *France-Observateur 297* (19 January 1956)," in *André Bazin's New Media*, ed. Dudley Andrew (Oakland: University of California Press, 2014), 120, 28.

"pure television."[19] Commenting on the broadcast, Painlevé himself described it as a shared medical-televisual accomplishment enabled by the "combined use of the universal endoscope, the brilliancy amplifier and the extremely sensitive television tube."[20]

Painlevé had orchestrated it so that the patient could follow his broadcasted operation on a nearby screen. The broadcast thus also allowed a "descent into yourself," according to Bazin—alluding to religious or psychological self-analysis—"thanks to TV, man has become his own Plato's cave."[21] He argued: "The honest human of this half-century can no longer ignore, for example, the miracles of modern surgery and the stupefying investigations of endoscopy ... A world where man can see the interior of his own living body, is no longer the same; we have to change with it."[22]

Painlevé's live endoscopy shows the emergence of a new form of liveness, a radical type of networked intimacy tied to the new possibilities of live television. Here, television functioned as an experimental space to present a new phenomenon—the endoscopic view—while also self-reflexively staging its own participation within this new system, negotiating television as both object and instrument of investigation. This experiment fueled discussions about the medium's unique possibilities, leading Bazin to hail live TV as enabling "a new cultural style." Two years later, this new style was also brought to the impenetrable realm of the human brain in *En direct du Cerveau Humain* ("Live from the human brain"). This episode most incisively shows how the "text" of the broadcast interacted with television as an apparatus to shape an experimental form of television that offered a live view of the brain at work.

The Brain "*En direct*" and the Epistemological Seductions of Television

In 1956, the brain literally became "live." That year, the popular French TV program *En direct de* ("Live from...") aired *En direct du Cerveau Humain*.

---

[19] Ibid., 129.
[20] Anon., "Television Report. Televised Endoscopy. Note about a Programme Broadcast on 14 December 1955 Presented by Jean Painlevé," *Science and Film* (1956). Painlevé remarked on the fact that nothing would escape the cameras' attention in this surgery studio with four different viewpoints; doctors wrote to him saying that they had been able to make a diagnosis from afar.
[21] Bazin, "La Télévision," 120.
[22] Bazin, *Radio-Cinéma-Télévision*, 1956, cited in Hervé Brusini and Francis James, *Voir La Vérité: Le Journalisme De Télévision* (Paris: Presses Universitaires de France, 1982), 117.

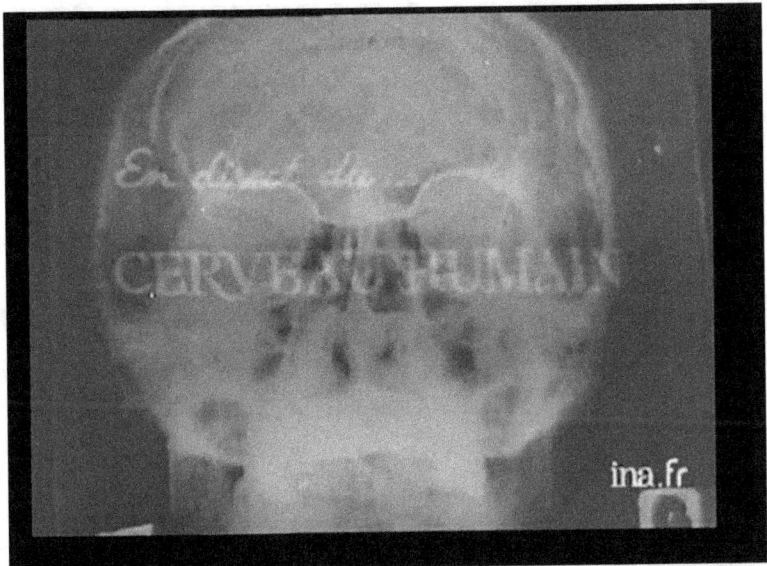

Figure 4.2.1 *En direct du Cerveau Humain*, 1956 (television broadcast still).

Every week, the program offered a different "live" experience, a televisual expedition to a remote and strange location like a coalmine, the bottom of the sea, an atomic reactor, or the inside of the human heart. The format emphasized television's spatiotemporal pyrotechnics of liveness. As one French commentator remarked in 1957, "The essence of television is direct reportage."[23]

Through reportage, the inside of the body could now be understood as an exotic, potentially visible location. At the start of *En direct du Cerveau Humain*, the host introduced that week's adventure as an extra special, perhaps impossible endeavor, framing the mission as venturing into one of the "great problems of mankind," that is, the desire to see inside the brain, "to hold it, to extract a sign from it" (Figure 4.2.1). Paris's Salpêtrière Hospital is the setting, and the narrative is centered around the battle against epilepsy, for which new views of the inside of the brain seemed very promising. Scene by scene, the dream of seeing inside the brain becomes fulfilled by modern science: viewers are shown a range of new techniques enabling scientists to approximate the brain. We cut between interviews with neurologists and

---

[23] Moureau, *Radio-Ciné*, Janvier 1957, cited in ibid., 46.

Figure 4.2.2 *En direct du Cerveau Humain*, 1956 (television broadcast still).

biologists at the host's table, medical procedures with complex machines, and a variety of views into the brain, such as an X-ray held up to the screen, a microscopic view of white matter, and the zigzagging lines on an EEG recording sheet.

Also taking part was eminent epilepsy researcher Henri Gastaut, who aimed to draw more media attention to the cause and the French League against Epilepsy. One scene cuts between Gastaut, standing in front of a large brain diagram with flashing lights to explain nerve cell activity, and a view of an EEG-fitted female subject whose brain activity fed into the brain diagram and caused the lights to change when she opened and closed her eyes (Figures 4.2.2 and 4.2.3). Through the succession of these various setups, the format of the television reportage enabled an assembly of different images, sites, and disciplines into a unified image of a new science, a science that could now be envisioned as having access to the invisible activity of the brain.

The "liveness" of the brain on TV becomes prominent in the episode's most daring intervention: the attempt to directly transmit a pneumoencephalography to the TV screens at home. First, the camera shows a darkened room filled with machines. A nurse adjusts the body of a patient who is hidden from view, in preparation for the procedure. The

Figure 4.2.3 *En direct du Cerveau Humain*, 1956 (television broadcast still).

voice of the host, who is positioned back at the machine's display, describes how a successful visualization depends on positioning the patient and fine-tuning the parameters. At one point, we cut to a direct transmission from the visualization apparatus. The screen turns black with a few spots of gray. The voice of a medical researcher suggests we can discern a jawbone and the eye-socket rim, yet the screen shows scarcely more than a black void (Figure 4.2.4). The host apologizes, explaining the procedure is delicate and has unfortunately failed to work on the spot. Quickly, he announces the next segment.

En direct du Cerveau Humain brought the invisible insides of the human head to living rooms in the form of a new image, a new presence. It introduced a new form of liveness, a networked intimacy, that was specifically related to this brainmedia assemblage. By interrupting the smoothness of the ongoing broadcast, the failed brain visualization ended up reinforcing liveness by emphasizing the broadcast's unscripted authenticity. Gaffes are thus part of the exciting potential of live science television; failing to switch to the right feed, a faltering machine, a missing voiceover: they all augment the medium's claim to "actuality." This oscillation between the smooth editing of a pliable medium and the anticipation of unexpected and unscripted events enhanced

Figure 4.2.4 *En direct du Cerveau Humain*, 1956 (television broadcast still).

the verisimilitude and also revealed the delicacy of the technological procedures. Though such novel visualizations were clearly feeble at times, the ultimate outcome of this broadcast was the idea of a live brain that had moved into the realm of the potentially visible.

En direct du Cerveau Humain's technological gaffes were subsumed in its narrative arc of technological progress, obscuring the pain, anxiety, and failure surrounding technomedical interventions. The episode thus fully emanates what José van Dijck has called the "myth of total transparency" in its framing of scientific visualizations.[24] The intermedial exchanges between medical and broadcast media shape a particular "epistemological seduction," portraying medical imaging as noninvasive and painlessly dissolving boundaries between "the television screen, the walls of the studio, the operating theatre, and the patient's skin."[25] This is not innocent or merely informative, Van Dijck argues, as it actually generates "false ideas about transparency and non-intervention."[26]

---

[24] van Dijck, *The Transparent Body*, 6.
[25] Ibid., 73.
[26] Ibid.

The medical gaze foregrounded in the above broadcasts engages in a familiar equalization of the human eye with technological visualization—and of the mediated body with the material body—and does so in a way that conjures a painless transparency. *Endoscopie en direct* celebrated a future of humanity's new medico-televisual intimacy with itself, yet omitted that this live endoscopy had to be performed on a sword swallower—the only person who could bear the pain of the bright TV lamp in his esophagus.[27] Similarly, viewers of *En direct du Cerveau Humain* were not told that pneumoencephalography was an extremely painful procedure for which air had to be injected into the base of the spinal column.[28] On the level of the broadcast text, these medico-scientific programs may thus be understood as celebrating both the sciences portrayed and the combined medico-televisual gaze that offered a myth of a transparent body, a painless total visibility. Exposing these myths is important, as is acknowledging that televisual celebrations of science often show little of the practical uncertainties, power imbalances, and bodily anguish of medico-scientific practice. Yet, my aim in recounting this history is to emphasize the way that such brain and body broadcasts also self-reflexively negotiated what television was as a new medium and how it brought new relations between studio and world into being. Television, in these cases, functioned as both object and instrument of investigation.

This history of brain (and body) broadcasts adds new substance to an analytical view emphasizing television's generative and self-reflexive qualities: the idea that the mechanism by which things appeared *on* television could become foregrounded and questioned *by* television, particularly in experimental broadcasts. Media scholar Lorenz Engell emphasizes this view by describing television as an experimental system, emphasizing its capacity to configure novel phenomena through continuous transformation, persistently reflecting on what counts as a successful broadcast, and redefining its instruments and technologies in the process.[29] This requires taking seriously the conditions that allow this complex, distributed, experimental system (broadcasting institutions and infrastructures, audiences, technical devices, etc.) to produce novelty, understanding it as an indefinite space where

---

[27] Hamery, *Jean Painlevé*.
[28] On the history of pneumoencephalography, see Bettyann H. Kevles, *Naked to the Bone: Medical Imaging in the Twentieth Century* (New Brunswick, NJ: Rutgers University Press, 1998), 101.
[29] Engell, "Fernsehen Mit Unbekannten." See also Judith Keilbach and Markus Stauff, "When Old Media Never Stopped Being New. Television's History as an Ongoing Experiment," in *After the Break: Television Theory Today*, ed. Marijke de Valck and Jan Teurlings (Amsterdam: Amsterdam University Press, 2013).

new relations and concepts can emerge. Within this "play of possibilities," television continuously shifts between foreground and background, between a well-defined instrument that shapes what it broadcasts and moments in which its self-constitution as a medium is experimentally altered and (self-reflexively) on display.[30]

Looking at these broadcasts through the lens of experimental television allows us to recognize their generative qualities. *Endoscope en direct* emphasized a live intimacy, the possibility of our descent into ourselves. *En direct du Cerveau Humain*, on the other hand, allowed for the assembling of medical technology (pneumoencephalography) with the TV screen at home, conjuring the strange sense of it becoming part of a wider network of technical visualization. The physical screen thus is invisible (a window into another world) but also gains new materiality as a technological surface and part of a larger amalgamation of machines.

One of the makers of *En direct de* would later describe the main goal of their medical broadcasts as "creating television," emphasizing instantaneity and intimacy: "See here a heart that beats, see here a man who has mastery over this beating heart. You are at home. The technology of television allows you to see, appreciate, not necessarily understand it."[31] This meant foregrounding its ability to generate new forms of liveness and captivate a viewer who could be amazed by a television of attractions without knowing exactly the workings of the phenomena on display. Here, "creating television" meant shaping both a new understanding of television itself and also producing new relations with the world.

By creating television with a live connection to the brain, the broadcasters created a new self-image of television and at the same time a new live brain. This live brain was part and parcel of a new networked intimacy, not only shaped *on* television but also enabled by an experimental technological network between TV sets at home, studios, and visualization machines. The black void—transmitted directly from the Salpêtrière's X-ray device—may have emanated a particularly strange presence that connected to a broader cultural imaginary of television's "portal" quality. As Jeffrey Sconce has argued, discourses about television in the 1950s were influenced by a broader and long-established imaginary of a potential "aliveness" (being animated, sentient, haunted, or possessed) of electronic technologies.[32] Television was attributed with uncanny, "occult" powers just like radio and telegraphy had been in the past. Yet, the "electronic elsewhere" produced by

---

[30] Keilbach and Stauff, "When Old Media Never Stopped Being New," 80.
[31] Barrère, host of *En direct de*, in Brusini and James, *Voir La Vérité*, 118.
[32] Sconce, *Haunted Media*, 126.

television was even more intense because of its new audio-visual quality: it was "at once visibly and materially 'real' even as viewers realized it was wholly electrical and absent."[33] In various cultural commentaries, the TV had been conceptualized as a type of gateway, as having a weird and eerie presence in the living room, and even as a sentient entity itself. These peculiar dimensions of television's liveness were now augmented by the particular brain–X-ray–television connection offered by *En direct du Cerveau Humain*: TV itself now seemed to possess a brain.

Toward the Brain as TV: The Cybernetic Living Brain

Following my analysis of how brains *on* television created a networked intimacy and turned TVs into cerebral machines, I now trace a simultaneous television–brain assemblage that led to conceptions of brains *as* television, turning to cybernetics-affiliated research that proposed a variety of resonances between the brain, TV studio, visual cortex, broadcasting, hallucinations, TV viewers, and distorted screens. These are most present in the work of cybernetics EEG researcher William Grey Walter.

Walter's research into electrical activity in the human brain (of both epilepsy patients and neurotypical subjects) was characterized by a particularly exuberant back and forth between theoretical hypotheses, small contraptions (working models), textual analogies, and tinkering with EEG recording machines, with cathode ray displays and TVs also playing key roles. Yet, he not only worked *with* media but also appeared *on* popular media. In the 1940s, he started to speak on the radio and became a welcome guest on UK TV (including the famous BBC show *The Brains Trust*) in the 1950s. His work thus is characteristic of cybernetics' particularly public life and of television's zigzagging resonances in this media-oriented body of scholarship.

Here I extend the existing (and now vast) body of historical scholarship on cybernetics (including Walter's models) with a media-analytical emphasis on the assembling of brainmedia.[34] Television took on specific significance

---

[33] Ibid.
[34] Several cybernetics historians have mentioned the perception-as-scanning hypothesis. In their studies of William Grey Walter's work on EEG, Andrew Pickering and Cornelius Borck elaborated on his working models for scan-like brain mechanisms and noted Walter's invention of a television-like EEG apparatus called the toposcope. Andrew Pickering, *The Cybernetic Brain: Sketches of Another Future* (Chicago: University of Chicago Press, 2010). Borck, *Brainwaves*. There is a vast library on the history of cybernetics starting in the mid-1940s and on its uptake in fields ranging from

within the material-discursive field of cybernetics, where it featured, in crisscrossing movements, as cultural emblem, lab instrument, and technical analogy. Two specific television–brain (brain *as* television) figures receive most attention in my analysis: the so-called cortical scanning hypothesis (the idea that perception worked as a scanning mechanism) and the related epilepsy research in which flicker lamps ignited new comparisons between flicker televisions and flicker brains. In the remainder of this chapter, I unpack the history of the cybernetic television brain to describe the way that brains and televisions were intertwined in complex material and discursive ways during TV's 1950s golden era.

Speaking on BBC Radio in 1951 for a three-part series called "Patterns in Your Head," Walter told his listeners about "brain prints" called EEG and how such records suggest a new understanding of the active brain as a "pattern-seeking and pattern-making instrument."[35] He offered an analogy to explain: "The radio you're listening to now is making a pattern in time," while the more complicated TV was able to change received signal patterns into "patterns in space" through a scanning mechanism that translated them into images on the screen. Starting in the 1940s, Walter and other researchers claimed that the procedure by which the eye and visual cortex allowed for perception might be similar to television's raster scanning. Walter argued that just like pictures on a TV screen could look distorted if there was a flickering light in the studio, exposing humans to flicker could also interfere "with the normal process of scanning" in the brain, resulting in altered EEG patterns and subjects seeing strange patterns, even when their eyes were closed.[36] Those flicker effects were usually hard to bear for his subjects, similarly to the flickering pictures on TV. Flicker could "produce a brain storm as wild as any distortion on the television screen."[37]

Though cybernetics research comprised only a small part of scientific research on the nervous system and mental processes (such as in

---

organization management and urban planning to couples therapy and group psychology. Claus Pias, "The Age of Cybernetics," in *Cybernetics: The Macy Conferences 1946-1953*, ed. Claus Pias (Zurich: Diaphanes, 2016). Within this literature, the cybernetic vision of a computing brain and the relation between cybernetics and neurophysiology, behavioral science, experimental psychology, and psychiatry have also been analyzed in a number of publications; for example, Kay, "From Logical Neurons to Poetic Embodiments of Mind"; Pickering, *The Cybernetic Brain*; Joseph Dumit, "Plastic Diagrams: Circuits in the Brain and How They Got There," in *Plasticity and Pathology: On the Formation of the Neural Subject*, ed. David Bates and Nima Bassiri (New York: Fordham University Press, 2016).

[35] W. Grey Walter, "Pattern-Making and Pattern-Seeking," in *Patterns in Your Head: A Weekly Programme about Work in the World of Science* (Great Britain: BBC Home Service, 1951).
[36] Walter, *The Living Brain*, 111–12.
[37] Walter, "Pattern-Making and Pattern-Seeking."

neurophysiology and neuropharmacology) in the mid-twentieth century, the story of the "cybernetic," information-processing brain has received much attention in the past decades. Historian Max Stadler, commenting on this vast literature, has cautioned that the narrative may have become all too dominant.[38] The prominence of this heterogeneous body of ideas about human brains and their relation to "thinking machines" and "electronic brains" (new information-processing machines) may be due to cybernetics' public visibility mid-century, as cybernetics researchers and popular science authors published successful books like *Giant Brains or Machines That Think* (1949), *Design for a Brain* (1952), *The Living Brain* (1953), *Minds and Machines* (1954), *La Pensée Artificielle* (1956), *The Human Brain* (1955), *The Computer and the Brain* (1958). Historian Ronald Kline calls it the "cybernetics craze" of the 1950 and 1960s, and Geoffrey Bowker notes that cybernetics also became a cult subject for a wider audience through newspapers, magazines, science exhibitions, and radio and TV shows.[39]

When analyzing the meaning of media within cybernetics discourses about brain science, it is important to consider cybernetics' status as a (pop-) cultural phenomenon that overlapped with infrastructures of (popular) mediation. Cyberneticists talked about television-like brains on radio and TV; they studied the mechanisms of communication devices while working on and with such devices; they theorized about demonstrative mechanisms of brain research while doing public demonstrations of such mechanisms; they were hailed as mediators of science while shaping new networks of mediating science. Cybernetic theorizing was characterized by these structures of performing knowledges, affecting the material-discursive loops through which the television–brain could be assembled.

In this fully media-saturated scientific discourse, Walter, as a scientist speaking on BBC Radio, is characteristic of the way cybernetics knowledge could be performed. Not only did he regularly appear on TV and radio, he also showed his mechanical models at London's Festival of Britain in 1951 and published a popular account of his research (*The Living Brain*, 1953).[40] Its overarching theme—the "living brain"—allowed Walter to frame his work as part of a long history of analyzing the brain as an active organ (starting as far

---

[38] Stadler, "The Neuromance of Cerebral History."
[39] Ronald Kline attributes cybernetics' popularity to public enthusiasm, ambivalence, and anxiety about the development of electronic computers and the rise of automation. Kline, *The Cybernetics Moment*, 69; Geoffrey Bowker, "How to Be Universal: Some Cybernetic Strategies, 1943–70," *Social Studies of Science* 23, no. 1 (1993): 108. Also Claus Pias, "Zeit Der Kybernetik," in *Cybernetics—Kybernetik. The Macy-Conferences 1946–1953. 2: Dokumente Und Reflexionen*, ed. Claus Pias (Zürich: Diaphanes Verlag, 2004).
[40] Walter, *The Living Brain*.

back as David Hartley's ideas of "brain vibrations" in 1749). In his narrative, the "living" of the living brain took on significance as the central tenet of cybernetic brain research: the idea that new theoretical and physical models and research technologies had brought researchers closer to accessing and understanding the workings of the active, living human brain.

The cybernetic idea of new access to the living brain arose not from one new technology but from an accumulation or networking of different hypotheses, analogies, instruments, and models. In *The Living Brain*, Walter expounded on his playful, lifelike models or automatons, which he created to better understand human cognition and perception; the toposcope, a (television-like) machine he devised for more comprehensive EEG visualizations; and new methods for EEG experimentation (flicker stimulation, for example) that might allow for more direct correlation between EEG and behavior. Together, these experimental and interlocking methodologies shaped the vision of new access to the active living brain. This way, brains did not need to literally be cracked open, but were tackled through different revelatory approaches, what Walter called "brain mirrors."[41] The EEG record provided "bits and pieces of the mirror," but so did his models, technical setups, and visualization apparatuses.[42]

Walter's writings on brain research evince cybernetics' sometimes dizzying analogical loops between technical instruments, working models, mechanical analogies, and hypotheses. A characteristic example from *The Living Brain* is the rhetorical juxtaposition of two images through which Walter argued for the similarity between the human brain and a "profile scanning device" he had developed to process EEG graphs (Figure 4.3).[43] Even though his argument is hard to follow (comparing various elements of his device to the visual cortex, fluctuating electrical activity, and receiving neurons), on a visual level the analogy seems to make perfect sense: the illustration shows how both produce steady zigzag records that form patterns depending on visual stimuli. In this example, the machine–brain comparison makes an intricate summersault: lab instruments developed to process EEG recordings may themselves be mechanically similar to the workings of the brain. Such comparisons were suggestive, partial, and sometimes even tautological. Yet, this did not matter for Walter's success, as Cornelius Borck has suggested: perhaps that was more due to "constructive productivity itself rather than the establishment of a theory."[44] It was his exuberant ongoing tinkering, his excitement for ambiguous analogies, that constituted its value.

[41] Ibid., 60.
[42] Ibid.
[43] Ibid., 110.
[44] Borck, *Brainwaves*, 263.

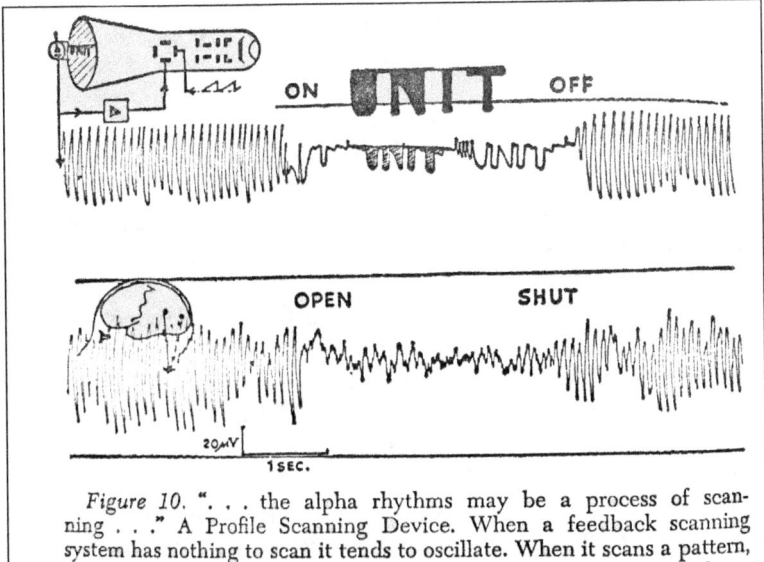

Figure 10. ". . . the alpha rhythms may be a process of scanning . . ." A Profile Scanning Device. When a feedback scanning system has nothing to scan it tends to oscillate. When it scans a pattern, its movements reproduce a facsimile of the pattern on a time base. Below, the alpha rhythms of a normal subject seem to suggest similarity of mechanism in model and brain.

Figure 4.3 Illustration of an analogy between a profile-scanning device and the human brain, 1953.

Taken together, what was most astonishing about Walter's cybernetic framing of the living brain was conceptually uniting this heterogeneous array of engineered devices, theoretical presuppositions, and research approaches. N. Katherine Hayles has aptly typified cybernetic thinking's bricolage of modeling and hypothesizing as "analogical linking."[45] This linking often went flexibly back and forth between objects and referents, as Claus Pias noted, with "productive fuzziness."[46] When researchers used machine–human comparisons, or when they designed mechanical constructions to mimic mental and behavioral processes, they created what Hayles describes as a discursive (or perhaps more suitably material-discursive) field through which "animals, humans, and machines can be treated as equivalent cybernetic systems."[47]

---

[45] N. Katherine Hayles, *How We Became Posthuman: Virtual Bodies in Cybernetics, Literature, and Informatics* (Chicago, IL: University of Chicago Press, 1999), 93. "Cybernetics as a discipline could not have been created without analogy" (91).
[46] Pias, "Zeit Der Kybernetik," 22.
[47] Hayles, *How We Became Posthuman*, 91, 93.

Cortical (Television) Scanning, the "Ineluctable Inference"

When television emerged in the 1930s, it was immediately compared to "human eye" cameras.[48] In the 1940s, these eye-television analogies became more elaborate and specific: cybernetics theorists observed that regular nerve activity patterns were involved in the process of visual perception and hypothesized that the relation between eye, brain, and environment could fruitfully be compared to a radar- or television-like scanning "sweep." Walter and his coresearcher Vivian Walter, for example, posited that alpha rhythms worked as "scanning generators" that "convert the spatial pattern of excitation on the visual cortex into a code of impulses on a time base for transmission to other regions."[49] Thus, (television) scanning offered a plausible model for perception that took hold of the cybernetics community in the late 1940s.[50]

Differing versions of the perception-as-scanning idea circulated, developed in parallel or resonating between researchers such as Kenneth Craick (1943), Warren McCullogh and Walter Pitts (1947), Norbert Wiener (1948), and William Grey Walter (1949).[51] Walter, who during wartime helped to invent the circular radar screen with the sweeping light beam (a "plan position

---

[48] Alden P. Armagnac, "'Human-Eye' Camera Opens New Way to Television," *Popular Science* (1933).

[49] V. J. Walter and W. Grey Walter, "The Central Effects of Rhythmic Sensory Stimulation," *Electroencephalography and Clinical Neurophysiology* 1, nos. 1–4 (1949): 82.

[50] While the idea of alpha as a "scanning sweep" for perception was discarded, hypotheses about the importance of nerve rhythms in the process of perception (including the idea of "discrete" perception) continue to be debated until the present. Rufin VanRullen, "Perceptual Rhythms," in *Stevens' Handbook of Experimental Psychology and Cognitive Neuroscience*, ed. John Serences (American Cancer Society/John Wiley, 2018).

[51] While in 1943 Wiener still remarked that "scanning is a process which seldom or never occurs in organisms" (the television receiver's detection rate (20 million per second) was far too fast for what the eye could achieve), by 1948 he was convinced that scanning presented the model system by which not only machines, but also humans were able to deal with a vastness of data that needed computing. Arturo Rosenblueth, Norbert Wiener, and Julian Bigelow, "Behavior, Purpose and Teleology," *Philosophy of Science* 10, no. 1 (1943): 23; Norbert Wiener, *Cybernetics: Or Control and Communication in the Animal and the Machine*, 2nd ed. (Cambridge, MA: MIT Press, [1948] 1961), 141. Kenneth J. W. Craik, *The Nature of Explanation* (London: Cambridge University Press, [1943] 1967), 74. Walter Pitts and Warren S. McCullogh, "How We Know Universals: The Perception of Auditory and Visual Forms," *Bulletin of Mathematical Biophysics* 9, no. 3 (1947). Walter and Walter, "The Central Effects of Rhythmic Sensory Stimulation," 82. For a discussion of some of these theories, see M. Russell Harter, "Excitability Cycles and Cortical Scanning: A Review of Two Hypotheses of Central Intermittency in Perception," *Psychological Bulletin* 68, no. 1 (1967).

indicator"), would look back at the 1940s in *The Living Brain*: "Goal-seeking missiles were literally much in the air in those days; so, in our minds, were scanning mechanisms."[52] What the theories had in common was that scanning (as present in radar or TV technology) was an efficient mechanism to relay spatiotemporal information, and that human perception might plausibly be based on a similar system. This also meant that perception was determined by a particular cycle time, as historian Andrew Pickering explains: "The idea of a brain that lives not quite in the instantaneous present, but instead scans its environment ten times a second to keep track of what is going on."[53] Researchers saw this corroborated by electrophysiological observations: alpha rhythms seemed to represent a periodic scanning process (a "time of sweep") and were already known to be involved in visual perception.[54]

Walter pursued various lines of research to test the scanning hypothesis. In tandem with research on nerve activity in living subjects, he started to examine rudimentary mechanisms of behavior (including perception) through simple automatons with light-sensitive cells, what he called "working models."[55] These automatons (lovingly dubbed "tortoises") were viewed as lifelike because they appeared to see with their rotating "eyes" (photoelectric cells) and search for particular positions in space.[56] Because they resembled the behavior of living creatures, they fortified the idea that human behavior might be based on a scanning principle too. Hence, these working models could support and strengthen the physiological theory of "internal scanning as a function of electrical brain rhythms in the final stage of sensory perception."[57] Even if the models did not prove the scanning theory, they at least fortified its plausibility.

In the 1950s, not everyone was convinced by cybernetics' analogies and models. Several historians have noted how, at the famous Macy cybernetics conferences, the analytical usefulness and rhetorical worth

---

[52] Walter, *The Living Brain*, 125.
[53] Pickering, *The Cybernetic Brain*, 48. Note here that the liveness associated with television is different from the cycle time associated with its scanning technology. On the notion of temporality in cortical scanning and a longer genealogy of the "brain clock hypothesis," see Henning Schmidgen, "Zeit Als Peripheres Zentrum: Psychologie Und Kybernetik," in *Cybernetics: The Macy Conferences, 1946-1953. Band 2 Cybernetics: Essays and Documents*, ed. Claus Pias (Berlin: Diaphanes, 2004). H. Schmidgen, "Cybernetic Times: Norbert Wiener, John Stroud, and the 'Brain Clock' Hypothesis," *History of the Human Sciences* 33, no. 1 (2020).
[54] Wiener, *Cybernetics*, 23.
[55] In *The Living Brain*, Walter explained this method as allowing "some elementary experience of the actual working" of a small number of brain units by constructing (limited though feasible) "working models." Walter, *The Living Brain*, 120. See Peter Asaro, "Working Models and the Synthetic Method: Electronic Brains as Mediators between Neurons and Behavior," *Science Studies* 19, no. 1 (2006).
[56] Hayward, "Our Friends Electric."
[57] Walter, *The Living Brain*, 131.

of the proposed analogies and hybrid models were a prominent topic of discussion.[58] Debates on the models' heuristic value (the nature and extent of their representativeness) were prevalent, particularly in the discussion surrounding the "digital model" for the functioning of (networks of) neurons.[59] Critiques also came from outside cybernetics' inner circle: neurosurgeon Geoffrey Jefferson, for example, writing in the *British Medical Journal* in 1949, argued that while "modern automata" such as Walter's were characteristic of the "ingenuity of invention at the present time," they were perhaps all too seductive.[60] The cortical scanning hypothesis was also criticized in 1954, as psychologist W. Sluckin wrote in his popular account of cybernetics, *Minds and Machines*, that it had become "somewhat doubtful."[61] While cybernetics researchers suggested analogies to understand "how the brain might be organized to act as it does," it was unclear "to what extent [the brain] resembles the theoretical mechanisms."[62] In the 1950s, researchers set out to experimentally test the scanning hypothesis related to EEG rhythms, finding no evidence to support it.[63] Nevertheless, Sluckin's account shows the hypothesis' lasting appeal and potency, as "the hypothetical picture of the mechanisms does not conflict with any known facts of neuroanatomy and neurophysiology" (on the contrary, it seemed positively supported by them), and "the explanation of certain features of perception offered by it is interesting, plausible, and almost certainly amenable to further testing."[64]

[58] Pias, "The Age of Cybernetics," 17.
[59] In Chapter 1, we already encountered Ralph Gerard's objections to the idea that nervous networks were akin to electrical circuits. On the debates surrounding the electrical switch analogy for nerve activity, see Kline, *The Cybernetics Moment*, 45–47. Paul N. Edwards, *The Closed World: Computers and the Politics of Discourse in Cold War America* (Cambridge, MA: MIT Press, 1997), 191–3. Jean-Pierre Dupuy, *On the Origins of Cognitive Science: The Mechanization of the Mind* (Cambridge, MA: MIT Press, 2009). Pias, "Elektronenhirn Und Verbotene Zone."
[60] Geoffrey Jefferson, "The Mind of Mechanical Man," *British Medical Journal* 1, no. 4616 (1949). Ultimately, Jefferson showed a certain disdain toward a new breed of scientists as model tinkerers (1110).
[61] W. Sluckin, *Minds and Machines* (Harmondsworth: Penguin Books, 1954), 130.
[62] Ibid., 132.
[63] D. M. MacKay, "Some Experiments on the Perception of Patterns Modulated at the Alpha Frequency," *Electroencephalography and Clinical Neurophysiology* 5, no. 4 (1953): 559–62; E. G. Walsh, "Visual Reaction Time and the Alpha-Rhythm, an Investigation of a Scanning Hypothesis," *Journal of Physiology* 118, no. 4 (1952): 500–508; H. B. Barlow, "Eye Movements during Fixation," *Journal of Physiology* 116, no. 3 (1952): 290–306.
[64] Sluckin, *Minds and Machines*, 132. See W. Grey Walter, "Features in the Electro-Physiology of Mental Mechanisms," in *Perspectives in Neuropsychiatry: Essays Presented to Professor Frederick Lucien Golla by Past Pupils and Associates*, ed. Frederick Lucien Golla and Derek Richter (London: H. K. Lewis, 1950), 77.

Sluckin's remarks illustrate the significant power of cybernetic modeling around 1950. An attractive theoretical model (television scanning) and corroborating mechanical models (such as Walter's tortoises) formed an analogical chain that allowed the hypothesis to persist despite lacking experimental neurophysiological evidence. In 1955, at a UK symposium on cerebral activity, Walter still mentioned the televisual "signal-in-time to signal-in-space" transformation as a key element of cognition, yet concluded that the newly understood complexity of rhythm formation in the brain meant that "no tenable hypothesis regarding the signals which we can observe passing to and fro in the brain can be based on any means of communication familiar to us in everyday life."[65]

Gradually, the scanning hypothesis and television analogy lost ground to the vocabulary of communication theory. At the 1955 symposium, Donald Mackay proposed describing brain function "in more fundamental and general terms, which may hope to outlast changes in engineering fashion" and cerebral activity in terms of "information-flow systems."[66] Looking back at the cortical scanning hypothesis in 1962, Walter would explain that a decade of comparing television-scanning mechanisms to brain activity rhythms had been based on an "ineluctable inference."[67] With recent evidence, he claimed, "comparisons with familiar artificial scanning mechanisms—television and radar, for example—though illustrative, need not be taken too seriously, since there is ample evidence that within the brain the intrinsic and evoked rhythms, whatever their functions, are extremely complex in their interactions with signals, with one another, and with brain topology."[68] But whereas the raster scanning of television did not persist as an analogy for perceiving and interpreting visual patterns, the medium of television continued to serve significant roles in the fundamental process of analogical linking within cybernetics discourse.

---

[65] "Electrical Signs of Mental Function (Part of Cerebral Activity: Report of a Symposium)," *Advancement of Science* 12, no. 1 (1956). In 1953, Walter "deliberately tried to avoid using the word 'scanning' because it has so many associations in ordinary speech and it has become rather tainted, perhaps, with too many simple-minded ideas" and suggested the word "abscission" to denote some of the time-dependent processes in nerve activity. "Studies on Activity of the Brain," in *Cybernetics. Circular Causal and Feedback Mechanism in Biological and Social Systems. Transactions of the Tenth Conference*, ed. Heinz von Foerster, Princeton, NJ, April 22–24, 1953.

[66] Donald MacKay, "Towards an Information-Flow Model of Cerebral Organisation," *Advancement of Science* 12, no. 1 (1956). Such as the idea of cerebral activity as an "information-flow system."

[67] W. Grey Walter, "Spontaneous Oscillatory Systems and Alterations in Stability," in *Neural Physiopathology*, ed. R. G. Grenell (New York: Harper & Row, 1962), 248.

[68] Ibid.

Too Close to the Screen: Television as Trigger and Mirror

While the cortical scanning hypothesis did not hold up as an articulation of the brain *as* television, television still interlaced with the material-discursive field constituting cybernetic engagement with the brain in other pervasive and recursive ways. A first clue to this is visible in *The Human Brain* (1955), a popular introduction to cybernetics by science writer John Pfeiffer. Pfeiffer described the scanning brain as part of the era's particularly media-saturated environment:

> Your mind wanders with a rhythm of ten pulses per second as it idly scans its own contents and those of the world outside. The instant the brainstem alerts the cortex and you focus your attention, however, the waves are blacked and fade or vanish entirely. The ripples on the chart flatten out when you watch a television broadcast, hear a loud noise, figure out your income taxes or become emotionally upset.[69]

In *The Human Brain*, watching TV becomes the reference task par excellence for focused attention while also being similar to emotional turmoil. Pfeiffer recounts how EEG was being tested as a method to diagnose behavioral disorders in children who "won't even sit still to look at television" and might turn out to be cerebrally damaged.[70] Television thus turns into a diagnostic instrument for subjects in media-saturated environments: its intensity (attention grabbing, emotionally upsetting) ensures that not attending to it ("won't sit still") must be a sign of an abnormally scanning mind.

Television's role thus is liminal: Pfeiffer signals the brain's relation to television as *the* contemporary medium of the time—a scanning brain preoccupied with *the* activity of the present—yet also makes clear that watching TV may lead to nervous breakdowns (presented as a particular condition of the present). In a chapter on electrical rhythms in the brain, he sketches a hospital scene in which the patient is none other than a TV executive. The man undergoes an EEG recording because he has trouble speaking under pressure. A normally functioning brain, Pfeiffer explains, can be compared to a complex organization in which part of the organ doles out "final decisions and executive orders," yet, this function was out of order in the executive's brain (abnormal waves indicated the presence of a brain

---

[69] John E. Pfeiffer, *The Human Brain* (New York: Harper, 1955), 108.
[70] Ibid., 114.

tumor).⁷¹ Here, the organ itself seems to morph into a busy TV studio where stress levels are too high. This comparison turns the brain-as-TV-studio into a boundary figure, on the verge of nervous collapse.

Pfeiffer is most astute on the liminal role of television when he turns to epilepsy research, explaining that a poorly tuned or designed TV set can literally cause mental breakdowns when its flickering screen triggers an epilepsy-like seizure. This was based on research, intensifying in the 1940s, on typical and atypical EEG patterns prompted by strobe lights.⁷² William Grey Walter was one of the principal researchers using this "fresh technique" that resulted in "strange patterns, new and significant ... from the swift scribbling of the pens in all channels of the EEG."⁷³ Walter and other researchers designed setups in which both epileptic and neurotypical subjects were exposed to flashes of blue-white xenon light (with a peak intensity of 88.000 candles).⁷⁴ Specific flash rates would affect the brains' normal rhythms, they found, causing behavioral and perceptual changes (illusory sensations, or even convulsions, for example). Since flicker could cause seizures in people who had never been diagnosed with epilepsy, Walter's research demonstrated the remarkable potential of all subjects to produce such irregular responses and peculiar EEG patterns. Epilepsy, as Pfeiffer put it, was "not something remote and strange" but rather "an exaggeration of tendencies that we know are built into the brain ... In a sense, we are all potential epileptics."⁷⁵

Since epilepsy-like convulsions and sensations were now understood to be latent in everyone, television's potential as a trigger became a pressing research topic. The 1950s saw the first case reports of seizures or altered behavior triggered by distortions on TV screens. Researchers started talking about "television epilepsy" to describe what they hypothesized was

---

[71] Ibid., 60.
[72] The directness that Walter attributed to the flicker stimulus plus EEG response was based on the idea that flicker caused EEG patterns that were "of the same kind as the spontaneous patterns in the brain." Walter, *The Living Brain*, 106; W. Grey Walter, V. J. Dovey, and H.Shipton, "Analysis of the Electrical Response of the Human Cortex to Photic Stimulation," *Nature* 158, no. 4016 (1946), cited in Hayward, "Our Friends Electric," 303; Walter and Walter, "The Central Effects of Rhythmic Sensory Stimulation," 57; Walter et al., "Analysis of the Electrical Response of the Human Cortex to Photic Stimulation"; W. G. Walter and V. J. Walter, "The Electrical Activity of the Brain," *Annual Review of Physiology* 11, no. 1 (1949): 209; Jimena Canales, "'A Number of Scenes in a Badly Cut Film': Observation in the Age of Strobe," in *Histories of Scientific Observation*, ed. Lorraine Daston and Elizabeth Lunbeck (Chicago: University of Chicago Press, 2011).
[73] Walter, *The Living Brain*, 91.
[74] Walter et al., "Analysis of the Electrical Response of the Human Cortex to Photic Stimulation."
[75] Pfeiffer, *The Human Brain*, 114.

a photosensitive epilepsy specifically related to television.[76] Such fits were reported to happen predominantly when viewers approached the screen to fine-tune a defective set or press a button, especially if they were in a dark room and when the screen showed jerky or skipping pictures (jumpy cartoons were thought to be particularly triggering).

By 1962, French epilepsy researcher Henri Gastaut (a participant in *Le Cerveau en direct*) wrote that the number of seizures in front of TVs had increased so much that "the question of a real television epilepsy" needed urgent attention.[77] With the help of French broadcaster RTF, he attempted to recreate the conditions for these fits, but did not succeed in finding precise parameters to delineate a specific television epilepsy.[78] Perhaps, some researchers speculated, television's emotional content was part of the trigger.[79] Whatever the cause, commentators stressed it was vital to continue analyzing the effects of television on the brain as part of a broader study on the effects of new stimuli in the modern-day environment on human behavior.[80]

By 1966, Gastaut no longer spoke of television epilepsy as a distinct syndrome, but listed it as one of the sources of sensory-triggered epilepsy. Yet, though television epilepsy turned out to be a short-lived and only tentative media-specific pathology in the 1950s, its existence points to the emergence of a new brainmedia assemblage in which dark rooms, television flicker, and triggerable brains fortified a conception of a subject with a brain that could be engrossed by the screen to such an extent that the scanning eye-plus-brain would produce atypical nervous rhythms. If brains were like distorted TVs, it was hypothesized, then distorting the television-like brain could be helpful to understand its (normal) scanning mechanism.

With flicker research, researchers aimed to probe what William Grey Walter and coresearcher Vivian Walter described as the "borderline between physiology and pathology": by inducing altered behavior at critical frequencies, thus "jamming" the device or "throwing the system out of gear," as they put it, these stimulation patterns were thought to help understand (typical and pathological) physiological mechanisms in the brain.[81] In 1946,

---

[76] J. Klapetek, "Photogenic Epileptic Seizures Provoked by Television," *Electroencephalography and Clinical Neurophysiology* 11, no. 4 (1959); H. Gastaut, H. Regis, and F. Bostem, "Attacks Provoked by Television, and Their Mechanism," *Epilepsia* 3, no. 3 (1962); S. N. Pantelakis, B. D. Bower, and H. Douglas Jones, "Convulsions and Television Viewing," *British Medical Journal* 2, no. 5305 (1962); M. H. Charlton and Paul F. A. Hoefer, "Television and Epilepsy," *Archives of Neurology* 11, no. 3 (1964).
[77] Gastaut et al., "Attacks Provoked by Television, and Their Mechanism."
[78] Ibid.
[79] Servit, in comment section, ibid., 444.
[80] Ibid.
[81] "Borderline," in Walter and Walter, "The Electrical Activity of the Brain," 209. "Jamming," in Walter and Walter, "The Central Effects of Rhythmic Sensory Stimulation," 83. An

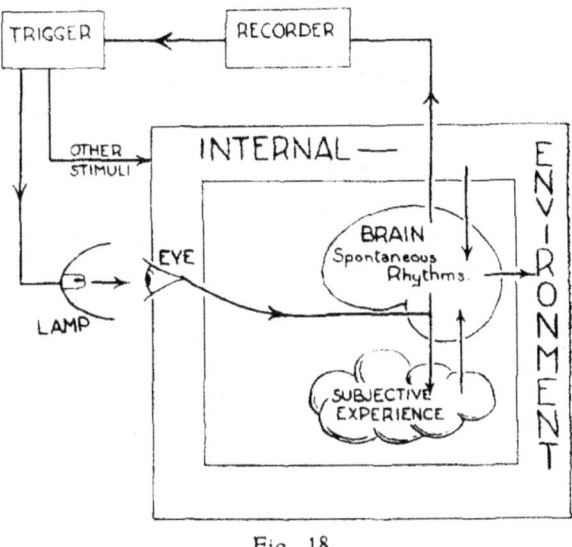

Fig. 18

Figure 4.4 Illustration of a trigger circuit, 1949.

Walter and coresearchers made an important contribution to this research by synchronizing the rate of the flash lamp to EEG measurements.[82] This "lamp>visual cortex>amplifier>trigger>lamp" trigger circuit effectively created a feedback system between brain and stroboscope that would settle at a "critical" frequency, revealing a particular individual "scanning" rhythm (Figure 4.4).[83] In the EEG record, Walter observed that feedback flicker could cause irregular rhythms in the visual area, which could in turn spill over to other areas (like "ripples ... overflowing") of the brain.[84]

---

attachment on the subjects' throat allowed for their descriptions to be precisely located on the temporal EEG record.

[82] Walter et al., "Analysis of the Electrical Response of the Human Cortex to Photic Stimulation," 541.

[83] Ibid. Shipton's 1949 explanation of the trigger circuit shows the influence of cybernetic vocabulary after the publication of Wiener's *Cybernetics* in 1948. Now, Shipton described the brain and human behavior in terms of feedback circuitry and emphasized the parallels between a physiologist and an electronic engineer. The trigger circuit diagram in Walter and Walter (1949) was now captioned "block diagram showing the internal and external feedback networks used in these experiments." H. W. Shipton, "An Electronic Trigger Circuit as an Aid to Neurophysiological Research," *Journal of the British Institution of Radio Engineers* 9, no. 10 (1949): 375.

[84] Walter, *The Living Brain*, 91.

In this cybernetic field of flicker research, television occupied a liminal position between physiology and pathology. While it could *evoke* epilepsy by sending out signals with a certain rhythm, a flickering broadcast also functioned as an analogy of the human brain when it was *expressive of* ("suffering from") disturbed rhythms if TV cameras had been subjected to flicker and displayed warped patterns and "TV ghosts" instead of a steady image. In turn, if a TV set was synchronized to the input frequency, the screen would start to resonate in such a way that it would become "a display of the scanning raster itself."[85]

Flicker-stimulated television presented a particularly revelatory ("mirroring") analogy: if stimulated in specific ways, the TV screen would reveal a pattern indicative of its inner mechanics. Flicker experimentalists like Walter and John Smythies hoped that, just as a distorted TV screen could be diagnosed with a particular mechanical problem by looking at the patterns, brains might be better understood by consulting their EEG patterns as well as the visual patterns subjects reported seeing under the influence of flicker.[86] When Walter described the effect of flicker as producing "a brain storm as wild as any distortion on the television screen," the brainstorm becomes a diagnostic message that ties together distorted patterns on the TV screen, distorted patterns in visual perception, and atypical EEG rhythms.[87] In cybernetic research into perception, comparing the flicker-stimulated brain and the flicker-stimulated TV set thus had an important rhetorical upshot: it strengthened the possibility of deducing which structure or mechanism caused an unusual output. In this material-discursive assemblage, seizures were like distorted TV screens, while distorted screens could recursively function as epileptic stimuli. Brains and TVs were gathered both in a new discursive field and in a very contemporary world in which viewers got a little too close to the screen.

---

[85] Walter and Walter, "The Central Effects of Rhythmic Sensory Stimulation," 83.

[86] "It is mechanically impossible for the televisual mechanism to give us a *direct* view of the events televised." J. R. Smythies, "The Stroboscope as Providing Empirical Confirmation of the Representative Theory of Perception," *British Journal for the Philosophy of Science* 6, no. 24 (1956): 333–4. The idea of revealing the mechanisms behind the distortion is also clear in the work of William Grey Walter and Vivian Walter. "The receiving station would see such a pattern and it would change with frequency, but the image would not be anything like the stimulus, in fact it would not be related to the stimulus at all; it would be a display of the scanning raster itself." "The Central Effects of Rhythmic Sensory Stimulation," 83.

[87] Walter, "Pattern-Making and Pattern-Seeking." The step from typical to pathological was an important element in cybernetics. Pathological mechanisms were indicative of typical workings, or it made such bifurcations more a question of a sliding scale. As Rodri Hayward explains, cybernetic theorists looked at mechanical mechanisms that went haywire (such as a frantic oscillatory process in a feedback mechanism) as models for psychopathology. "Our Friends Electric," 303.

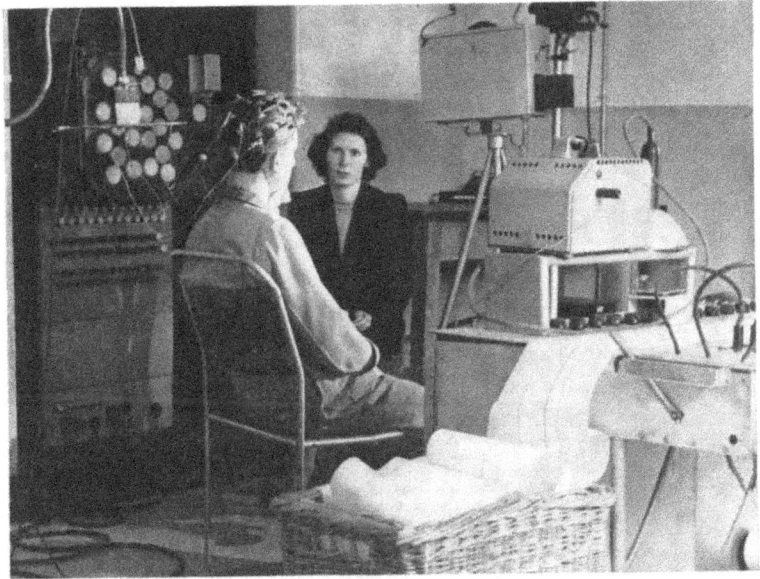

Figure 4.5 EEG-set-up at the Burden Neurological Institute in Bristol, England. Toposcope display visible left from the subject and researcher Vivian Walter. c. 1954.

The Toposcope: The Brain Displays Itself

The TV studio brain was emotionally taxed, potentially "overflowing," a brain that might work through scanning beams that resonated with particular stimuli, and, importantly, a brain that could reveal its structures if triggered and jammed in the right way and with certain "brain mirrors." At the end of the 1940s, Walter and his coresearcher Shipton designed a new type of "brain mirror" to enhance the epistemic authority of EEG recording (Figure 4.5). This new visualization apparatus for EEG measurements was called the toposcope ("Topsy" for short), a device made up of a series of cathode ray tubes, "like 22 small television or radar sets," as Walter explained in the popular journal *Scientific American*.[88] Each tube was connected to an electrode placed on the

---

[88] W. Grey Walter, "The Electrical Activity of the Brain," *Scientific American* 190, no. 6 (1954): 62; W. Grey Walter and H. W. Shipton, "A New Toposcopic Display System," *Electroencephalography and Clinical Neurophysiology* 3, no. 3 (1951): 283. Around 1950, several brain researchers had started to develop topographic displays of EEG activity; see

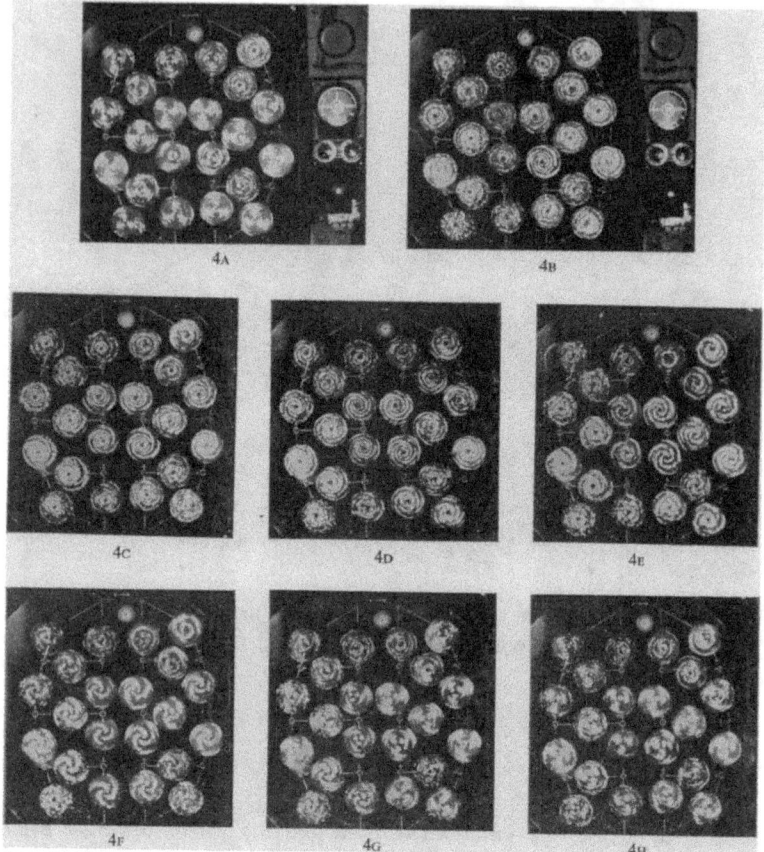

FIG. 4.—The effect of mental activity on alpha frequency and distribution in normal subject of Fig. 2A. The duration of each exposure was 8 seconds.

Figure 4.6 Toposcope display, c. 1957 (photographs).

scalp and, taken together, these TV tubes thus showed the distribution of nerve cell activity mapped onto the head (Figure 4.6). The toposcope's new skill was not only showing brain activity over time but also revealing its spatial distribution mapped onto the oval of the skull, "as though you were looking down on the head of a patient from above."[89] Its twenty-two screens with

---

H. Petsche, "From Graphein to Topos: Past and Future of Brain Mapping," in *Topographic Brain Mapping of EEG and Evoked Potentials*, ed. Konrad Maurer (Berlin: Springer, 1989), 14.

[89] Walter, "Studies on Activity of the Brain," 21.

swirling light spots allowed what Walter called "toposcopic observations" of nerve activity patterns that had a "greater dramatic effect" than regular EEG recordings with rows of zigzag lines.[90] With characteristic rhetorical lyricism, Walter portrayed the toposcope as the realization of Sherrington's enchanted loom (discussed in Chapter 1): finally, a technology with "millions of flashing shuttles weave a dissolving pattern" seemed to have been materialized.[91]

While the toposcopic methods developed by Walter and other research groups showed differences between subjects during different mental tasks, they ultimately did not generate patterns that could be meaningfully interpreted and never came into widespread research practice. Yet, they are evidence of the desire to create more comprehensive (spatiotemporal) and persuasive data visualizations mapped on the anatomical space of the brain in the 1950s. The toposcope, Walter explained, allowed "the illusion that the observer is looking at the head of the patient and can see the significant discharges where they are actually occurring."[92] Not only that, he argued, "for demonstration purposes [it] was most pleasing."[93] In EEG records, he had seen how flicker caused irregular rhythms to spill over from the visual cortex to other areas, and now the toposcope made this "broadcasting" (in Walter's words) of signals to other areas in the brain—that is, other screens—visible.[94] This epistemic promise of visibility through broadcasting was strengthened by the growing ubiquity of cathode ray tubes in the lab. With improved tube technologies in the 1940s, researchers had started to develop experimental setups with more complex visual displays that employed custom or multiple cathode ray tubes. (In visual perception research, TV screens were now frequently employed to offer subjects visual tasks that could be altered in frequency, brightness, and hue.[95]) This pervasiveness of display tubes resonated with television's symbolic purchase as a preeminently "live" medium, strengthening Topsy's promise of directness.

Moreover, Walter's multiscreen toposcope also shows how an experimental setup strengthened a theoretical concept. Topsy fortified the deductive logic of a jammed brain television by integrating feedback as a means to have the brain "display itself." Walter cited preliminary experiments that showed Topsy

---

[90] Walter, *The Living Brain*, 111.
[91] Walter, citing Sherrington, in "Studies on Activity of the Brain," 22.
[92] "Discussion on the Electro-Encephalogram in Organic Cerebral Disease," *Proceedings of the Royal Society of Medicine* 41, no. 4 (1948): 237.
[93] Walter and Shipton, "A New Toposcopic Display System," 282.
[94] Walter, "The Electrical Activity of the Brain," 62.
[95] For example, R. M. Hanes, "A Scale of Subjective Brightness," *Journal of Experimental Psychology* 39, no. 4 (1949); MacKay, "Some Experiments on the Perception of Patterns Modulated at the Alpha Frequency."

was able to detect mental or nervous disorders hitherto "undetectable in conventional brainprints."[96] By showing nerve cell activity on single screens as well migrating from screen to screen, Topsy enabled intricate new aggregate patterns whose full comprehension was projected to the future. Its dramatic effect was most striking in a flicker feedback setup, as the toposcope showed peculiar forms and geometry, going wild just as a distorted TV.[97] "When flicker is used," he explained, the display "comes near to being a moving picture of a mind."[98] Similar to the way the stroboscope's flashing light was coupled to a particular oscillation of the brain, so the timing of the toposcope's swirling screens could be made dependent on the oscillation of nerve cell activity. As the lights circling on the screens could be conditioned by "what is happening in the person's brain at that time and that place," brain activity could "display itself" with its own, ever-changing scanning rhythm.[99] Topsy displayed what Walter described as "the brain driving the machine," and, coupled to a stroboscope, it could become a "machine driving the brain."[100] At the 1953 cybernetic Macy conference, Walter argued that the toposcope, with its feedback mechanism and television-like display, could help acquire "a direct impression of coincidence, congruence, and contingency" in brain activity.[101] Though he noted that determining causality was perhaps still the root problem of understanding the living brain, in his (promissory) view, Topsy had elevated research into stimulus and response and the relation between nerve physiology and human behavior through a particularly intimate coupled system, a televisual setup between brains and screens.

These technological promises of new revelatory circuits emerged, from the 1940s until at least the 1960s, as part of a research culture in which experimental opportunism sometimes prevailed over proper patient care.[102] William Grey Walter and other psychophysiologists conducted experiments on people with epilepsy and psychiatric patients without (what we now see as) appropriate ethical consent and treatment protocols. Newly created "direct impressions" of the brain's activity were the result of painful methods

[96] Walter, "The Electrical Activity of the Brain," 62.
[97] Walter, *The Living Brain*, 111.
[98] Walter, "The Electrical Activity of the Brain," 63.
[99] Walter, "Studies on Activity of the Brain," 22.
[100] Ibid., 23.
[101] Ibid.
[102] On these "risky experiments of uncertain value" performed on epilepsy suffers, see, for example, Ellen Dwyer, "Neurological Patients as Experimental Subjects: Epilepsy Studies in the United States," in *The Neurological Patient in History*, ed. L. Stephen Jacyna and Stephen T. Casper (Rochester, NY: University of Rochester Press, 2014), 44–60. Cited in Rachel Elder, "Speaking Secrets: Epilepsy, Neurosurgery, and Patient Testimony in the Age of the Explorable Brain, 1934–1960," *Bulletin of the History of Medicine* 89, no. 4 (2015): 761–89.

of forced seizures in patients. Moreover, by the mid-1950s, light and flicker stimulation experiments were also conducted on patients diagnosed as schizophrenics or psychotics who had "depth electrodes" inserted into their brains, a procedure with a high chance of death.[103] Researchers sat vulnerable patients in front of television projection tubes and carousel slide projectors to "reproduce" "all kind of wave forms" in painful and dangerous new brainmedia circuits.[104]

Conclusions

In this chapter's first part, my analysis of brains on TV in the 1950s described new brainmedia assemblages that used television to connect medical visualization apparatuses, institutional broadcasting structures, discourses on liveness, and portrayals of brain science to shape a new image of the "live brain." Interpreting brain broadcasts as experimental television opens up a study of brainmedia that moves beyond the spectacular representation of brain science in science broadcasts, and includes the reciprocity with which views of the "brain at work" were shaped with and through television's quest for liveness in the 1950s.

---

[103] Researchers legitimized experimentation with patients who had depth electrodes inserted in their brains on the grounds of the necessity of these electrodes to evaluate and improve the efficacy of a scheduled psychosurgery for these patients (lobotomy or leukotomy). H. W. Dodge et al., "Technics and Potentialities of Intracranial Electrography," *Postgraduate Medicine* 15, no. 4 (1954). On this periods of "unrivalled" "empirical human experimentation," see John Gardner, "A History of Deep Brain Stimulation: Technological Innovation and the Role of Clinical Assessment Tools," *Social Studies of Science* 43, no. 5 (2013). More research is necessary to trace the influence of US military funding on Grey Walter's research at the Burden Neurological Institute.

[104] "All kind of waveforms," in A. Kamp, C. W. Sem-Jacobsen, W. Storm van Leeuwen, and L. H. van der Tweel, "Cortical Responses to Modulated Light in the Human Subject," *Acta Physiologica Scandinavica* 48, no. 1 (1960): 1–12. A. Kamp et al., "Cortical Responses to Modulated Light in the Human Subject," *Acta Physiologica Scandinavica* 48, no. 1 (1960). The most astonishing example of a deep-brain trigger loop was an experiment allegedly carried out by Grey Walter himself. Philosopher Daniel Dennett recounts a presentation by Grey Walter to an Oxford University audience in 1963, in which he talked about an (unpublished) experiment for which he created a trigger circuit between a carousel slide projector and a patient implanted with a depth electrode. Activity in the motor cortex would trigger a next slide on the projector in front of the patient's eyes. Daniel Dennett, *Consciousness Explained* (Boston: Little, Brown, 1991), 167–8. As Dennett reports, "he told us that he told his patients that he needed to do follow up testing on their epilepsy post-surgically (but in fact, he was simply eager to use them as long-term experimental subjects)." E-mail by Daniel Dennett to Dirk Hartmann, May 10, 1999, cited in Dirk Hartmann, "Neurophysiology and Freedom of the Will," *Poiesis & Praxis* 2 (2001): 283.

In the second part, I zoomed in on a parallel history of cybernetic explorations of the living brain to discuss heterogeneous material-discursive movements between televisions and brains around 1950. Paying particular attention to the relations drawn between a flicker brain circuit (lamp>visual cortex>amplifier>trigger>lamp) and a flickering TV receiver, I argued that fusions of televisions and brains opened up a revelatory structure and deductive logic. By jamming the (television–brain) system in coordinated ways, the structural working of the brain was envisioned as deducible from a variety of "brain mirrors." The upshot of this was that the brain could ultimately "display itself" (like a televisual scanning raster) through a particular flicker feedback loop that was both television-like and (in the case of Walter's toposcope) relied on TV technologies. At different stages of this intricate rhetorical bricolage, television—both conceptually and materially part of the flicker feedback system—helped fortify the epistemic efficacy of EEG flicker research.

Ultimately, the fusion of EEG with television epilepsy research changed brain research practice in this cybernetic era. This becomes most visible in Walter's *The Living Brain*, when he compared his experiments with the "communication engineer's method of the Black Box."[105] While flicker stimulation and EEG did not "completely expose" the contents of the brain, they did provide "a more direct means of enquiry."[106] Now, with feedback flicker research, the black box brain had become television-like, a seizure-prone apparatus that could be part of experimental systems where it could be made to drive visualizations mirroring its inner workings. Just what the patterns on these displays revealed was hard to understand, but the medium of television helped establish a deductive logic that allowed toposcopic patterns to be understood as directly expressing physiological and anatomical mechanisms. Though flicker feedback research had not opened the black box, its walls had now become screens displaying potentially meaningful expressions of invisible structures. My analysis of this media-saturated history also suggests that the genealogy of the cybernetic concept of the black box is more complex than previously thought.[107]

---

[105] Walter, *The Living Brain*, 104.
[106] Ibid.
[107] In canonical accounts, the genealogy of the black box comprises a complex historical lineage, including behaviorist ideas on inferentiality (B. F. Skinner's "boxes") and anti-aircraft predictors (boxes with buttons) that responded to the enemy's "active oppositional intelligence" through the combined behavior of pilots and airplanes. P. Galison, "The Ontology of the Enemy: Norbert Wiener and the Cybernetic Vision," *Critical Inquiry* 21, no. 1 (1994).

Within these brainmedia, the nature of the black box brain changed through resonances with television and the constitution of the flicker trigger circuit. Literally encased in a black box (a structure placed over the head) during flicker research, the human subject became part of a material-discursive field in which comparisons with television shifted the importance of the black box's "terminals for output," which were now conceived as mirrors of the active brain.[108] More than any black box, the televisual flicker brain was prone to showing its internal structures as patterns on the outside via external "brain mirrors": EEG rhythms, toposcopic patterns, hallucinatory observations. Moreover, because of flicker circuit's feedback nature and the toposcopic brain mirror that filtered out EEG noise, the black box brain was now thought to "display itself." This self-revelatory capacity was symbolically strengthened through the comparison with television. In the cybernetic universe, the living brain would be accessible because it could express its insides.

Even as literal television analogies (scanning) waned, TV technology remained closely associated with the brain. At the end of the 1950s, hospitals started experimenting with closed-circuit television EEG (CCTV-EEG), by which measuring an atypical rhythm would trigger a video recording matching the EEG record with images of a patient's behavior.[109] TVs became common in hospital and psychiatric wards, and CCTV became a method of teaching psychiatry and a tool for behavioral research in the mid-1950s.[110] Simultaneously, the concept of feedback gained new meaning in the domains of psychiatry and psychophysiology. If cyberneticists influenced a vision of psychiatric patients as having faulty feedback loops, now patients, psychiatrists, and brain researchers

---

[108] Black box, for example, in E. G. Walsh, "Experimental Investigation of the Scanning Hypothesis," paper presented at the The EEG Society, Maudsley Hospital, London, April 26, 1952. Cited in D. A. Pond, "The EEG Society: Maudsley Hospital, London April 26th, 1952. Society Proceedings," *Electroencephalography and Clinical Neurophysiology* 4, no. 3 (1952): 371.

[109] By Goldensohn and Koehle in 1962 in New York; see reported in E. S. Goldensohn, "Simultaneous Recording of EEG and Clinical Seizures Using Kinescope," *Electroencephalography and Clinical Neurophysiology* 21 (1966); T. F. Collura, "History and Evolution of Electroencephalographic Instruments and Techniques," *Journal of Clinical Neurophysiology* 10, no. 4 (1993).

[110] Milton M. Berger et al., "The Use of Videotape with Psychotherapy Groups in a Community Mental Health Service Program," *International Journal of Group Psychotherapy* 18, no. 4 (1968); Donald S. Kornfeld and Lawrence C. Kolb, "The Use of Closed-Circuit Television in the Teaching of Psychiatry," *Journal of Nervous and Mental Disease* 138, no. 5 (1964).

were advised to experiment with video therapy as one of various forms of biofeedback.[111] The 1960s, as I recount in the next chapter, saw brains newly circuited in feedback systems and a vision of the circuited self that was pitted directly against the hegemonic structures of broadcast television, beyond the brain *as* and *on* TV.

---

[111] Andrew Pickering, "Psychiatry, Synthetic Brains and Cybernetics in the Work of W. Ross Ashby," *International Journal of General Systems* 38, no. 2 (2009); J. H. Clark, "Adaptive Machines in Psychiatry," in *Progress in Brain Research*, ed. N. Wiener and J. P. Schadé (Amsterdam: Elsevier, 1963).

5

Interfacing the Real-Time Brain: EEG Feedback in Art and Science, 1964–77

In 1965, the live brain entered the musical stage. Sitting center stage, composer Alvin Lucier was wired up with an EEG headband that measured his brain activity, which was subsequently routed to amplifiers setting off various sound-making devices, including gongs, cymbals, metal ashcans, cardboard boxes, and tape-recorders (Figure 5.1). A decade after performing his now-famous EEG composition *Music for Solo Performer*, Lucier reflected on the piece in an interview by Robert Ashley circa 1975:

> In a way art has always been dealing with a state that you find yourself in. Here we are, we're born in this time, and these things are here. I mean, the EEG was here without me doing anything about it. See, I don't think of technology as technology. I think of it as the landscape. We're born and brought up in a landscape and there's not much I can do about the fact that there are EEG amplifiers. I mean you could hardly pass a law ending it, right? I mean, there's nothing I can do to invent it or to make it go away. So, I see it as an environ …—it's a landscape. If you worked in a medical center EEG would be just like a tree, it's what you see every day. And it's touching. I mean, a composer in the nineteenth century, or in this century, they are talking about the landscape that they are in … I'm just doing that, I'm doing a very simple thing.[1]

Lucier described his musical EEG setup as a reflection on contemporary technology, on the way technology was omnipresent—inherent to the landscape or environment—and thus also involved him, the composer, as

---

[1] Alvin Lucier interviewed by Robert Ashley: Robert Ashley, *Alvin Lucier—Music with Roots in the Aether*, c. 1975, https://www.youtube.com/watch?v=nRa5x6j26Is, June 7, 2019. Chapter 5 is a revised and expanded version of Flora Lysen, "The Interface Is the (Art)Work: EEG-Feedback, Circuited Selves and the Rise of Real-Time Brainmedia (1964–1977)," in *Brain Art: Brain-Computer Interfaces for Artistic Expression*, ed. Anton Nijholt (Cham: Springer, 2019).

Figure 5.1 Alvin Lucier (left) and John Cage (right) preparing a performance of *Music for Solo Performer* at the festival "John Cage at Wesleyan," 1988 (photograph).

part of that landscape.[2] According to the piece's 1965 score, an "alteration of thought content" would change the signals fed into the amplifiers.[3] It also mentioned that brain activity might be used "to activate radios, television sets, lights, alarms, and other audio-visual devices" in the future.[4] By linking humans and technology in an electrical circuit, and by showing EEG outside a scientific context, Lucier staged a contemporary musical environment of transmutable energy flows. He deemed the behavioral categorization of brainwaves rather insignificant: "I really didn't care what the brainwaves were ... I just never had the answer."[5] What mattered for the performance

---

[2] Performed on stage in 1965 at the Rose Art Museum of Brandeis University, scheduled right after John Cage's performance of *4'33"* No. 2. As Douglas Kahn mentions in Douglas Kahn, *Earth Sound Earth Signal. Energies and Earth Magnitude in the Arts* (Berkeley: University of California Press, 2013), 88.
[3] Alvin Lucier, "Music for Solo Performer (1965) for Enormously Amplified Brain Waves and Percussion," in *Chambers: Interviews with the Composer by Douglas Simon* (Middletown: Wesleyan University Press, 1980).
[4] Ibid., 69.
[5] Alvin Lucier in a videotaped interview with Robert Ashley: Ashley, *Alvin Lucier—Music with Roots in the Aether, c. 1975.*

was the fact that living, active brains were intimately and evidently part of a contemporary environment brimming with new amplification technologies.

Lucier thus effectively created the first artistic EEG feedback circuit. Any change in thought was reflected in the EEG measurements, thus changing sounds in the environment, which would in turn impact the thoughts of the observer, and so on. The work sprung from a collaboration with a Brandeis University colleague, physicist Edmond Dewan, who had provided the amplifying devices and know-how for the (technically rather complicated) piece.[6] Dewan researched EEG feedback for the US Airforce (inspired by Norbert Wiener's cybernetics) and had, one year prior, received widespread public attention from national newspapers and television for inventing what was rather hyperbolically called a "thought-controlled device," a simple lamp that could be switched on or off according to a subject's measured levels of alpha brainwave activity.[7]

Lucier's and Dewan's mid-1960s collaboration indicates the rise of a new attention to brainwaves in resourcefully devised feedback circuits as well as new exchanges between artistic and scientific spheres of experimentation. In this chapter, I analyze such EEG feedback research between 1964 and 1977, tracing its rise from Dewan's media-hyped feedback setup to the waning enthusiasm for EEG feedback and the concurrent emergence of so-called brain–computer interfaces (BCIs) in the early 1970s. Based on my historical sources on artistic and experimental EEG feedback experimentation, I make two central observations about EEG research in this period: first, there is a new epistemological commitment to the "interface" in EEG feedback research, and, secondly, the EEG vocabulary of "feedback" and "communication" persists in later BCI research despite these concepts changing meanings in machine-oriented, closed-loop systems that emphasized the notion of "real-time" transmission.

In the first part of this chapter, I trace EEG feedback's emergence in the 1960s as part of a heterogeneous array of experimental conditioning techniques that would come to be called "biofeedback" and included EEG measurements (predominantly brainwaves measured in the frequency range called "alpha") but also heart rates, skin conduction, or muscle activity, for example. In this context, I examine the rise of an interest in alpha feedback in relation to a cultural interest in the self, as well as a new field of research

---

[6] Kahn, *Earth Sound Earth Signal*.
[7] Edmond M. Dewan, "Communication by Electroencephalography. Experiment at the Stanley Cobb Laboratories at Massachusetts General Hospital," *Air Force Cambridge Research Laboratories, United States Air Force* (1964). (In fact, Dewan admitted that EEG was imprecise. "The use of EEG for communication purposes ... seems to have very little likelihood of practical application at this time."). Ibid.

that included both scientific and artistic approaches. I claim that alpha research created new ways of performing knowledges about the live brain—new forms of brainmedia—in which the approaches of artists resonated with those of other experimenters. At the time, a number of artists created elaborate installations in which physiological signals functioned as input for intricate technological circuits of flickering displays, closed-circuit television, blinking lights, alternating sounds, and colorful projections. These media environments produced what I call visions of a "circuited self." Experimental research into alpha feedback offered similar visions, leading to an interest in designing evocative and playful interfaces through which alpha could be made sensible to participants. I argue that such emphasis on interfaces and interfacing constituted a new epistemological commitment in which interfacing took precedence over understanding precise mental correlates of measured brainwaves, an attention shift that I describe with the idea that the "interface is the work."

In the second part, I turn to the emergence of a related type of EEG interface research dubbed "brain–computer interfaces." I explain how the proximity of EEG feedback research to BCI research (particularly in an American context) in 1970 meant that new BCI developments still used the vocabulary of "feedback" and "communication" that had been prevalent in EEG feedback research. Yet, while EEG feedback had projected a "communicating with the self" and new types of self-insight, BCI research employed feedback and communication to speak about the way computers could measure and interpret brain activity, resulting in real-time updates through BCIs. In the early 1970s, research into conditioning alpha brainwaves was increasingly criticized. I describe the contested field of feedback science, a conflict between a cultural interest in exploring the self and an increasing criticism of EEG feedback's experimental methods. Ultimately, even though EEG feedback gave way to real-time BCIs in the 1970s, what remained was the central importance of the interface.

The Groovy Science of Alpha

In this chapter, I predominantly focus on EEG feedback research in the historical context of the United States, where a substantial number of people experimented with EEG feedback and where the borders between artistic and scientific experimentation were particularly blurry.[8] However, it is

---

[8] For example, artist David Rosenboom's early scientific informers were neurophysiologists Neil Miller (Rockefeller University), Lester Fehmi (Stony Brook), and Edgar Coons

important to note that EEG feedback experimentation was an international phenomenon: after the aforementioned experiments by Dewan and Lucier at Brandeis University in the mid-1960s, work with EEG feedback also appeared in a number of different institutional and experimental sites in the 1960s and 1970s, including in Italy and France.[9] In the Netherlands, brainwave feedback first appeared in 1970, when Dutch countercultural newspaper *Paradiso Fox* interviewed American physiologist Peter Crown, who had been researching brainwave feedback as a postdoc at Rotterdam's Dijkzigt Hospital, "the only scientist in Holland researching this new and still very vague area of cerebration."[10]

When I emailed Crown (now based in New York) to ask him about his experiences, he told me his work had been "appropriately criticized" for its poorly executed methodology by clinical neurophysiology professor Willem Storm van Leeuwen (who had studied with Edgar Adrian in Cambridge and with William Grey Walter in Bristol and had reported about tentative EEG conditioning in 1957).[11] When Crown returned to the United States in 1970, he continued to study brainwaves and feedback phenomena both artistically and in clinical, experimental settings. In the Netherlands, EEG feedback did not appear to find further resonance within the research community (at least not until the 1990s). This signals the existence of variable local research cultures and historical lineages through which EEG was enacted and employed as a clinical and experimental research tool, which in turn also affected the materialization of feedback experimentation at different locations.[12]

(NYU); the artist Richard Teitelbaum worked on alpha-wave feedback with psychologist Lloyd Gilden (Queens College, New York), and artists Woody Vasulka and Richard Lowenberg collaborated with Peter Crown (at the EEG computer laboratory at the New York Medical College).

[9] An influential crossroads, for example, was the 1968 First International Electronic Music Congress in Florence, Italy. Here, Finnish artist Erkki Kurenniemi was introduced to the biomusic of the American artist Manford L. Eaton, for example, which later resulted in his own development of a brainwave-modulating musical instrument. Members of the Rome-based "Musica Elettronica Viva" (including Richard Teitelbaum) spread information about their 1967 piece *Spacecraft*, which employed EEG and ECG signals. Another line runs from Florence to Bordeaux (where Roger Lafosse developed his Corticalart device) and Paris, where electronic musician Pierre Henry performed an EEG music installation at MNAM (Musée d'Art Moderne de la Ville de Paris) in 1971.

[10] "Alpha Brain Waves—New Key to Inner Peace?" *Paradiso Fox*, no. 3 (1970): 11.

[11] Peter Crown, email communication, December 8, 2018. H. A. Gastaut et al., "Étude Topographique Des Réactions Électroencéphalographiques Conditionnées Chez L'homme: (Essai D'interprétation Neurophysiologique)," *Electroencephalography and Clinical Neurophysiology* 9, no. 1 (1957).

[12] Borck, "Between Local Cultures and National Styles," 457.

Focusing on EEG feedback practices in the United States, I trace the various spaces of encounter and lines of research of alpha-interested experimenters to show the emergence of newly interacting domains: established academic laboratories for neurophysiology, new EEG laboratories focused on alpha feedback, investigations associated with military research programs, as well as experimental art, music, and performance spaces. As will become evident, California is a particularly important site, as cybernetic-countercultural exchanges met what Foucault would later scornfully call the "Californian cult of the self."[13] The interdisciplinarity of these collaborations was also part of EEG feedback's self-image: researchers, practitioners, and popular media accounts stressed cross-communication and exchange as a way to overcome the so-called scattered state of the field of mind–brain research.[14] EEG feedback was hailed as "a child of the time," "a wedding between the psychologists, engineers, physicists, meditators and psychologists," "born from the Zeitgeist of 'state-of-the-art electronics, psycho-physiology, operant conditioning procedures, and a desire to explore inner space.'"[15]

The rise of EEG feedback across varying research spaces starting in the mid-1960s must also be understood as imbricated with the emergence of New Age, an inchoate body of thoughts and practices (variably intersecting, for example, with the Human Potential movement and the Holistic Health movement) that claimed an alternative worldview, including alternative approaches to science.[16] This imbrication meant that brainwaves were appropriated into discourses on energies and "potential." The electric force of the biological brainwave gained a cultural association with a person's "life potential," a move from "quantity into quality," as anthropologist Stefan Helmreich describes.[17] In a similar way, I note, the term "feedback" gained an expanded meaning, both as a technical term to denote the return from input to output and also as a cultural expression of communication: informing a person about his or her performance, offering a means of self-improvement

---

[13] Foucault, Michel. *Ethics: Subjectivity and Truth.* Edited by Paul Rabinow. (New York: New Press, 1997), 271.
[14] Barbara B. Brown, *New Mind, New Body; Bio-Feedback: New Directions for the Mind* (New York: Harper & Row, 1974), 303.
[15] Thomas Budzynski, cited in Jodi Lawrence, *Alpha Brain Waves* (New York: Avon Books, 1972), 18.
[16] On New Age's (varying) relations to established science, see Andrew Ross, "New Age Technoculture," in *Cultural Studies*, ed. L. Grossberg, C. Nelson, and P. Treichler (New York: Routledge, 1992).
[17] Stefan Helmreich, "Potential Energy and the Body Electric: Cardiac Waves, Brain Waves, and the Making of Quantities into Qualities," *Current Anthropology* 54, S7 (2013): S142.

and (self-)insight.[18] This ambiguity of brainwaves and feedback as both technobiological and cultural phenomena was present, in varying ways, in institutional research spaces of EEG research as well as in artistic, experimental spheres of EEG experimentation (though various researchers aimed to demarcate serious feedback research from the false promises of "technological gypsies").[19]

Clearly, the possibility of alpha feedback struck a note with particular white middle-class countercultural communities in the United States. In fact, alpha conditioning was part of a broader self-proclaimed "progressive" brain science: the idea that the brain could be trained for different mental states resonated with other research areas on things like LSD (and other altered states of consciousness) and the emerging notion of a "split" or "bisected brain" with which researchers distinguished between left-brain (variably associated with logical thinking, language, and Western rationality) and right-brain functionality (variably associated with intuitive, mystical, creative, and Eastern consciousness). In this discussion, the emerging science of EEG feedback offered strategic potential because it contained a dimension of the brain's mutability and trainability, which strengthened a vision of improving either left- or right-brain functionality.[20]

Discussing the cultural conditions that allowed for the bisected brain paradigm to flourish around 1970, historian Anne Harrington notes that it was suited to the ideological needs of the counterculture movement. Extending this observation, historian Michael Staub argues that the scientific theory was popular because it offered something for everyone, resonating with both people who had racist notions of essential differences and those who believed in transformation and improvement; a business elite searching for "creativity" as well as "do-gooders and romantics of all stripes."[21] Harrington's and Staub's remarks on the bisected brain are important, as they point to the need for a careful consideration of cultural discourses and situated structures of the research in which EEG feedback took place.

Taking heed of their observations, I argue that describing EEG feedback experiments in terms of fringe science (or describing some practices as more

---

[18] We may note that this "cultural" meaning of feedback was already implied in cybernetic research approaches to human psychology and abnormal behavior. Hayles, *How We Became Posthuman*, 65. Dumit, "Plastic Diagrams."
[19] Lawrence, *Alpha Brain Waves*, 196.
[20] Erik Peper, "Localized EEG Alpha Feedback Training: A Possible Technique for Mapping Subjective, Conscious, and Behavioral Experiences," *Kybernetik* 11, no. 3 (1972): 168.
[21] Harrington, *Medicine, Mind, and the Double Brain*, 283; Michael E. Staub, "The Other Side of the Brain: The Politics of Split-Brain Research in the 1970s–1980s," *History of Psychology* 19, no. 4 (2016): 269.

fringe than others) does not aptly capture the across-the-board pervasiveness of the promise of feedback and the desire to train the brain in this period. Instead, the emergence of EEG feedback can perhaps best be described as examples of what historians David Kaiser and W. Patrick McCray have called the "groovy science" that typified the "long 1970s": a media-savvy research culture that foregrounded alternative ways of organizing institutions, posed different questions with hands-on tools, nurtured charismatic scientific personalities, and found new exchanges with a broader public.[22] The challenge is to see how the "grooviness" of this new brainmedia—the particular assemblages of brains and forms of performing knowledges—fundamentally impacted modes of reasoning and the way the active brain could be thought. My main argument is that, in the context of growing concern about the methodological validity of EEG feedback research, a performative research culture foregrounded the importance of developing feedback interfaces. In turn, the way the concepts interface and interfacing were shaped also impacted the interpretation of BCI research.

"Broadening Horizons" and "The Golden Age of Man"

Arguably, alpha feedback had been around since the very beginning of EEG. In their seminal 1934 work on electrical brain activity, Edgar Adrian and Bryan Matthews designed circuits through which alpha wave measurements (the "Berger rhythm") triggered a sound, a feature that was particularly helpful when researchers were measuring their own alpha activity.[23] In the 1940s, an EEG trigger circuit had also been envisioned as a practical warning system for pilots and motor cyclists who were at risk of losing their focus, and in 1949, William Grey Walter and Edward Shipton (as described in Chapter 5) designed a flicker lamp EEG trigger circuit to examine and influence the nature of nervous rhythms in the visual cortex of both epileptic and neurotypical subjects.[24]

Beyond devising such EEG trigger circuits, researchers had long been interested to know whether giving subjects feedback on their EEG states

---

[22] David Kaiser and W. Patrick McCray, eds., *Groovy Science: Knowledge, Innovation, and American Counterculture* (Chicago: University of Chicago Press, 2016).
[23] Adrian and Matthews, "The Berger Rhythm."
[24] Alois Kornmüller, "Signalisierung Der Langsamen Wellen Des EEG Im Sauerstoffmangel" (MPG-A III/16/411945). Cited in Borck, *Brainwaves*, 243; Shipton, "An Electronic Trigger Circuit as an Aid to Neurophysiological Research."

offered a way to actively condition the suppressing or stimulation of brain activity and, in turn, whether this meant that the onset of particular mental states could also be trained. This research was predominantly concerned with blocking or attenuating alpha activity, a physiological state that had already been closely linked to the modulation of attention and visual imagery in early publications on EEG. Since the mid-1930s, however, reports about successful alpha conditioning (starting with the work of Durup and Fessard in France) were tentative and disputed; the effects observed seemed highly individual, unstable, and weak.[25]

In 1962, however, psychologist Joe Kamiya was one of the first to report on successful alpha experiments (which he had carried out in 1958), and reports of other successful experiments (notably those of researchers in the Soviet Union) subsequently started to circulate in the mid-1960s.[26] This turn to EEG feedback must be viewed as tied to contemporaneous developments: several studies in the early 1960s had tackled conditioning for other psychophysiological signals (heart rate, for example); a number of publications covered successful cortical-conditioning experiments in animals; and perhaps most significantly, EEG feedback research was part of an attempt by various researchers to combine established behaviorist methods with a new interest in mental events and introspective, verbal reports by experimental subjects.[27]

[25] For an overview of the situation, see Albino and Burnand (1964), who at that point do not seem to be aware of the results by Kamiya. R. Albino and G. Burnand, "Conditioning of the Alpha Rhythm in Man," *Journal of Experimental Psychology* 67, no. 6 (1964). Early sources mentioned in Gastaut et al., "Étude Topographique Des Réactions Électroencéphalographiques Conditionnées Chez L'homme." Gastaut et al. mention how early work on alpha conditioning by Durup and Fessard had not been followed up in France but was taken up in the United States. Subsequently, in the 1940s and 1950s, a number of researchers in Russia and Japan took up this line of research. See ibid., 1 ff1.

[26] Joe Kamiya, "Conditioned Discrimination of the EEG Alpha Rhythm in Humans," presentation at the Western Psychological Association, San Fransciso, April 19–21, 1962, listed in Thomas W. Harrell, "Proceedings of the Forty-Second Annual Meeting of the Western Psychological Association," *American Psychologist* 17, no. 9 (1962): 602; Joe Kamiya, "Conscious Control of Brain Waves," *Psychology Today* 1 (1968); P. V. Bundzen, "Autoregulation of Functional State of the Brain: An Investigation Using Photostimulation with Feedback," *Federation Proceedings. Translation Supplement; Selected Translations from Medical-Related Science* 25, no. 4 (1965); Joseph T. Hart, "Autocontrol of EEG Alpha," *Psychophysiology* 4, no. 4 (1968). For a description of early developments, see Joe Kamiya, "The First Communications about Operant Conditioning of the EEG," *Journal of Neurotherapy* 15, no. 1 (2011).

[27] On physiological conditioning and cortical conditioning in the early 1960s, see references mentioned in David P. Nowlis and Joe Kamiya, "The Control of Electroencephalographic Alpha Rhythms through Auditory Feedback and the Associated Mental Activity," *Psychophysiology* 6, no. 4 (1970). In his 1962 presentation, Kamiya envisioned his alpha conditioning experiment "as a demonstration of the behavioristic equivalent of an introspective act." Abstract of Kamiya Harrell, *Proceedings of the Forty-Second*

In light of this mixed behaviorist-introspective experimentation, Stoyva and Kamiya unfolded their approach to alpha control as part of a novel proposal for studying mental states they flamboyantly called a "new study in the strategy of consciousness" that depended on the superimposition of "intersecting" evidence from multiple sources.[28] To study internal states, they proposed a triangulation strategy cross-referencing operant conditioning (through "information feedback procedures"), physiological measures (brain activity measurements), and verbal reports (through post-experimental interviews). Subjects had told Kamiya that when alpha was "on" (when alpha EEG activity had been measured), they had felt relaxed or "not experiencing any visual imagery."[29] Such verbal reports were important, the authors noted, yet should not be taken entirely at face value. Similar to the uncertain relation between subjective accounts of dreams and occurrences of mental states, they theorized, researchers would "feel most secure in inferring" about mental states if they could observe (multiple) congruences between the verbal reports and physiological (EEG) measurements.[30]

Hence, the crux of the 1968 proposal for EEG feedback, one that defined the field of study, was the idea that information feedback procedures (the ring of a bell with the onset of alpha, for example) would allow a subject to train a strong cooccurrence between a mental state and the alpha rhythm. If feedback training caused the cooccurrence to become as strong to suggest an association between an alpha and a "mental event X," then such a hypothesis could be further corroborated by observing more cooccurrences. The proposed method, Stoyva and Kamiya explained, was a type of "convergent operationalism" that allowed for "strong inference."[31] Even though Kamiya admitted in a follow-up study that intermediary processes between physiological measurements and mental states were not well understood, he claimed strong evidence in favor of correlation.[32]

Ultimately, the researchers suggested, strong correlations also meant that subjects were thus acquiring some control over not only a physiological

---

Annual Meeting of the Western Psychological Association, cited in Vincent Nowlis, "Mood: Behavior and Experience," in *Feelings and Emotions: The Loyola Symposium*, ed. Magda B. Arnold (New York: Academic Press, 1970), 268.

[28] Johann Stoyva and Joe Kamiya, "Electrophysiological Studies of Dreaming as the Prototype of a New Strategy in the Study of Consciousness," *Psychological Review* 75, no. 3 (1968): 193, 95.

[29] Ibid., 202.

[30] Ibid.

[31] Ibid., 195.

[32] Nowlis and Kamiya, "The Control of Electroencephalographic Alpha Rhythms through Auditory Feedback and the Associated Mental Activity," 483.

(alpha) event but also the associated mental event.³³ In this way, the EEG feedback triangulation not only meant upgrading the psychophysiological principle with an element of operant conditioning, but it also opened up the notion that subjects could learn to trigger a certain mental state through a type of brain (or mind) training. In this way, it was not only introspection that appeared to have been recuperated, but also, as feedback researcher Elmer Green remarked, a certain type of "volition" (i.e., the subject's will) now seemed introduced into experimental psychology.³⁴

Around 1968, news about alpha conditioning and brain training rapidly spread, both in the popular press and in scientific commentary. In 1969, the famous psychologist Abraham Maslow excitedly reported on Kamiya's research (which he had encountered in a much-read *Psychology Today* article) and recounted how Kamiya had allowed his subjects to establish voluntary control over their EEGs so as to produce "a state of serenity, meditativeness, even happiness" (Figure 5.2).³⁵ For Maslow, the new operant-conditioning technique had revolutionary consequences: "The mind-body problem, until now considered insoluble, does appear to be workable after all."³⁶ Others similarly heralded the intersecting evidence provided by EEG feedback as a new scientific approach to subjective states. As researchers Robert Kantor and Dan Brown put it in 1970: "The internal state now has both verbal and physiological outputs that can be cross validated. When these traditional divisions of public/private data or of objective/subjective data are broken down, the horizons of science broaden out in large ways."³⁷

The mention of "broadening horizons" is characteristic of the revolutionary rhetoric surrounding EEG feedback. In the early 1970s, for example, EEG researcher Barbara Brown described the emergence of alpha feedback as the start of the "Golden Age of Man," an era in which "for the first time the mental self can communicate intelligently with the physical self."³⁸ Brown's statement indicates that new attention and proof for the self-regulation of brain activity depended on a broader cultural interest in

---

[33] Stoyva and Kamiya, "Electrophysiological Studies of Dreaming as the Prototype of a New Strategy in the Study of Consciousness," 192, 201.
[34] Elmer Green, Alyce M. Green, and E. Dale Walters, "Self-Regulation of Internal States," *Proceedings of the International Congress of Cybernetics 1969* (1970). Cited in John M. Grossberg, "Brain Wave Feedback Experiments and the Concept of Mental Mechanisms," *Journal of Behavior Therapy and Experimental Psychiatry* 3, no. 4 (1972): 246.
[35] A. H. Maslow, "Toward a Humanistic Biology," *American Psychologist* 24, no. 8 (1969): 728.
[36] Ibid.
[37] Robert E. Kantor and Dean Brown, "On-Line Computer Augmentation of Bio-Feedback Processes," *International Journal of Bio-Medical Computing* 1, no. 4 (1970).
[38] Brown, *New Mind, New Body; Bio-Feedback*, 14.

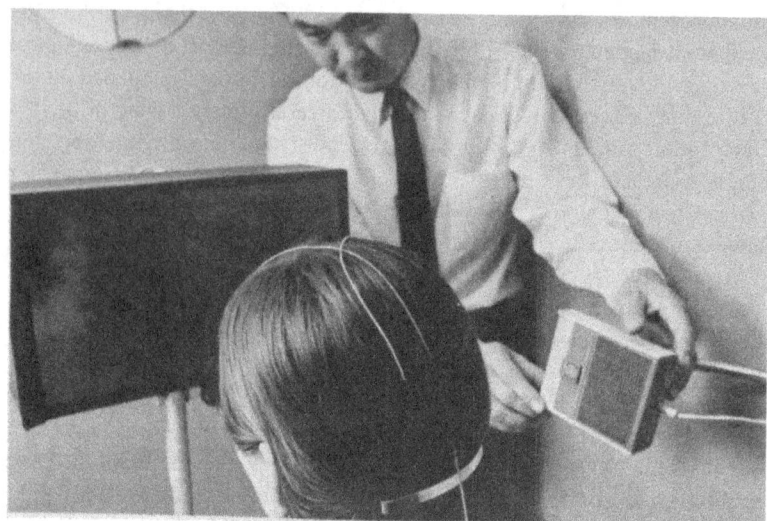

Figure 5.2 Joe Kamiya with EEG feedback setup and subject, c. 1968 (photograph).

knowing and transforming an inner self and on the proposition that EEG feedback offered new ways of accessing it.[39] Increased alpha states had been observed in practicing yogis, and alpha was also increasingly associated with meditative and spiritual states.[40] Some researchers proposed feedback could help develop a "nonverbal" communication method. This was part of a broader paradigm of approaches associated with a more intuitive way of interconnecting with others.[41] Accordingly, feedback research turned into

---

[39] Jackson Beatty, "Effects of Initial Alpha Wave Abundance and Operant Training Procedures on Occipital Alpha and Beta Wave Activity," *Psychonomic Science* 23, no. 3 (1971); Barbara B. Brown, "Recognition of Aspects of Consciousness through Association with the EEG Alpha Activity Represented by a Light Signal," *Psychophysiology* 6, no. 4 (1970); Elmer Green, Alyce M. Green, and E. Dale Walters, "Voluntary Control of Internal States: Psychological and Physiological," *Journal of Transpersonal Psychology* 2, no. 1 (1970); Hart, "Autocontrol of EEG Alpha"; Nowlis and Kamiya, "The Control of Electroencephalographic Alpha Rhythms through Auditory Feedback and the Associated Mental Activity," 6.

[40] Various researchers studied the EEG of practicing yogis and made claims about the relation between alpha waves and specific states of meditation, which in turn led to an enthusiasm for alpha conditioning as a way to train the brain for Zen relaxation and providing insight to an inner, spiritual self. Anne Harrington, *The Cure Within: A History of Mind-Body Medicine* (New York: W. W. Norton, 2009).

[41] See Starkey Duncan Jr., "Nonverbal Communication," *Psychological Bulletin* 72, no. 2 (1969).

what some have called an "experiential physiology": alpha conditioning reports had mentioned potential correlations with a variety of experiences of letting go, floating, feeling pleasure, not focusing, and increased awareness.[42]

"Disproportionate Excitement" and the Alpha Fad

Not everyone was convinced of EEG feedback's success. Soon after the surge of interest in 1968, researchers voiced two different kinds of objection: concern about its dualist or materialist approach to the mind and behavior, and increasing methodological doubts about the scientific validity of its experimental designs. In 1972, for example, psychologist John Grossberg described alpha feedback as old wine in new bottles: a "sophisticated modern version" of a fraught mind–brain dualism that thought it had closed the gap between psychology and biology.[43] With their narrow focus on alpha correlations to experience, Grossberg argued, feedback researchers had not considered crucial situational and historical components of the subjects' experiences, such as the setting, the presence and nature of the audience, and a person's past experiences. Additionally, a large number of researchers questioned and disproved the reliability of the reported correlations between alpha feedback and mental states, suggesting that many intermediary factors were involved.[44]

And yet, popular attention to biofeedback methods, including alpha feedback, remained steady. The practice enjoyed an "unusually long period of publicity," as feedback researcher Barbara Brown noted in 1974.[45] Considering the mounting contestation of EEG feedback, Grossberg dryly concluded that "the disproportionate excitement generated by such modest evidence is intriguing."[46] To understand this prolonged enthusiasm for the

[42] "Experiential physiology," in Nowlis, "Mood: Behavior and Experience." Examples of various descriptions of alpha in Kamiya, "Conscious Control of Brain Waves"; Brown, "Recognition of Aspects of Consciousness through Association with the EEG Alpha Activity Represented by a Light Signal."
[43] As Grossberg argued, "biological factors *participate to varying degrees in psychological activities.*" "Brain Wave Feedback Experiments and the Concept of Mental Mechanisms," 248.
[44] For an overview of critique and a response by Kamiya, see Sonia Ancoli and Joe Kamiya, "Methodological Issues in Alpha Biofeedback Training," *Biofeedback and Self-regulation* 3, no. 2 (1978). Publications on other types of biofeedback were also under increasing scrutiny in the early 1970s. See Neal E. Miller and Barry R. Dworkin, "Visceral Learning: Recent Difficulties with Curarized Rats and Significant Problems for Human Research," in *Cardiovascular Psychophysiology*, ed. Paul Obrist et al. (2017).
[45] Brown, *New Mind, New Body; Bio-Feedback*, 43.
[46] Grossberg, "Brain Wave Feedback Experiments and the Concept of Mental Mechanisms," 247.

potential of EEG feedback as part of biofeedback research, it is important to take note of the way this new field could emerge through new networks of research, new platforms for scientific publication, and new relations with an interested (lay) public for biofeedback.

Starting with the emergence of laboratories dedicated to biofeedback in the late 1960s, researchers began sharing regular newsletters to exchange information (followed by professional biofeedback societies and journals). They organized new types of public meetings and were in close contact with (popular) science writers and journalists.[47] This drew a motley crowd to EEG research. Kamiya characterizes the hybrid public at the first official biofeedback conference in Santa Monica, California, in 1969, as follows: "It was a mixture of uptight scientific types of all types, and people barefooted, wearing white robes, with long hair."[48] Accounts by alpha feedback researchers in the 1970s sketch an image of EEG laboratories beset by hundreds of feedback volunteers interested in exploring their inner selves and sharing their experiences at other laboratories and cultural platforms, all of which shaped informal networks of exchange.[49] While subjects were initially sometimes paid for their participation in experiments, Vincent Nowlis recounts how they often actively sought further chances to serve as subjects, "sometimes offering to do so without pay or even to pay for the opportunity!"[50]

The popular bestseller *New Mind, New Body* (1974) by biofeedback researcher Barbara Brown characteristically demonstrates the self-proclaimed antiauthoritarian status of biofeedback research. Brown framed biofeedback as a science of self-taught practitioners and experimenters, distinct from an older scientific elite, who actively sought to demonstrate their science to interested laypeople and did not shy away from popular media. On the contrary, for Brown, the reverberation of scientific insights to a wider audience was at the heart of feedback research and helped to shape (what were hailed as) more communitarian infrastructures of knowledge production.[51]

---

[47] Brown, *New Mind, New Body*; *Bio-Feedback*, 31.
[48] Kamiya, cited in Jim Robbins, *A Symphony in the Brain: The Evolution of the New Brain Wave Biofeedback* (New York: Grove Press, 2008), 65.
[49] Steven Fahrion and Patricia Norris, "Biofeedback & Self-Regulation: Introduction to a Reprint of 'Self-Regulation of Internal States' by Elmer Green, Alyce M. Green & E. Dale Walters (Originally Appeared in the Proceedings of the International Congress of Cybernetics 1969)," *Subtle Energies & Energy Medicine* 10, no. 1 (1999): 18.
[50] Nowlis and Kamiya, "The Control of Electroencephalographic Alpha Rhythms through Auditory Feedback and the Associated Mental Activity."
[51] Brown, *New Mind, New Body*; *Bio-Feedback*, 44. On communitarian conceptions of (scientific) knowledge production in the New Age movement, see Andrew Ross, *Strange Weather: Culture, Science, and Technology in the Age of Limits* (London: Verso, 1991).

Alpha feedback quickly surged in the popular press. After a 1968 *Psychology Today* article featured photographs of EEG researcher Joe Kamiya examining a subject seated in front of a monitor (a young woman, as would be the case in most feedback imagery), a mass of popular reports and self-help manuals appeared with titles such as *Alpha Brain Waves: The Startling New Way Thousands of Americans Are Using to Reach Peace of Mind* (1972), *Biofeedback: Turning on the Power of Your Mind. The First Comprehensive Book on a New Technique That Places the Power for Change and Control in the Hands of the Individual, and Not with an External Authority* (1972), *Alphagenics: How to Use Your Brainwaves to Improve Your Life* (1974), *The Power of Alpha-Thinking: Miracle of the Mind* (1976), *Biofeedback: A Partnership of Mind and Body* (1976) (Figures 5.3.1–5.3.4).

Various EEG researchers and alpha practitioners started producing and selling portable alpha-measuring instruments such as the Alphaphone, the Alpha Pacer, and the Mood Ring (a 1972 report counted forty-eight different US manufacturers).[52] A *New York Times* article remarked in 1971 that these alpha devices were "hot sellers among an introspective generation," and in 1973, *Popular Electronics* magazine instructed its readers how to build your own "brain wave feedback monitor."[53] In turn, a number of researchers scorned the alpha fad and distanced themselves from the health and rejuvenation claims of the "alpha sellers" while at the same time emphasizing the potential of serious alpha feedback practice.[54] At the annual meeting of the Biofeedback Society in 1972, researchers expressed their dismay over the "biofeedback boom" that had been created by "cultists, clinicians and the media," which had made researchers and clinicians uneasy about exaggerated claims.[55] Demarcating genuine feedback from sham versions and exaggerated claims, that is, boundary work between different spheres of feedback practice, became increasingly important.

---

[52] Lawrence, *Alpha Brain Waves*, 190.
[53] Gay Luce and Erik Peper, "Mind over Body, Mind over Mind," *New York Times*, September 12, 1971; Mitchell Waite, "Build an Alpha Brain Wave Feedback Monitor," *Popular Electronics* (1973).
[54] Lawrence, *Alpha Brain Waves*, 196–204. Thomas Budzynski, "Tuning in on the Twilight Zone," *Psychology Today* 11, no. 43 (1977).
[55] H. Stephen Leff, "A Case Study of Scientists' Opinions about the Regulations of Their Work: Opinions of the Members of the Society for Psychophysiological Research about the Regulation of Biofeedback Research and Technology," *Psychophysiology* 10, no. 5 (1973).

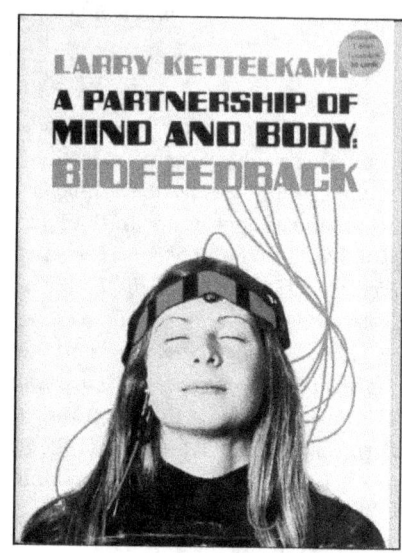
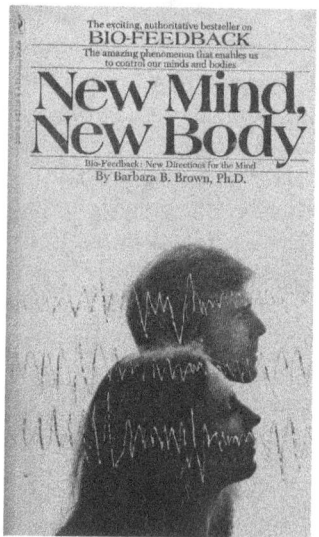

Figures 5.3.1–5.3.4 Covers of popular books on biofeedback and EEG feedback, 1973–6 (book covers, clockwise from the top left figure: Marvin Karlins and Lewis M. Andrews, *Biofeedback: Turning on the Power of Your Mind* [London: Garnstone Press, 1973]; Larry Kettelkamp, *A Partnership of Mind and Body, Biofeedback* [New York: Morrow, 1976]; Anthony A. Zaffuto, *Alphagenics: How to Use Your Brain Waves to Improve Your Life* [Garden City, NY: Doubleday, 1974]; Barbara B. Brown, *New Mind, New Body; Bio-Feedback: New Directions for the Mind* [New York: Harper & Row, 1974]).

Circuited Selves, Media Environments, and Radical Software

Grossberg's observation about the "disproportionate excitement" over EEG feedback, despite emerging scientific critique, can be better understood through biofeedback's relation to the 1960s increasing emphasis on the "self"—one based on the premise of new self-insight, and one that could be modulated through different techniques and technologies.[56] Science studies scholars Andrew Pickering and Jonna Brenninkmeijer have aptly suggested we understand EEG feedback as a particularly 1960s inflection of what Foucault termed a "technology of the self."[57] Like other 1960s techniques such as flotation tanks, stroboscopic devices, LSD, mutable architectures, flicker, video therapy, and breathing techniques, feedback was shaped by specific ideas of the self and in turn also shaped particular ways of being a self. As Brenninkmeijer has argued, the operationalization of the self in EEG feedback afforded a way of "working on the self by working on the brain" that simultaneously allowed for a brain-based self and a kind of spiritual self.[58] I argue that this conjunction of a new emphasis on the self and performances of particular EEG feedback technologies was key to the enthusiasm for EEG feedback research.

Understanding the technologically performed dimension of self-oriented EEG feedback helps to explain how a considerable number of scholars and practitioners continued to experiment with EEG feedback until the mid-1970s, despite intensifying objections to this type of research. Tinkering and playing with "technologies of the self" in experimental communities around 1970 shaped a particularly distributed notion of the self, a self that could not only be captured through introspection, but one that was envisioned as connected with, or plugged into, broader circuits or systems. EEG feedback setups generated opportunities that resonated with this new interest: recording and amplification technologies were viewed as offering self-insight but also abilities for the self to connect to wider circuiting systems, to become part of what was envisioned as a unitary whole or an all-encompassing environment. Artists' engagement with this was particularly indicative of what I call a vision

---

[56] Micki McGee, *Self-Help, Inc.: Makeover Culture in American Life* (Oxford: Oxford University Press, 2005). Karlyn Crowley, *Feminism's New Age: Gender, Appropriation, and the Afterlife of Essentialism* (Albany, NY: SUNY Press, 2011). Nikolas Rose, *Inventing Our Selves: Psychology, Power, and Personhood* (Cambridge: Cambridge University Press, 1998).

[57] Michel Foucault, *Technologies of the Self: A Seminar with Michel Foucault* (Amherst: University of Massachusetts Press, 1988). See Pickering, *The Cybernetic Brain*; Jonna Brenninkmeijer, *Neurotechnologies of the Self: Mind, Brain and Subjectivity* (London: Palgrave Macmillan, 2016).

[58] Brenninkmeijer, *Neurotechnologies of the Self*, 46.

of a "circuited self": one that was dispersed in media installations with various assembled, circuited elements that together modeled a vision of the world as a total environment. These artistic installations—often dubbed "environments" or "ecologies"—incorporated EEG as one of a variety of ways to meld technological media and "selves" to create circuited selves.

Particularly indicative of the circuited self were EEG feedback works such as Nina Sobell's *Interactive Brainwave Drawing: EEG Telemetry Environment* (1975), which conjured an imaginary of electricity flows through which selves (including brains, minds, and bodies) became part of a broader media circuitry that opened up new horizons of (group) communication and notions of "synchronization." Rosenboom's 1970 feedback work *Ecology of the Skin* combined alpha feedback headbands, synthesizers, a closed-circuit television system, "phosphene stations," and oscilloscope displays of brainwave activity to create a "group encounter brain biofeedback performance system" for ten participants, which he regarded as a "systems procedure" functioning as a "general monitor of internal balance," which he envisioned could also "be 'plugged in' to a central source of information" (Figure 5.4).[59] Ultimately, Rosenboom's brainmedia–self circuit aimed to create "information-energy exchange rituals," a "mediational language, a coherent energy" that "could reach the point where a group could almost hold conversations without talking," thus creating "more profound interactions so needed in our mechanistically functional world."[60]

Similarly interested in systems procedures, Nina Sobell started to include EEG in 1974 in her video installations with elements of time delay and closed-circuit systems.[61] She experimented with feedback video images in the EEG laboratory of Barry Sterman and subsequently created *Interactive Brainwave Drawing: EEG Telemetry Environment* (1975) (Figure 5.5).[62] In this piece, two subjects placed in a living room setting could watch direct video feedback of themselves on a monitor, while their EEG activity was recorded and translated into a zigzagging line visualizing the combined brain activity of both participants. Outside the room, five monitors displayed their EEG

---

[59] David Rosenboom, "Method for Producing Sounds or Light Flashes with Alpha Brain Waves for Artistic Purposes," *Leonardo* 5, no. 2 (1972): 143; Donal Henahan, "Music Draws Strains Direct from Brains," *New York Times*, November 25, 1970. Rosenboom mentions plugging into Buckminster Fuller's envisioned "National Resource Information System."

[60] Rosenboom, "Method for Producing Sounds or Light Flashes with Alpha Brain Waves for Artistic Purposes," 141.

[61] Nina Sobell, interview by Evelin Stermitz, August 22, 2007.

[62] Emily Hartzell and Nina Sobell, "Sculpting in Time and Space: Interactive Work," *Leonardo* 34, no. 2 (2001).

Figure 5.4 Photograph of David Rosenboom, *Ecology of the Skin*, 1970.

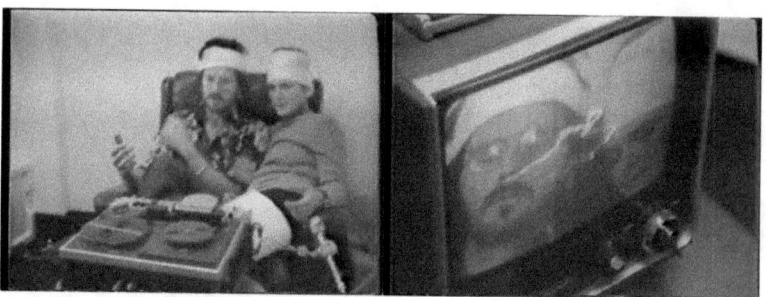

Figure 5.5 Nina Sobell and Michael Trivich, interactive electroencephalographic video drawings, 1972–4 (video stills).

recordings as well as previous participants' activity, which was superimposed on the live feed.

Rosenboom's and Sobell's EEG feedback work must be understood as part of a new (media) ecological framework that became influential in artistic discourses in the 1970s, specifically due to Gregory Bateson's writings on a cybernetics of the self.[63] Bateson's ideas were prominently circulated in the arts and culture journal *Radical Software* (published between 1970 and 1974).[64] This brand of media ecology dissolved the brain and the mind into an idea of the self unbounded by the body, a circuited self that stood in a continuous relation to the world and arose through changing flows of communication.[65] In *Radical Software*, these visions of a circuited, cybernetic self constituted the underlying philosophy for artistic experiments with portable video equipment, computers, and closed-circuit monitoring networks, DIY technologies that were envisioned as diametrically opposed to mass forms of communication, such as broadcast television, and ultimately connected multiple selves into what one author in *Radical Software* called a "world brain."[66]

Within this (video) art universe, feedback denoted both the literal real-time experimentation that new equipment enabled—direct and delayed audio and video playback, for example—but it also resonated with broader notions of feedback as the basis for an environment of reverberating, communicative flows. EEG feedback perfectly fitted these notions of circulating communication

---

[63] Gregory Bateson, "The Cybernetics of 'Self': A Theory of Alcoholism," *Psychiatry: Journal for the Study of Interpersonal Processes* 34, no. 1 (1971).

[64] See Peter Collopy, "The Revolution Will Be Videotaped: Making a Technology of Consciousness in the Long 1960s" (PhD dissertation, University of Pennsylvania, 2015).

[65] William Kaizen, "Steps to an Ecology of Communication: Radical Software, Dan Graham, and the Legacy of Gregory Bateson," *Art Journal* 67, no. 3 (2008): 87.

[66] Don Benson, "Neurone Cluster Grope," *Radical Software* 1, no. 2 (1970).

Interfacing the Real-Time Brain

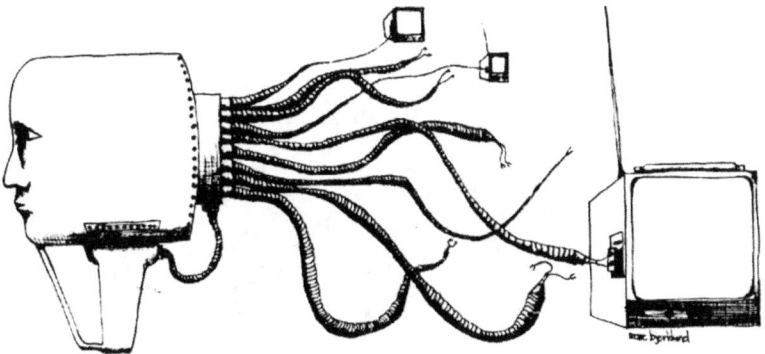

Figure 5.6 Marc Bjorlund, no title, 1971 (illustration).

and introduced a novel dimension of flow into this larger circuited world. Notions of the "synchrony" of brainwaves became intuitively aligned with cooperation in person-to-person communication, so the circuited self was envisioned as enabling communicative harmony, as circuits allowed for the playing, connecting, tuning, and merging of different selves.

By the end of the 1960s, it seemed that every systems-loving artist had become interested in alpha: Stan VanderBeek, for example, envisioned brainwave measurements as part of his *Movie Drome* experiment, and Nam June Paik spoke of relaying brainwaves via videophone lines, elevating people's mood to the point of "Electronic Zen."[67] In the pages of *Radical Software*, EEG feedback devices and training were discussed as "providing people with a chance to explore the internal, and in a socially constructive way."[68] Artists such as Woody Vasulka and Richard Lowenberg, working together with psychophysiologist Peter Crown, conceptualized new "Techno-Sensory Interface Projects" that included audio-video systems triggered by brainwave alpha rhythm readings to study, for example, "human control of purely contemplative creative processes."[69] Drawings in *Radical Software* showed heads sprouting meandering nerves connected to television screens (Figures 5.6 and 5.7).

[67] Jud Yalkut, "Electronic Zen: The Alternate Video Generation," http://vasulka.org/archive/Artists10/Yalkut,Jud/ElectronicZen.pdf1984. The filmmaker as video artist: Stan Vanderbeek (interview undated), 10, Paik (1966), cited in Kahn, *Earth Sound Earth Signal. Energies and Earth Magnitude in the Arts*, 278.
[68] Ralph Ezios, "Implications of Physiological Feedback Training," *Radical Software* 1, no. 4 (1971).
[69] Woody Vasulka and Richard Lowenberg, "Environetic Synthesizer," *Radical Software* 1, no. 2 (1970); Richard Lowenberg and Peter Crown, "Environetic Synthesis. Techno-Sensory Interface Projects" (Typoscript 1971).

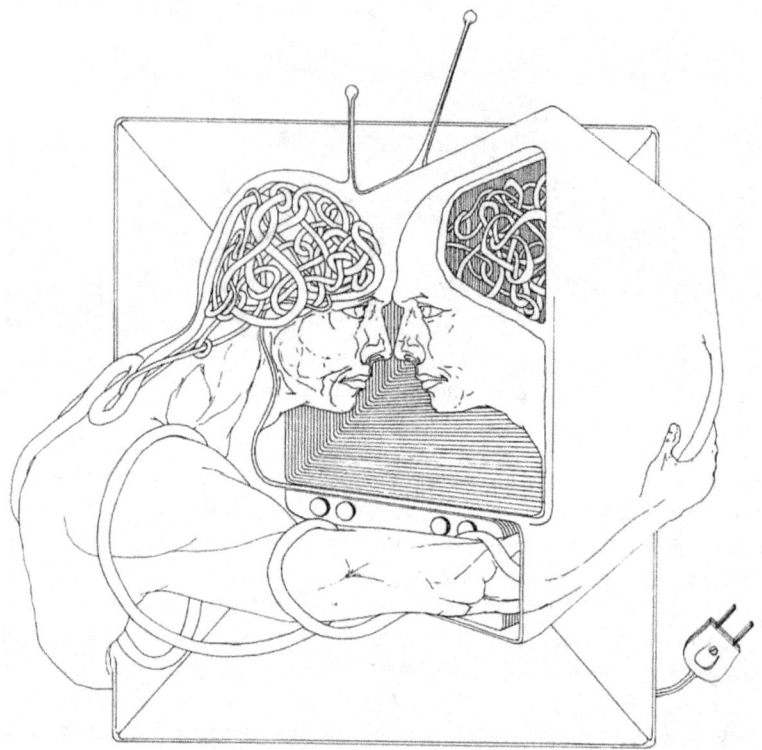

Figure 5.7 Richard Lowenberg, *Environetic Synthesis*, 1972 (illustration).

The fact that psychophysiological signals could also be harnessed for other purposes—not only self-regulation but also (governmental, dominating, nonvoluntary) regulation—did not go unnoticed by artists. Some artists' investigations into biofeedback (including EEG) seemed to align themselves all too smoothly with the possibility that sensory feedback could be employed to force behavioral change. Historian Joseph Brandon describes, for example, the affiliations between Manford Eaton's early 1970s *Bio-Music* and CIA research into interrogation and sensory deprivation. Eaton's work thus strengthened visions of the ultimate controllability and programmability of individual and group behavior.[70]

In contrast, in 1974, Richard Teitelbaum warned against such potential applications, concerned that EEG research allowed brains to be opened

---

[70] Branden W. Joseph, "Biomusic," *Grey Room* 45 (2011): 143.

up for others to command: "With some of the most technically 'advanced' psychology work currently being carried out in our prisons [under] the guise of aversion therapy and the like, there is clearly great cause for concern."[71] Similarly, a commentator in *Radical Software* commented on the "Big Brother" potential of biofeedback devices: malevolent governments might add it to an existing arsenal of governmental control such as propaganda, responsive environments, surveillance, and wiretapping. And yet, the author shrugged, "it's really nothing new. In the same way that so many of us have pushed aside the bullshit of broadcast television, we'll deal with what comes. Free universities will offer courses on cortical jamming techniques and *Radical Software* will be a hologram of How to Build an Alternative Brain Wave Network."[72] As this remark demonstrates, while the man–machine circuits created in and through EEG feedback art were potentially dangerous, they were ultimately envisioned as positive technologies of the self or promising ways of interconnecting with machines or other subjects. The emphasis, in this case as in many others, was on making tools and techniques accessible by building a network.

The creation of circuited selves with EEG feedback as part of elaborate media environments demonstrates a new approach to experimentation: here, emphasis shifted from operant conditioning through feedback toward situations in which participants' physiological measures were envisioned as being in flux and constant interaction with connected displays and circuits. While Kamiya had envisioned feedback research as a calibrating back and forth between "information feedback procedures," brain activity measurements and verbal reports, media artworks mixed these elements in a technophilic whirl. In this situation, the meaning of alpha decidedly shifted. Artistic EEG experiments shaped the idea of an alpha experience as a particularly distributed mode of being, a state of mind that was always already part of a media-techno-cultural-biological assemblage. Because experimental subjects had flocked to feedback rooms to find new technologies of the self through alpha, they also described feeling like an alpha-technological self.

This ontologization of a particular type of alpha—a peculiar hybrid between a physiological signal and an experience of a techno-cultural present—is particularly evident in the systems-interested artworks described here. Yet, this techno-environmental dimension also reverberated with other feedback

---

[71] Richard Teitelbaum, "In Tune: Some Early Experiments in Biofeedback Music (1966–74)," in *Biofeedback and the Arts: Results of Early Experiments*, ed. David Rosenboom (Vancouver: A. R. C. Publications, 1976).

[72] S.S., "Not Surprisingly, You Can Now Own Your Own Brain Feedback Device…," *Radical Software* 1, no. 4 (1971).

research communities in the borderlands. Barbara Brown, for example, discussing alpha feedback in 1970, noted that subjects with higher levels of alpha had reported feeling "dissolved into the environment."[73] Similarly, describing their alpha experiences in the experiments of researcher Elmer Green, subjects used phrases such as "the off-conscious" and "the threshold of consciousness."[74] Green envisioned a creativity program through which subjects might voluntarily induce the state "associated with such descriptive statements." Hence, while experimental EEG feedback had set out to explore the mental correlate of alpha, it had effectively produced a new object, the mental state of "feeling alpha" that wholly depended on the media-environmental alpha interface.

Techno-Sensory Interface Projects and New Modes of Communication

Artistic visions of feedback systems and circuited selves employed the term "interface" (such as the aforementioned technosensory interface projects) as a notion that intuitively fitted (the vocabulary of) new visions of reciprocity and new forms of communication between selves, brains, and machines. At present, the notion has become more narrowly associated with graphic user interfaces or even simply with the screen, that is, with a visual structure that allows users to interact with a digital environment. Yet, as scholars have argued, the notion emerged as a significant new concept in the 1950s, with the rapid rise of human–computer communications and more complex black-boxed computing systems that necessitated mediators between humans and machines.[75] For artists as well as EEG technology researchers, interfaces were envisioned and framed as crucial nodes for the interaction between humans, bodies, brains, and machines, an interaction that was often framed with a vocabulary of "communication" or "dialogue." Importantly, the interfaces discussed in this emerging computing era had a speculative

---

[73] Brown, "Recognition of Aspects of Consciousness through Association with the EEG Alpha Activity Represented by a Light Signal," 449.

[74] Green et al., "Self-Regulation of Internal States," reprinted in Steven Fahrion and Patricia Norris, "Biofeedback & Self-Regulation," *Subtle Energies & Energy Medicine* 10, no. 1 (1999): 33.

[75] J. Drucker, "Humanities Approaches to Interface Theory," *Culture Machine* 12 (2011); Florian Cramer, "What Is Interface Aesthetics, or What Could It Be (Not)?" in *Interface Critism: Aesthetics beyond Buttons*, ed. Christian Ulrik Andersen and Søren Bro Pold (Aarhus: Aarhus University Press, 2011); Branden Hookway, *Interface* (Cambridge, MA: MIT Press, 2014); Florian Hadler and Daniel Irrgang, "Instant Sensemaking, Immersion and Invisibility. Notes on the Genealogy of Interface Paradigms," *Punctum. International Journal of Semiotics* 1, no. 1 (2015).

and futurist dimension. As historian Jonathan Grudin notes, publications on human–computer interaction such as "Man–Computer Symbiosis" (Licklider 1960), "Augmenting Human Intellect" (Engelbart 1962), and "A Conceptual Framework for Man–Machine Everything" (Nelson 1973) were future-oriented; their interfaces were "inspiring visions and prototypes," shaping "a world that did not exist" (yet).[76]

From the mid-1960s onwards, intricate setups—dark rooms with brain-controlled switches, differently colored lights, or changing sounds– were a principal element of EEG feedback experimentation. While researchers did not always explicitly use the word interface to denote the structures that allowed for a feedback between humans and technology, the carefully designed feedback circuit and the performative interactions of the human with elements of this circuit became central to EEG feedback research.[77]

In 1970, Brown devised an EEG feedback interface that used three EEG frequency ranges of theta, alpha, and beta to operate red, yellow, and green lights.[78] Building on this research, she subsequently experimented with ways to move away from the abstraction of such signals to create cues that had "some symbolic meaning or interest" to subjects.[79] To do so, she designed two playful devices to make alpha waves more vivid to participants: the alpha train and the alpha wave racetrack, which received much press attention (Figure 5.8).[80] The alpha train would start or stop according to the alpha waves a subject produced, while the alpha wave racetrack could be played by two subjects competing for alpha wave control. New interfaces were part of the framing of EEG as enabling new types of dialogue, as, one newspaper explained, communication was at the basis of alpha feedback not only between "man and his inner being" but also between scientists and laypeople.[81]

---

[76] Jonathan Grudin, "Three Faces of Human-Computer Interaction," *IEEE Annals of the History of Computing* 27, no. 4 (2005): 48.
[77] See, for example, Joe Kamiya's notes on his improvement of the feedback sounds and presentation of the subject's performance in Mark S. Schwartz et al., "The History and Definitions of Biofeedback and Applied Psychophysiology," in *Biofeedback, Fourth Edition: A Practitioner's Guide*, ed. Mark S. Schwartz and Frank Andrasik (New York: Guilford, 2017).
[78] Brown, "Recognition of Aspects of Consciousness through Association with the EEG Alpha Activity Represented by a Light Signal."
[79] Brown, *New Mind, New Body; Bio-Feedback*, 42.
[80] Footage of Brown's train is included in the documentary *Dialogue on Biofeedback*, https://www.youtube.com/watch?v=3gGQF2ltH8c, Veterans Administration, 1974.
[81] Robert Kirsch, "Biofeedback: In the Beginning Was Alpha," *Los Angeles Times*, August 18, 1974.

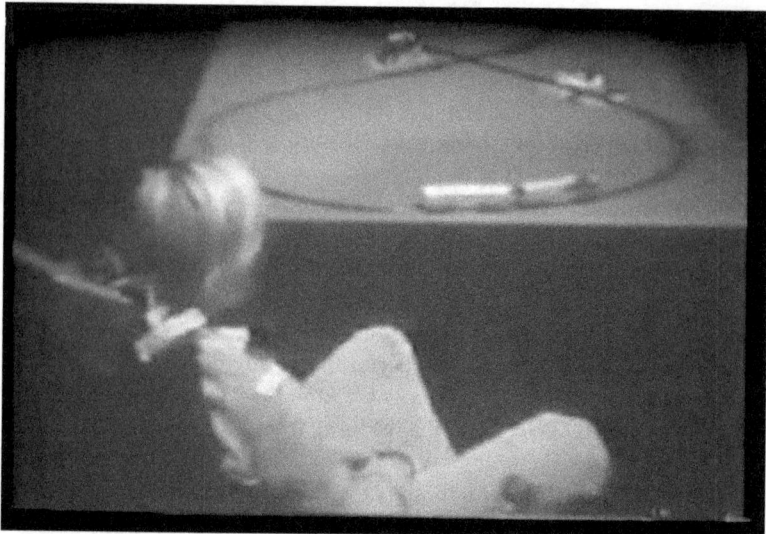

Figure 5.8 Feedback training with alpha train in Barbara Brown's lab, c. 1974 (video still).

The visually attractive, interface-oriented setups of EEG feedback facilitated this public attention. As Brown recounts, "the laboratories were invaded by television film crews and reporters, newspaper, and magazine writers. The settings were ideal for visual media: brain wave recordings, sound and colored light displays, and even tiny trains or racing cars which could be hooked up and energized by the subject's alpha waves."[82] Edmond Dewan's earlier brain-controlled electric lamp had been a similar popular favorite. Already in the lab, the setup was demonstrated in front of audiences who could give orders to the performing subject. The performative potential of this experiment became even more potent when CBS News featured Dewan's setup and concluded their report with a scene in which the subject's EEG signals supposedly switched on a television set and returned the viewer to the CBS studio (Figures 5.9.1 and 5.9.2).

These popular imaginaries of brains plugged into everyday (media-) technological circuits were common; as Barbara Brown recounts, "scientists and tv crews alike began plotting brain wave electric companies. Almost everyone decided that in the future it would be possible to have the brain

---

[82] Ibid.

Interfacing the Real-Time Brain 181

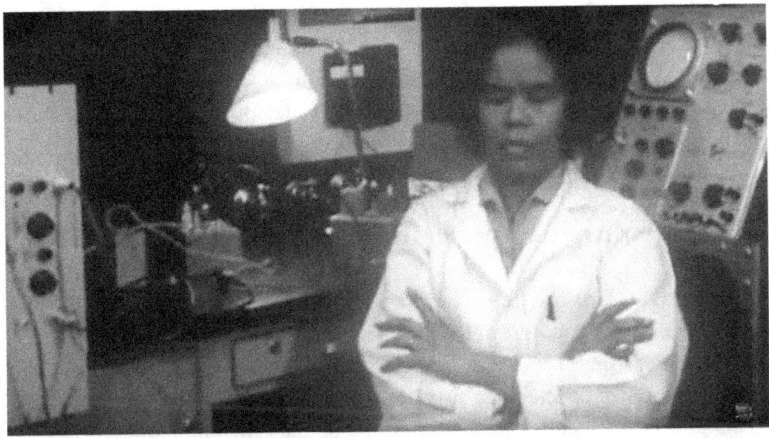

Figures 5.9.1 and 5.9.2 Edmond Dewan, brain-controlled lamp setup, 1964 (television broadcast stills).

start the coffee pot."[83] As a humorous response to the media hype of alpha feedback, artist Richard Teitelbaum and his collaborators created *Alpha Bean Lima Brain* (1971), in which alpha waves in California were transmitted to New York to activate water sprinklers in pots containing dry lima beans. "Naturally," Teitelbaum recounts, a "film of this performance was featured

[83] Brown, *New Mind, New Body; Bio-Feedback*, 42.

on the NBC evening news in Los Angeles."[84] Whether ironic or poetic, both scientific and artistic installations emphasized the performative, playful, interesting, or even humorous aesthetic of the designed interface between EEG measurements and the measured subjects. Both in art and in EEG science, the interface became the (art)work.

New Micro-Temporalities of the Brain in Real Time

My analysis so far shows how an interest in EEG feedback in both scientific and artistic communities resulted in experiments with new types of interfaces. This emphasis on interface design also resonated with public interest and media reports of spectacular situations of brains interacting with blinking lights and other technologically circuited signals. In tandem, artistic EEG works offered a vision of alpha activity as an intricate flow of energy that was part of a larger (technologically) circuited whole. A major reason for the popularity of alpha feedback was the notion that these interfaces appeared to allow people to communicate with themselves, enabling people to become aware of their own brain activity on a moment-to-moment basis and train their activity accordingly.

However, such appeals to alpha feedback as a technology of the self, while popular, became increasingly contested. Certain experimental paradigms were questioned, as were the reliability of EEG feedback (including alpha feedback), as well as the meaning of EEG measurements (its status as a psychophysiological measure, that is, the relation between alpha and mental states). In the beginning of the 1970s, EEG feedback was dismissed by a growing number of researchers; the standard for measuring the active brain was rapidly shifting.

Already in the 1960s, researchers had sought to define smaller temporal components of the measured EEG activity and relate them to sensory, cognitive, or motor events through evoked responses (using a sound or flash, for example). If measuring and averaging would be granular enough, researchers thought, consistent, minute patterns of electrical changes could be extracted from the messy and noisy brainwave recordings. The tedious work of averaging such evoked potential trials could be sped up with the

---

[84] Teitelbaum, "In Tune: Some Early Experiments in Biofeedback Music (1966–74)." Another example is Jacqueline Humbert's 1973 proposal for *Alpha Garden*, an installation in which a "synchrony detector" (for brainwave synchrony between two subjects) would control a sprinkler system. David Rosenboom, *Biofeedback and the Arts: Results of Early Experiments* (Vancouver: A. R. C. Publications, 1976).

introduction of commercially manufactured computing devices in the 1960s.[85] This resulted in measuring consistent patterns of minute spikes called "event-related potentials" (ERPs).[86] In 1964 and 1965, three separate research teams reported discovering three such minute wave patterns (peaks and valleys occurring at particular moments) that consistently correlated with specific mental activities.[87] The precise psychophysiological meanings of these ERPs were a vivid topic of discussion—and remain so today.[88]

The ultimate consequence of this shift in temporalities was a major transformation in the conception of interfaces between brains and machines and of feedback. This was a shift between the history of EEG feedback thus far—which appeared to allow self-regulation and communication with the self—and new research, starting around 1970, into the microtemporalities of the active brain examined through "real-time" communication between brains and computers (BCIs). In both types of research, the interface, feedback, and communication between man and machine were central, yet understood in radically different ways.

While alpha self-regulation proposed interfaces and feedback as way to give subjects insight into themselves, new man–machine communications proposed in the 1970s gave the computer feedback about a subject's performance in real time. There is an important conceptual difference

---

[85] Clyne (1962), cited in Charles Shagass, *Evoked Brain Potentials in Psychiatry* (New York: Plenum Press, 1972). History section, 1–3.

[86] The first so-called ERP was published in W. Grey Walter et al., "Contingent Negative Variation: An Electric Sign of Sensori-Motor Association and Expectancy in the Human Brain," *Nature* 203, no. 4943 (1964). On the history and use of ERPs, see Geoffrey F. Woodman, "A Brief Introduction to the Use of Event-Related Potentials (ERPs) in Studies of Perception and Attention," *Attention, Perception & Psychophysics* 72, no. 8 (2010); Jacques J. Vidal et al., "Biocybernetic Control in Man-Machine Interaction: Final Technical Report 1973–1974" (PhD dissertation, Computer Science Department, School of Engineering and Applied Science, University of California, 1974); Jacques J. Vidal, "Real-Time Detection of Brain Events in EEG," *Proceedings of the IEEE* 65, no. 5 (1977).

[87] The 1960s saw the publication of the "contingent negative variation" (CNV, associated with "expectancy," discovered by a team of researchers including William Grey Walter), the readiness potential (*Bereitschaftspotential*), and the P300, originally related to "stimulus uncertainty." Walter et al., "Contingent Negative Variation"; Hans H. Kornhuber and Lüder Deecke, "Hirnpotentialänderungen Bei Willkürbewegungen Und Passiven Bewegungen Des Menschen: Bereitschaftspotential Und Reafferente Potentiale," *Pflüger's Archiv für die gesamte Physiologie des Menschen und der Tiere* 284, no. 1 (1965); S. Sutton et al., "Evoked-Potential Correlates of Stimulus Uncertainty," *Science* 150, no. 3700 (1965); Kornhuber and Deecke, "Hirnpotentialänderungen Bei Willkürbewegungen Und Passiven Bewegungen Des Menschen."

[88] John Russell Knott and William Cheyne MacCallum, *The Responsive Brain: Proceedings of the Third International Congress of Event-Related Slow Potentials of the Brain. Bristol, England 12–18 August 1973* (Bristol: John Wright and Sons, 1976). For a general historical review, see Frank Rösler, "From Single-Channel Recordings to Brain-Mapping

between these two approaches in the 1970s that may not be immediately apparent: both EEG feedback and BCIs used EEG measurements, and both drew on a vocabulary of interfaces and feedback. Yet, while the main aim of EEG feedback was to train the brain through observable, meaningful feedback on performance, the goal of BCIs was to create a situation in which tasks could be continuously updated in relation to EEG activity (not necessarily with meaningful feedback to the subject), resulting in EEG data that was primarily used for fundamental research into psychophysiological correlations. This type of man–machine interfaces worked through a "closed-loop" feedback, based on the incremental interpretation of the EEG and task performance data by computers. Thus, while a vocabulary of feedback and communication and an emphasis on interface design persisted, the meaning of those terms fundamentally changed.

A particularly clear and critical juncture in this regard occurred at the beginning of the 1970s in the context of two major research grant schemes on "Biocybernetics," developed by the US Defense's Advanced Research Projects Agency (DARPA), which examined EEG's potential for military applications.[89] In 1970, in the wake of the enthusiasm for alpha feedback, DARPA had initiated a first research grant to assess "self-regulation as an aid to human effectiveness." A number of well-known alpha feedback researchers (such as Joe Kamiya and Johann Stoyva) were funded through the grant, which resulted in a final DARPA report (not published until 1976) that left little doubt about the prospects of self-regulation: "Biofeedback offers much less powerful and robust self-control over certain internal physiological events than many researchers believed on the strength of early anecdotal evidence," "alpha enhancement or suppression have not been found to affect performance," and "there is no easy and powerful key available in biofeedback for significant enhancement of performance."[90]

Devices: The Impact of Electroencephalography on Experimental Psychology," *History of Psychology* 8, no. 1 (2005).

[89] One part of the program researched the "voluntary regulation of physiological events to enhance performance" (self-regulation as an aid to human effectiveness, part A, 1970–5), while the other was interested in developing "closely coupled man/machine interfaces for the extension and enhancement of human performance" (biocybernetics technology and behavior, part B, 1973–9). See Robert L. Benshoff, "Self-Regulation as an Aid to Human Effectiveness and Biocybernetics Technology and Behavior" (PhD dissertation, San Diego State University, 1976), https://apps.dtic.mil/docs/citations/ADA021105. Judith Daly, "Close-Coupled Man/Machine Systems Research (Biocybernetics) Completion Report" (Cybernetics Technology Division Program, DARPA, 1981).

[90] Benshoff, "Self-Regulation as an Aid to Human Effectiveness and Biocybernetics Technology and Behavior."

Even though the first DARPA report crushed any hope for self-regulation through biofeedback, it stimulated a second biocybernetics grant scheme that aimed to employ psychophysiological signals not as a means for humans to regulate their own behavior but to supply such signals to a computer that would thus allow the man–machine system to take over some of the human's functions, a goal that was called a "prosthetic" approach.[91] Five of the DARPA-funded research teams used brain activity measurements to work toward this ideal of man–machine communication.[92] Tracing these EEG projects reveals the shift in approach to the active brain under the biocybernetics paradigm. Exemplary in this respect is the (partially) DARPA-funded work of computer engineer Jacques Vidal, who, working at UCLA's Brain Research Institute, famously coined the term "brain–computer interface" in his 1973 paper "Toward Direct Brain–Computer Communication."[93] (In fact, in 1966, UCLA's Thelma Estrin had already reported on the use of a "closed loop" or "feedback mode" at Mary Brazier's neurophysiology laboratory, describing how visual stimuli could be "delivered by the computer to a subject based on computations from his on-going recorded brain waves."[94]) Vidal's BCI consisted of three computers and a considerable number of circuited displays, control terminals, A–D converters, buffer controllers, and other components.[95] The intricate system measured its subjects' ERPs and used two rudimentary game-playing situations (a figure ground visual task and a space shuttle shooting computer game drawing on the existing game *Spacewar!*, popular among laboratory workers) to examine how ERPs related to specific visual and cognitive tasks. Vidal described his BCI as a "real-time system" that aimed for a true "man–computer dialogue" that would turn the computer into a "genuine prosthetic extension of the brain."

---

[91] Daly, "Close-Coupled Man/Machine Systems Research (Biocybernetics) Completion Report," 1. A wide range of research programs received support: at Harvard University, for example, researchers tried to use EMG to develop a silent, fast typewriter; at the Naval Health Research Center (San Diego), researchers employed computers to monitor physiological states; and in Arlington, researchers tested how computers could create "realistic environments" in biocybernetics research.

[92] Ibid., 4–5.

[93] Jacques J. Vidal, "Toward Direct Brain-Computer Communication," *Annual Review of Biophysics and Bioengineering* 2, no. 1 (1973).

[94] Thelma Estrin, "Neurophysiological Research Using a Remote Time-Shared Computer," in *Data Acquisition and Processing in Biology and Medicine: Proceedings of the 1966 Rochester Conference*, ed. Kurt Enslein (Rochester, NY: Elsevier, 1968); "On-Line Electroencephalographic Digital Computing System," *Electroencephalography and Clinical Neurophysiology* 19, no. 5 (1965).

[95] Vidal et al., "Biocybernetic Control in Man-Machine Interaction: Final Technical Report 1973–1974."

Vidal's BCI had introduced the term "real time" into the circuit of brain–machine interfacing. "Real time" was a technical term that emerged in the 1940s with the engineering of automatic analyzers for computing. It was first used by electrical engineer J. Eckert in 1946, who explained that a machine worked in "real time" (printed in scare quotes) if it exhibited a "sufficient speed to fit its time variable with real time" (without scare quotes).[96] "Real time" meant a sufficient correspondence to the time of "the human element," to use Eckert's words, that is, the time in which a human could perceive and react.[97] Importantly, with this definition, technical processes could thus potentially be too slow or too fast to function in real time, that is, in relation to (human-related) real time. With new attempts to use computing in biofeedback interfaces around 1970, the ideal of "real-time" communication between man's physiological signals (EMG, EEG, ECG, etc.) and machines could be advanced as a significant line of investigation.

The concept of real time was important for ERP research because the minute temporal events that constituted ERPs shifted over time (e.g., because a subject gradually adjusted to a task), and a computer needed to calculate these ERPs during the experiment, adjusting the task on the fly. Vidal defined his real-time system "as one which controls an environment by receiving data, processing it and returning the results sufficiently quickly to affect the environment at that time."[98] To return measurements sufficiently quickly, Vidal argued that a round trip of less than 0.5 second was necessary to "directly communicate brain messages" (i.e., to update the task in a way that the subject would experience a change sufficiently fast).[99] Practically, this also meant that real-time processing of EEG feedback constrained the depth of the analysis. More precise and differentiating procedures of batch processing and averaging (which usually take place offline) would slow the analysis, so Vidal's BCI employed a reference set of typical EEG waveforms for various responses.

In this setup, "real time" referred to adjusting the tasks (visible on the interface) to the measured ERPs from moment to moment. The notion was

---

[96] First-time use cited in the *Oxford English Dictionary*. J. P. Eckert, Jr., "Continuous Variable Input and Output Devices," in *The Moore School Lectures: Theory and Techniques for Design of Electronic Digital Computers*, ed. Martin Campbell-Kelly and Michael R. Williams, The Charles Babbage Institute Reprint Series for the History of Computing (Cambridge, MA: MIT Press and Tomash, 1946; reprint 1985); ibid., 394.
[97] Ibid., 397.
[98] Vidal, "Toward Direct Brain-Computer Communication," 169. Vidal cited a definition of "real-time systems" from James Martin, *Design of Real-Time Computer Systems* (Englewood Cliffs, NJ: Prentice Hall, 1967).
[99] Ibid., 169.

less important when it came to overt feedback for subjects. Even though Vidal also spoke of relaying a "real-time score of performance" to the subject (who could thus evaluate their score), this was difficult to accomplish in practice and was not the goal of the system. Whatever the feedback to the subject was, it would always be lagging with respect to the measured microtemporalities of the brain. More important was the real-time feedback to the machine; the fact that the subject's responses "were automatically incorporated" into the computer to adjust the computer's evaluation of the relation between the ERP and the task.[100]

What feedback meant in this circuit between EEG measurement, computer display, and subject had thus shifted. Even though Vidal used the vocabulary associated with earlier EEG (alpha) feedback, in his setup, the EEG signals flowing from man to machine effectively were much more important than those running in the other direction.[101] The significant information flow was that of the measured ERPs, the subject's task responses to the machine, and the continuous (real-time) update of the task interface, not any "real-time score" presented to the subject.

Vidal proposed that this might change in the future. Once ERPs could be "read in context, and translated by the computer into a perceivable change in the environment" (e.g., a signal on a screen), he argued, "the specific brain activity becomes a 'behavior,' open to conscious validation."[102] In other words, Vidal projected that with the accumulation of research on the communication and calculation running from the subject's ERPs to the machine, eventually the ideal result would be meaningful real-time feedback of ERPs to the subject. For biocybernetics, humans fundamentally constituted a form of latency in the man–machine system: because they could not react within a short enough interval, BCI was envisioned as a means of reducing latency.[103] Even though the subject—the human element in Eckert's definition of real time—was a central part, eventually it was the microtemporality of the subject's brain that mattered most in the BCI.

---

[100] Ibid., 177.
[101] As Vidal's collaborator Robert J. Hickman put it in 1977: "Clearly the input-output protocol between man and machine is a key aspect of this concept. The emphasis here has been centered on the input channel of the man-machine system, that is the communication channel from man to the computer ... Signals collected directly from the brain are processed for entry into the computer so that, in effect, a neurological prosthetic efferent is created." Robert J. Hickman, "A Neurocybernetic Approach to Man-Machine Communication," *Proceedings of the San Diego Biomedical Symposium* 16, no. 1 (1977): 475.
[102] Vidal, "Real-Time Detection of Brain Events in EEG," 636.
[103] Hickman, "A Neurocybernetic Approach to Man-Machine Communication," 475.

This new directionality in man–machine communication becomes poignantly visible in another 1970s project funded by the same biocybernetics program. At the University of Illinois, Emanuel Donchin used ERP measurements (P300) to give the computer information about the possibility that subjects might have erred in their decision. In Donchin's research, the directionality of the "biocybernetics channel" (a closed-loop application) clearly ran from man to machine. He proposed that when a subject's P300 ERP was shorter than their reaction time, this meant that they "acted before 'thinking'" and therefore tended to make mistakes.[104] Incorporating a "biocybernetics error correction algorithm" would "allow subjects to respond as fast as they wished while allowing the computer to correct for errors using an algorithm based on P300."[105] Hence, the experiment did not aim to give subjects feedback about their functioning (or only minimally), but to allow computers to assess the quality of the subject's performance (such as a word categorization task) on the basis of their brain activity and task performance. Hopefully, Donchin speculated in 1977, this research would result in marketable products, though he did remark that managers should not "hasten to purchase small, handheld error-correcting biocybernetic gadgetry," as the error correction method might not yet be reliable enough.[106] In Vidal's lab, researchers projected that man–machine communication could be applied to controlling prosthetics, commanding systems too complex and fast for human control, and for investigative interrogations (lie detection).[107]

What counted as the (circuited) "self" for a community of EEG feedback researchers was decidedly different than the subject who became part of a biocybernetics BCI in "real time." I have described the former as a hybrid science that emerged through interacting artistic and experimental entanglements fundamentally focused on self-insight, and the latter as an approach to the active brain through closed-loop research that was closely imbricated with military research into error correction, warning mechanisms,

---

[104] In a *Science* paper based on the P300 experiment, Donchin (together with coresearchers Marta Kutas and Gregory McCarthy) argued that studying ERPs allowed researchers to go beyond subject's overt responses and offered an index for mental chronometry. M. Kutas, G. McCarthy, and E. Donchin, "Augmenting Mental Chronometry: The P300 as a Measure of Stimulus Evaluation Time," *Science* 197, no. 4305 (1977).

[105] Emanuel Donchin, "Event-Related Brain Potentials: A Tool in the Study of Human Information Processing," in *Evoked Brain Potentials and Behavior. Proceedings of a Conference on Evoked Brain Potentials and Behavior Held at the Downstate Medical Center, State University of New York, Brooklyn, New York, May 19–21, 1977*, ed. Henri Begleiter (New York: Plenum Press, 1979), 72.

[106] Ibid., 73–4.

[107] Hickman, "A Neurocybernetic Approach to Man-Machine Communication," 475.

and machine learning. Moreover, zooming in on this experimental and ideological juncture in the beginning of the 1970s, I have also noted a simultaneous import in BCI research of a vocabulary of "communications," "channels," and "feedback" between man and machine.

1970s ERP researchers found remarkable ways of rhetorically performing (and thereby invoking) machine–man communication, as evidenced by the emphasis on "two-way communication" at the International Congress of Event-Related Slow Potentials in Bristol in 1973. On the invitation of William Grey Walter, meetings took place in the Burden Neurological Institute's lab spaces, where a closed-circuit television system allowed for presenting "live on-line experimentation with human subjects" in smaller lab rooms to the main body of congress participants.[108] "Live on-line experimentation" and communal feedback were framed as a new type of collective knowledge exchange as well as a way to show experimental methods in action; a procedure, the editors of the conference proceedings admitted, that was rarely "smooth, clear and unambiguous."[109] This closed-loop experiment constituted a way of performing knowledge that enacted a special interest in creating systems of circuited flows, hence an imaginary of a media environment that was similar to contemporaneous artistic experiments with systems installations. Nevertheless, feedback and communication had decidedly different meanings in these contexts: while alpha feedback was about giving insights into the self, this 1973 experiment could only relay an EEG measurement "live"; it could not make visible, in real time, the types of microtemporal measurements that mattered for assessing task performance and incrementally improving the BCI's algorithms. This crucial difference between the two domains is, I argue, obscured by a similar vocabulary, a shared interest in interface design, and—as becomes clear in the 1973 example—a practice of performing technological circuits.

Two-way communication, the idea that brains could communicate not only with computers but also with subjects, remained an ideological horizon based on the hope that the precise mental correlates of ERPs could be deduced

[108] John Russell Knott and William Cheyne McCallum, "An Experimental Approach to Brain Slow Potentials," in *The Responsive Brain*, ed. John Russell Knott and William Cheyne McCallum, *The Proceedings of the Third International Congress of Event-Related Slow Potentials of the Brain. Bristol, England 12–18 August 1973* (Bristol: John Wright and Sons, 1976), 212.

[109] Viewers could see ERP experiments in action and immediately relay suggestions and questions back to the lab. One special "live" experiment constituted the EEG measurement of the closed-circuit coordinator himself, whose actions and ERP data were visible on screen. Ibid., 213.

and rendered meaningful for the subject. In a paper on "Real-Time Detection of Brain Events in EEG" in 1977, Vidal projected that measuring ERPs and task performance through the paradigm of biocybernetics feedback, that is, "to bring about real-time discrimination" using computers, "would provide a quantum increase in the power of the psychophysiological method."[110] Yet, during the 1970s, the correlation between the minute wave patterns of the ERPs and mental behavior was contested. Donchin, for example, admitted that the functional significance of P300 was "controversial," but claimed that it was sufficiently clear that it was "a real component," as there were "reliable, systematic, relations between P300 and psychological processes."[111] The fundamental question underlying ERP experimentation was not: do these minute waves in the EEG correspond to a behavioral or mental phenomenon? But: to which psychological function do ERP components correlate?[112] As such, during the explosion of ERP research, the fundamental uncertainty about its meaning could be momentarily deferred in favor of a focus on improving technology for ever-finer measurements and faster transmissions of information to the computer.[113] This shift of attention and the possibility to *suspend* theoretical understanding of ERPs were facilitated, I argue, by a historical affiliation between EEG feedback research and BCI research, which allowed biocybernetics approaches to be read in proximity to the feedback paradigm of subjective insight.

Not many artists pursued the new paradigm of BCI and ERP research in the 1970s; BCI's research communities were decidedly different from those that had shaped around alpha research, as was BCI's experimental direction (focusing on machine learning and mapping different ERPs). An exception is the artist David Rosenboom, who, inspired by Donchin's work in the mid-1970s, set out to experiment with ERP feedback at the Experimental Aesthetics Laboratory at Toronto's York University, creating an electronic instrument that

---

[110] Vidal, "Real-Time Detection of Brain Events in EEG," 636.
[111] Donchin, "Event-Related Brain Potentials: A Tool in the Study of Human Information Processing," 66.
[112] Here, I paraphrase a remark from a 1981 review of ERP research in which Walter S. Pritchard notes a shift of emphasis from examining the possibility of psychophysiological correlation to a new, more confident position: "P300 exists; what psychological function correlates with *it*?" Walter S. Pritchard, "Psychophysiology of P300," *Psychological Bulletin* 89, no. 3 (1981): 508.
[113] Donchin, "Event-Related Brain Potentials," 66. The controversy of P300 becomes clear in "Appendix: The Relationship between P300 and the Cnv. A Correspondence Conducted by Dr. Emanuel Donchin," in *The Responsive Brain*, ed. John Russell Knott and William Cheyne MacCallum, *The Proceedings of the Third International Congress of Event-Related Slow Potentials of the Brain. Bristol, England 12–18 August 1973* (Bristol: John Wright and Sons, 1976).

would adapt to ERP components measured while subjects were listening.[114] Reporting on the results in the short-lived *Journal of Experimental Aesthetics* in 1977, Rosenboom's coworker Christopher Nunn remarked that "at the present time, the characterization of patterns and brainwaves which encode unique responses to various stimuli is vague and incomplete," which made it impossible to pursue a reliable experimental application.[115] Conversely, writing about these experiments in 1997, Rosenboom shifted attention from ERPs' lack of correlative meaning to the issue of technology: "At that time, results were limited by available instrumentation, most importantly the lack of portable, inexpensive computing power with the speed necessary to provide meaningful results to a subject in a real-time feedback paradigm."[116] Similar to Vidal's hope for real-time feedback from machines to humans, Rosenboom was confident, in the 1970s and again in the 1990s, that new microtemporal measurements would eventually be made meaningful to the subject.[117]

The 1970s saw a bustling practice of measuring and categorizing ERPs and their proposed psychophysiological correlates, a practice that was also criticized toward the end of 1970s for its feeble theoretical and empirical basis.[118] While a large number of researchers confidently strode forward, mapping and measuring ERPs, some researchers shifted their attention. Around 1980, Jacques Vidal, the "inventor" of the BCI, would move away from researching the active human brain toward designing new computer interfaces (a "virtual workspace") and computer architectures that imitated neural networks ("silicon brains").[119] Beyond the brain, the interface now primarily constituted his work.

[114] Christopher Mark Nunn, "Biofeedback with Cerebral Evoked Potentials and Perceptual Fine-Tuning in Humans," *Journal of Experimental Aesthetics* 1, no. 1 (1977).
[115] Ibid., 68.
[116] David Rosenboom, *Extended Musical Interface with the Human Nervous System* (San Francisco, CA: International Society for the Arts, Sciences and Technology, 1997), 42.
[117] Several researchers doubted whether this method of conditioning by means of feedback based on ERPs would work. Nunn solicited reviews on his work from ERP researchers Sutton, Shagass, and Donchin, who questioned his experimental design and the possibility that P300 could ever be used for consistent, meaningful feedback to subjects. Nunn, "Biofeedback with Cerebral Evoked Potentials and Perceptual Fine-Tuning in Humans," 68.
[118] Steven J. Luck, *An Introduction to the Event-Related Potential Technique* (Cambridge, MA: MIT Press, 2014), 5.
[119] Jacques. J. Vidal, L. H. Miller, and G. Devillez, "Adaptive User Interfaces for Distributed Information Management," in *Human-Computer Interaction: Proceedings of the First U.S.A.-Japan Conference on Human-Computer Interaction, Honolulu, Hawaii, August 18–20, 1984* (Amsterdam: Elsevier, 1984). See Jacques J. Vidal, "Silicon Brains: Whither Neuromimetic Computer Architectures," in *Proceedings IEEE International Conference on Computer Design: VLSI in Computers (ICCD '83)* (Port Chester, NY: 1983).

Conclusions

EEG feedback is still controversial today. In a 2015 systematic review, Robert Thibault and his coresearchers conclude that, while there is evidence that subjects partaking in EEG neurofeedback experiments display changes both in behavior and brain activity patterns, "the current literature does not support a direct connection between the specific feedback and the observed alterations."[120] In other words, EEG neurofeedback has various (positive) effects, but these effects are—as far as the past fifty years of research shows—not clearly caused by feedback about specific brain activities.[121] A neurofeedback practitioner would probably be puzzled by the confidence of Thibault's statements. Starting with the new biofeedback institutions established around 1970, researchers and clinicians have established a number of journals specialized in bio- and neurofeedback, publishing articles that generally do claim positive outcomes.[122] Notwithstanding this controversy, most prominent in the past half-century is the development of powerful new computing methods, visualization techniques, display designs, and new repertoires for (playful) feedback situations. Interfaces and interfacing still constitute the main gist of the work.

Today, as in the 1960s, the true fulfillment of successful neurofeedback is projected to the (technological) future: with faster transmission speeds and higher imaging resolutions, ultimately, training the self through the brain is thought to be attainable. Joseph Dumit describes this promissory practice as one in which the future is "hedged," in the sense that scientists are certain that technology has the potential to eventually generate the satisfactory data and answers, thereby "expressing confidence that neuroscience is posing the question properly even if the results are not yet in."[123] This position of

---

[120] Robert T. Thibault et al., "Neurofeedback, Self-Regulation, and Brain Imaging: Clinical Science and Fad in the Service of Mental Disorders," *Psychotherapy and Psychosomatics* 84, no. 4 (2015): 196. See Jimmy Ghaziri and Robert T. Thibault, "Neurofeedback: An Inside Perspective," in *Casting Light on the Dark Side of Brain Imaging*, ed. Amir Raz and Robert T. Thibault (Cambridge, MA: Academic Press, 2019).

[121] Very recently, a 2018 systematic review of research on neurofeedback with fMRI (fMRI-nf) concludes that, while it is still unclear how fMRI feedback approaches might lead to behavioral improvements in the clinical domain, it is certainly possible to change brain activity with fMRI-nf. Robert T. Thibault et al., "Neurofeedback with Fmri: A Critical Systematic Review," *NeuroImage* 172 (2018).

[122] Robert T. Thibault and Amir Raz, "The Psychology of Neurofeedback: Clinical Intervention Even If Applied Placebo," *American Psychologist* 72, no. 7 (2017).

[123] Joseph Dumit, "How (Not) to Do Things with Brain Images," in *Representation in Scientific Practice Revisited*, ed. Catelijne Coopmans et al. (Cambridge, MA: MIT Press, 2014), 305.

technology as a dominant vector for scientific advancement is particularly prominent in EEG research, as Borck has remarked, because we can only know the active, electrical brain by means of "circuitry between brain, mind and machine."[124]

My analysis of *how* technological advancement could be positioned in varying research contexts by tracing the histories of EEG feedback through heterogeneous experimental spaces contributes to an understanding how the active, electrical brain is constituted through a circuitry between brain, mind, machine, and media—brainmedia. The entanglement of scientific and cultural spheres, as well as artistic and scientific EEG experiments, effected a particular research emphasis on the designed setup that provides feedback, a prominence of interface over EEG data—the idea that the interface is the work. My expanded perspective on the art of EEG feedback and BCI demonstrates how new ways of performing knowledges are shaped in hybrid, intersecting spheres of artistic, experimental, and popular sites of knowledge production.

In this chapter, I have sketched the emergence of new experimental research communities in which both artists and other experimenters participated. By outlining these interchanges, my aim has been to steer away from a depiction of artists "appropriating" EEG research. Instead, researchers from different backgrounds—psychologists, physiologists, *and* artists, particularly in hybrid collaborations—were interested in insights into the self and self-optimization, nonverbal communication (both with machines and with the self), and visions of brain-to-brain synchronization within media-technological environments, resulting, as I have argued, in a move toward the centrality of the interface.

My historical description of art and science engagement with EEG feedback also shows how there ultimately was more enthusiasm than anxiety about the possibility of feedback in experimental communities. Although I have cited interlocutors' concerns about the possibility that psychophysiological signals may be employed to commercial or domineering ends, the eagerness to transform and circuit the self seems to have emphasized the progressive potential of EEG experiments *over and above* the possibility of directing human behavior and regulatory applications of EEG measurements. While media artists in the 1960s and 1970s were famously critical of the monodirectional, hegemonic broadcasting structure of mass television, alpha equipment fell into the category of potent tools for new forms of circuiting and interfacing.

---

[124] Borck, *Brainwaves*, 16.

Nevertheless, my account of the proximity of EEG feedback research to an artistic-experimental practice interested in media environments should also acknowledge potential critiques of dominant structures of feedback and circuiting. Nam June Paik, Stan VanderBeek, Hans Haacke, Dan Graham, and Alvin Lucier are but a few well-known examples of artists and art critics who questioned and subverted the way contemporary lived experience was increasingly enveloped and saturated by hegemonic media-technological systems.[125] Art critic Jack Burnham, for example, argued in 1969 that new types of data-gathering "real-time systems" were in need of serious investigation by the arts.[126] He mentioned recent examples of large digital control systems that had emerged in the United States, such as SAGE (for computer control), Telefile (for online banking), and SABRE (for airline reservations); systems that now have the power to gather data from environments, "in time to effect future events within those environments."[127] Burnham observed that the feedback logic of real-time data processing effectively meant that these systems "are *built into* and *become part of* the events they monitor."[128] Precisely because of this technological enmeshing, Burnham proposed what was to be done: "With increasing aggressiveness, one of the artist's functions, I believe, is to specify how technology uses us."[129] This 1969 imperative continues to ring urgently true today.

---

[125] Hannah Higgins and Douglas Kahn, *Mainframe Experimentalism: Early Computing and the Foundations of the Digital Arts* (Berkeley: University of California Press, 2012); Pamela M. Lee, *Chronophobia: On Time in the Art of the 1960s* (Cambridge, MA: MIT Press, 2004); Charlie Gere, *Art, Time and Technology* (Oxford: Berg, 2006); David Joselit, *Feedback: Television against Democracy* (Cambridge, MA: MIT Press, 2007).
[126] Jack Burnham, "Real Time Systems," *Artforum* 8, no. 1 (1969).
[127] Ibid., 51.
[128] Ibid.
[129] Ibid., 55.

6

Synchronizing Two Dynamic Brains: Art–Science Experiments and Neuroscience in the Wild, 2013–19

On a November night in 2013, a friend (subject P.) and I (F.) participated in an experiment. In Amsterdam's EYE Filmmuseum, our heads were fitted with EEG-measuring caps, and we stepped into a darkened screening room. Positioned in the middle of the space was an illuminated dome with two chairs facing each other inside. We were invited to take a seat. P. and I had been told that each side of the dome would light up with a real-time projection of our faces (she would see me, I would see her), but only if the EEG-measuring software had detected "brainwave synchronization." Our task was to play and experiment with our mutual interaction to see if we could turn the visualizations on or off. Stepping into the dome, I could not remember the exact assignment: something about "being on the same wavelength" or "feeling in sync." I felt a bit giddy. Facing P., I tried not to move too much, as we had been told this could interfere with the EEG recording. Sometimes I would see a flash of her projected face on the half dome in front of me, accompanied by loud electronic buzzes. We tried closing our eyes only to realize we would not be able to see the visual feedback. We tried thinking of the same things: blue, the sea. We carefully held hands. The visualization did not seem to correspond to anything we tried. Aware that other friends were in attendance, I felt joyfully competitive even though I was "failing," and I anticipated telling them about the experience. I felt slightly awkward. I wondered if she felt the same. Her face told me she did, delightedly uncomfortable, but the visualization did not appear.

Titled *Mutual Wave Machine*, this "crowd-sourcing neuroscience experiment" was conducted by cognitive neuroscientist Suzanne Dikker and digital media artist Matthias Oostrik as part of the 2013 Museumnacht in Amsterdam (Figure 6.1).[1] Its main question was: "What can the brain tell us about what it means to be 'on the same wave length' with another person?"

---

[1] Suzanne Dikker, *Mutual Wave Machine*, 2013.

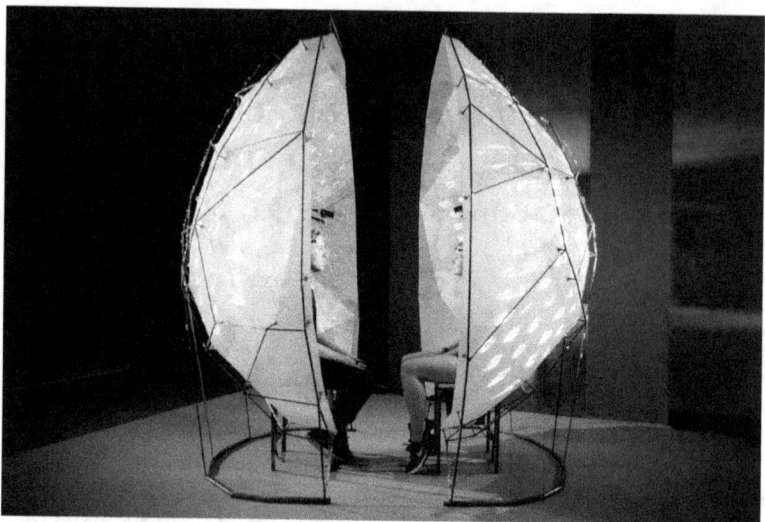

Figure 6.1 Suzanne Dikker and Matthias Oostrik, *Mutual Wave Machine*, 2014 (photography by Oleg Borodin).

hypothesizing that "your brainwaves are more in sync with those of another person when you are feeling connected to that person."[2] By 2019, Dikker and Oostrik have performed the experiment at fourteen locations worldwide (including the Silicon Valley Contemporary Arts Festival, Athens's Benaki museum, and Lowlands Festival in the Netherlands), and over five thousand people have participated. Each pair of subjects, after sitting in the dome together, is asked to fill out a questionnaire recording their age, gender, and handedness and asking about their relationship (how long they know each other, what their relation is). Additionally, each participant is asked to evaluate whether they felt a strong connection with their "brainwave partner" and to describe what strategies of connecting supposedly worked. Cross-correlating the results of hundreds of measured pairs, the results of the study were published on a special website dedicated to the project and a scientific article based on the measured data was published in 2021.[3] Dikker

---

[2] Suzanne Dikker, Karina Kirnos, and Siena Oristaglio, "Neuroscience Exclusives: Digital Mai Exclusives Package" (Marina Abramovic Institute, 2014), n.p.

[3] Dikker, *Mutual Wave Machine*, http://www.suzannedikker.net/mutualwavemachine#results. Dikker, Suzanne, Georgios Michalareas, Matthias Oostrik, Amalia Serafi maki, Hasibe Melda Kahraman, Marijn E. Struiksma, and David Poeppel." Crowdsourcing Neuroscience: Inter-Brain Coupling during Face-to-Face Interactions Outside the Laboratory." *NeuroImage* 227 (2021): 117436.

(based at New York University and Utrecht University) and other researchers have also implemented the technical and experimental knowhow gained during this experiment in other research, such as in a study of the behavior of multiple EEG-wearing high school students in a classroom setting, published in *Current Biology* (2017) and the *Journal of Cognitive Neuroscience* (2018).[4]

Mutual Wave Machine connects very different temporalities: the microtemporality of moment-to-moment changes in the brain and the lifetime duration of personal relationships. In order to examine the neural basis of social interaction, the researchers hypothesize about the EEG activity patterns of two subjects engaged in mutual interaction, as this activity potentially correlates with feelings of "being in sync," as well as with the qualities and quantities of familiarity: knowing each other as a friend, lover, or enemy, for example, or knowing each other for over ten years, for less than an hour, or not knowing each other at all. With its audacious proposal of connecting the neural and the social, *Mutual Wave Machine* is an example of research in the field of social neuroscience, which emerged around 2000. Particularly, the experiment is part of a recent subfield of social neuroscience that uses new computational approaches to measure the brain activity of two or more interacting people at the same time, what is called "two-body neuroscience." As Dikker and colleagues explained in their 2019 paper about *Mutual Wave Machine*, the novelty and promise of this field is that it allows "measuring communication 'live.'"[5]

In this context, as the authors explained, measuring "live" means an approach that allows for a more comprehensive understanding of the neural basis of social interaction by examining realistic human interactions that cannot be captured in conventional laboratory experiments. *Mutual Wave Machine* introduces this new form of liveness as part of an emerging—and, as I will explain, also contested—field that examines the phenomenon of "brain-to-brain synchronization." Focusing on the contemporary field of brain-to-brain synchronization, this chapter shows how new ways of performing knowledges and novel forms of liveness result in new and also tendentious forms of "live brains" today.

Mutual Wave Machine is but one example of the many and varying new manifestations of the active brain outside the scientific laboratory since the shift from the "Decade of the Brain" in the 1990s to what some researchers

---

[4] Suzanne Dikker et al., "Brain-to-Brain Synchrony Tracks Real-World Dynamic Group Interactions in the Classroom," *Current Biology* 27, no. 9 (2017); Dana Bevilacqua et al., "Brain-to-Brain Synchrony and Learning Outcomes Vary by Student–Teacher Dynamics: Evidence from a Real-World Classroom Electroencephalography Study," *Journal of Cognitive Neuroscience* 31, no. 3 (2018).

[5] Pawel J. Matusz et al., "Are We Ready for Real-World Neuroscience?" *Journal of Cognitive Neuroscience* 31, no. 3 (2019).

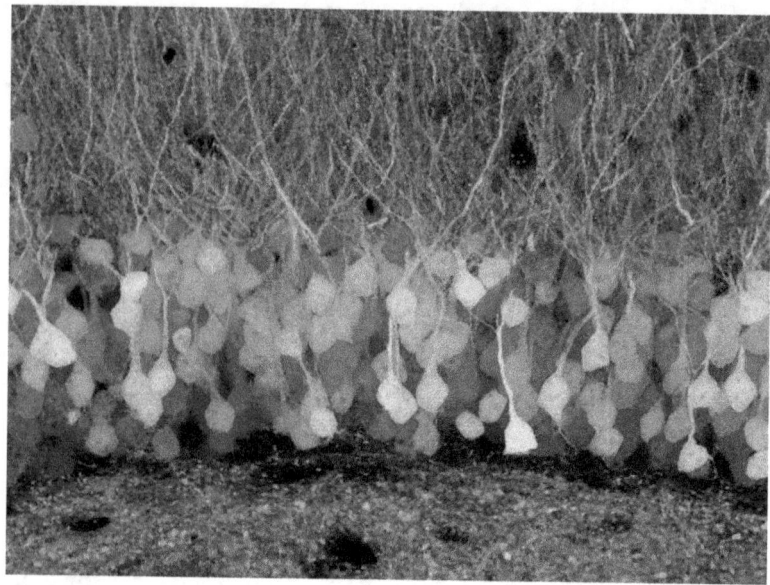

Figure 6.2 Jean Livet et al., *Brainbow* image of dentate gyrus, 2007.

have called the new, twenty-first "Century of the Brain."[6] In the first decade of this new century, perhaps the most circulated brain image was the colorful *Brainbow* (first published in 2007): a microscopic cross-section of a mouse brain showing individual neurons lighting up in ninety different fluorescent shades (Figure 6.2).[7] Scientists and science writers have praised the "technicolor" aesthetics of *Brainbow*, which allowed it to win a number of imaging competitions.[8] The emergence, in the past two decades, of such scientific imaging competitions and image exhibition platforms—in which scientists and artists often share the stage—demonstrates the continuing importance of image-oriented research and visual mediations in twenty-first-century neuroscience.[9]

---

[6] Rafael Yuste and George M. Church, "The New Century of the Brain," *Scientific American* 310, no. 3 (2014).
[7] Flora Lysen, "Technokittens and Brainbow Mice. Negotiating New Neural Imaginaries," in *Speculations on Anonymous Materials; Nature after Nature; Inhuman: [Fridericianum, Kassel, September 29, 2013–June 14, 2015]*, ed. Susanne Pfeffer (London: Koenig Books, 2018).
[8] Alison Abbott, "Colours Light Up Brain Structure," *Nature News*, October 31, 2007.
[9] Andrea Gawrylewski, Jennifer Leman, and Liz Tormes, "The Brain in Images: Top Entries in the Art of Neuroscience," *Scientific American* (2019).

Yet, from the 1990s until today, the public circulation and performance of the active brain also continues to change shape in relation to new media platforms and new forms of mediation that impact the dynamics of conducting and performing brain research. A striking example, in this respect, is the 2011 publication of research into human perception by Jack Gallant and coresearchers at the University of Berkeley, California, who asked subjects to watch Hollywood movie trailers in an fMRI scanner to subsequently predict what subjects had viewed based on their measured brain activity, creating blurry, dream-like image predictions based on a database of randomly selected YouTube clips.[10] Whether this research heralds the possibility of a type of "mind reading" or "brain decoding"— opening up what Gallant calls a "window into the movies in our minds"— researchers are currently debating.[11] Notwithstanding, the operationalization and circulation of Gallant's research certainly reveals new and mindboggling metaphorical circuits of contemporary brainmedia by which a YouTube-powered experiment of an envisioned YouTube-like mind actually circulates on YouTube.[12]

In this chapter, *Mutual Wave Machine* functions as an evocative crossroads for my examination of new assemblages of brainmedia in contemporary neuroscience. As a public art–science installation, the work aptly exemplifies new ways of performing knowledges in neuroscience as it developed since the 2000s, representing the new relations between artists and scientists, new ways of conducting experiments ("crowdsourcing neuroscience," for example), new platforms and forms to present science, and new spheres for science reporting.

First I explain why and how an analysis of contemporary neuroscience through the lens of art–science practice is fruitful and why such an examination should steer away from a simplistic critique of neuro-hype. Second, I situate *Mutual Wave Machine* in the emerging field of brain-to-brain synchronization research and as part of a broader discourse on what researchers call "real-world" neuroscience, "real-life"

---

[10] Shinji Nishimoto et al., "Reconstructing Visual Experiences from Brain Activity Evoked by Natural Movies," *Current Biology* 21, no. 19 (2011).

[11] Yasmin Anwar, "Scientists Use Brain Imaging to Reveal the Movies in Our Mind," *Berkeley News*, September 22, 2011. "Brain Reading," in *Neuroforensics: Exploring the Legal Implications of Emerging Neurotechnologies: Proceedings of a Workshop*, ed. Lisa Bain et al. (Washington, DC: National Academies Press, 2018). For a skeptical view of mind reading research, including Gallant's image prediction experiment, see Nikolas Rose, "Reading the Human Brain How the Mind Became Legible," *Body & Society* 22, no. 2 (2016).

[12] UC Berkeley, "Vision Reconstruction." Laura Matthews, "Scientists Turn Brain Images into Youtube Videos," *International Business Times*, November 24, 2011.

neuroscience, or "neuroscience in the wild": a recent movement toward a more ecologically valid and "naturalistic" experimental practice. As a crowdsourcing neuroscience experiment that allows for measuring many brains, *Mutual Wave Machine* is an example of how this vision of real-world social neuroscience pairs with the promise of exploratory big data science. I argue that public practices of crowdsourcing neuroscience allow scientists to link crowdsourced (big data) neuroscience and "citizen science." This prompts questions about what type of citizens—citizens with brain data—are interpellated by such artistic, "live," crowdsourced performances. In the third part, I analyze the relation between *Mutual Wave Machine* as an art–science experiment and as a scientific experiment. I observe a disjunction between the multilayered media experiences this "real-world" neuroscience experiment affords and the way mediation is ultimately treated in the scientific analysis of brain-to-brain experiments. In the fourth and final part, I turn to the notion of "synchronization" that is prevalent and often even central in two-body neuroscience experiments. By offering a media-historical analysis of the concept of synchronization in social interaction research, I explain how issues of harmonization, affect, and operability are sedimented in the conceptualization and practice of current social neuroscience research. My plea here is for a new sensitivity to media histories when conceptualizing live brains in reflections on current neuroscience.

Investigating New Forms of Neuroscientific Life

Mutual Wave Machine is but one recent example of a large number of EEG-based artworks that have started to populate galleries, tech conferences, and a variety of art-meets-science events in the past decade.[13] Because consumer-grade EEG headsets (with lyrical names such as Neurosky, Emotiv, and Muse) have become more affordable and widely distributed, mindfulness practitioners, gamers, and artists have started using them to experiment with sonifications and visualizations of the EEG data.[14] Consider, for example,

---

[13] For an overview, see Mirjana Prpa and Philippe Pasquier, "Brain-Computer Interfaces in Contemporary Art: A State of the Art and Taxonomy," in *Brain Art: Brain-Computer Interfaces for Artistic Expression*, ed. Anton Nijholt (Cham: Springer, 2019).

[14] For examples and further analysis, see Flora Lysen, "Kissing and Staring in Times of Neuromania: The Social Brain in Art-Science Experiments," in *Artful Ways of Knowing, Dialogues between Artistic Research and Science & Technology Studies*, ed. Trevor Pinch, Henk Borgdorff, and Peter Peters, Routledge Advances in Art and Visual Studies (London: Routledge, 2019).

Stelarc's performance *Spectacle of Mind* (2010), in which the performance artist wore an EEG headset to create a "brainwave-generated" art piece, a surrealist-looking animation of his own face projected on a huge screen.[15] Or Lisa Park's much-publicized installation *Eunoia* (2013), in which EEG measurements cause buzzing noises and ripples on the surfaces of a number of water basins, depending on the level of the performer's concentration.[16] Or the installation *E.E.G. KISS* (2014–ongoing) by artists Hermen Maat and Karen Lancel, which stages two kissing subjects while their measured EEG activity is made audible and perceptible.[17] Quite a number of these works conjure meditative atmospheres and intricate chains of technical mediation that remind us of the media environments and visions of circuited selves discussed in Chapter 5. Yet, the current field of EEG experimentation is more diverse. While some artists use existing EEG devices to create evocative displays of scientific data, others employ EEG headbands to ultimately question and expose the black-boxed assumptions in (patented) consumer EEG software that sort brain activity according to (contested) labels such as "bored," "excited," "meditation," or "frustration."[18]

In relation to this broader field of EEG art, *Mutual Wave Machine* stands out in the way its dimensions of spectacle, reflection, and scientific experimentation continuously intersect. Its creators emphasize the work's heterogeneous status: part scientific experimentation meant to result in publishable research; part vehicle for public engagement with science that shows the messy work of scientific experimentation; part attractive, immersive experience that allows for exploration and play.[19] The experimenters aim to *do* and *show* research into brain-to-brain synchronization within an evocative, aesthetic setting. Though I question notions of engagement and experience as they are attached to performances

[15] Stelarc, *Spectacle of the Mind*, 2010.
[16] Lisa Park, *Eunoia*, 2013.
[17] Karen Lancel and Hermen Maat, *E.E.G. KISS*, 2014–ongoing, https://www.lancelmaat.nl/work/e.e.g-kiss/. Flora Lysen, "Neurofutures of Love," *Baltan Quaterly* (2016).
[18] See "Kissing and Staring in Times of Neuromania: The Social Brain in Art-Science Experiments." About contested data measuring of consumer-grade headsets, see N. N. Y. Chu, "Brain-Computer Interface Technology and Development: The Emergence of Imprecise Brainwave Headsets in the Commercial World," *IEEE Consumer Electronics Magazine* 4, no. 3 (2015).
[19] Suzanne Dikker, Suzan Tunca, and Sean Montgomery, "Using Synchrony-Based Neurofeedback in Search of Human Connectedness," in *Brain Art: Brain-Computer Interfaces for Artistic Expression*, ed. Anton Nijholt (Cham: Springer, 2019). On showing science in action, see *Crowdsourcing Neuroscience* (Imagine Science Films, 2014), https://www.youtube.com/watch?v=pyIs6Syvf_k, interview with exhibition footage and brain activity animation.

of crowdsourced neuroscience in this chapter, *Mutual Wave Machine* (2013–19) introduces a productive site for reflection as an art–science experiment, because it is both part of (public) neuroscientific research and a work of art that offers potential reconfigurations of, or reflections on, this very research.

I emphasize this duality because current critical accounts of neuroculture have dismissed the public presentation of the iconic, beautifully exhibited active brain all too easily. Since the 1990s "decade of the brain," a number of authors have voiced concerns about what Nikolas Rose and Joelle Abi-Rached call "pedagogies of brain awareness"[20]—that is, the multifarious ways in which media events, exhibitions, workshops, and public installations may prime laypeople to become, in Martyn Pickersgill's term, "ready to talk" about the brain and make them inclined to understand everyday behavior from a neurological perspective.[21] In a similar vein, Davi Thornton has argued that exhibitions of the brain help saturate public discourse with biological and neurological ways of thinking: "visitors are interpellated as neurobiological citizens, understanding their thoughts and behaviors as neurobiologically correlated."[22] The curators of the 2012 exhibition *Brain: The Mind as Matter* (Wellcome Collection, London) also point to these potential governing powers: brain displays will instantly become part of a "political economy of brains," they argue, whether "the curator goes with the ideological grain or against it."[23] These dynamics of neuro-governmentality and neurocentrism that come with exhibited brains are facilitated, some contend, by a particular beguilement on the part of the public that Sabrina Ali and colleagues have called "neuroenchantment," that is, laypeople's "sub-judicious fascination with brain science."[24] Nonexperts may be too easily persuaded to overestimate the present state of neuroscience knowledge and take up a neurocentrist perspective on behavior and the self. The authors propose the public should be protected against such neuroenchantment through education in critical thinking.

---

[20] Nikolas Rose and Joelle M. Abi-Rached, *Neuro: The New Brain Sciences and the Management of the Mind* (Princeton, NJ: Princeton University Press, 2013), 22.

[21] Ibid.; Martyn Pickersgill, Sarah Cunningham-Burley, and Paul Martin, "Constituting Neurologic Subjects: Neuroscience, Subjectivity and the Mundane Significance of the Brain," *Subjectivity* 4, no. 3 (2011): 357. See also Ortega and Vidal, *Neurocultures*.

[22] Thornton, *Brain Culture*, 70.

[23] Marius Kwint and Richard Wingate, "Curating the Brain," *Interdisciplinary Science Reviews* 38, no. 3 (2013): 196.

[24] Sabrina S. Ali, Michael Lifshitz, and Amir Raz, "Empirical Neuroenchantment: From Reading Minds to Thinking Critically," *Frontiers in Human Neuroscience* 8 (2014).

While I think a consideration of governing brains and fascinated publics is crucial, my analysis in this chapter goes against an all-too-sweeping critique of the enchanting powers of exhibited neuroscience or the seduction of the layperson. Instead, my aim in analyzing *Mutual Wave Machine* and other art–science works is to contribute to a deeper understanding of the modes of reasoning in current synchronization research (as part of social neuroscience) that are intricately tied to the (art and media) forms through which research is performed.

Art–science work is particularly suitable for this analysis, as it may be emblematic of contemporary forms of "scientific life" (to use Steven Shapin's term), that is, new structures of researching, experimenting, funding, and demonstrating in scientific practice today.[25] Shapin points to a new form of industrial-scientific entrepreneurship (exemplified in his research by San Diego's venture capital–funded biotech sector) in which whether a project will yield results is increasingly uncertain. He explains that a central element of this new scientific life, characteristic of late modernity, is the preoccupation with creating "pictures of possible worlds-to-come" and "technoscientific and economic future making."[26] In order to build (new forms of) trust and authority, entrepreneurial science frames research sites as creative playgrounds and places new emphasis on the charisma of individual researchers. It is through the creative personalities of researchers—the scientist-entrepreneur and scientists as performance artists—that these fields can best embody and build visions of the future, accentuating play and fun in research praxis.

Art–science installations and performances have emerged as a key form of future-making and are especially prevalent in cognitive neuroscience as it interfaces with developers of EEG devices and BCI software producers. On occasion, academically trained entrepreneurs and entrepreneurial neuroscientists also work as media artists, developing their visualizations and sonifications of brain activity in tandem with artistic installations and demonstrating their work at hybrid academic-, public-, and industry-oriented art–science events.[27] Hence artists are an integral part of these future-making practices. The question, then, is which futures such art–science experiments project and how specific forms of future-making allow for some futures over others.

---

[25] Steven Shapin, *The Scientific Life: A Moral History of a Late Modern Vocation* (Chicago: University of Chicago Press, 2008).
[26] Ibid., 309, xv.
[27] For an example of this form of "neuroscientific life" in the field of neuro-gaming, see Lysen, "The Interface Is the (Art)Work."

A Real-World Neuroscience with Hyper-Stakes

Research into brainwave synchronization such as pursued with the *Mutual Wave Machine* is still exploratory. To date, little is known about the significance of what is variably called "neural coupling," "interbrain synchrony," or "brain-to-brain synchronization."[28] Between 2010 and 2019, over 140 studies of brain-to-brain activity were published, with evocative titles like "'Stay Tuned': Inter-Individual Neural Synchronization During Mutual Gaze and Joint Attention," "On the Same Wavelength: Face-to-Face Communication Increases Interpersonal Neural Synchronization," "Teams on the Same Wavelength Perform Better: Inter-Brain Phase Synchronization Constitutes a Neural Substrate for Social Facilitation," "Getting into Sync: Data-Driven Analyses Reveal Patterns of Neural Coupling That Distinguish Among Different Social Exchanges."[29] While scholars have long investigated synchronization processes in social behavior ("being in sync" in the sense of walking at the same pace or speaking at the same speed), research correlating these observations and the phenomenon of brain-to-brain synchronization is now picking up.

"Brain-to-brain" measurements were sparked by new technological and computational inventions: the ability to scan two brains simultaneously, a process called "hyperscanning."[30] Starting around 2010, hyperscanning (both with EEG and fMRI) has been hailed as a paradigm shift in studying social cognition, opening up a whole new field that is (with different inflections) called "two-body neuroscience," "second-person neuroscience," or "2P neuroscience."[31] EEG hyperscanning, which uses portable headsets that allow

---

[28] Suzanne Dikker et al., "On the Same Wavelength: Predictable Language Enhances Speaker–Listener Brain-to-Brain Synchrony in Posterior Superior Temporal Gyrus," *Journal of Neuroscience* 34, no. 18 (2014); Guillaume Dumas et al., "Inter-Brain Synchronization During Social Interaction," *Plos One* 5, no. 8 (2010); Johanna Sänger, Ulman Lindenberger, and Viktor Müller, "Interactive Brains, Social Minds," *Communicative & Integrative Biology* 4, no. 6 (2011). As of yet, there is no shared definition for measuring brain-to-brain activity. Synchronization or "hyperconnectivity" means that there are "statistically significant correlations or covariances between different brain signals." Fabio Babiloni and Laura Astolfi, "Social Neuroscience and Hyperscanning Techniques: Past, Present and Future," *Neuroscience & Biobehavioral Reviews* 44 (2014): 80. About the lack of a shared definition, see Difei Liu et al., "Interactive Brain Activity: Review and Progress on EEG-Based Hyperscanning in Social Interactions," *Frontiers in Psychology* 9, 1862 (2018): 4.

[29] According to my own 2019 PubMed search for articles including the term "hyperscanning." See two review papers in 2014 and 2018: Babiloni and Astolfi, "Social Neuroscience and Hyperscanning Techniques," 80; Liu et al., "Interactive Brain Activity."

[30] Babiloni and Astolfi, "Social Neuroscience and Hyperscanning Techniques."

[31] Leonhard Schilbach et al., "Toward a Second-Person Neuroscience," *Behavioral and Brain Sciences* 36, no. 4 (2013): 404; Riitta Hari and Miiamaaria V. Kujala, "Brain Basis of Human Social Interaction: From Concepts to Brain Imaging," *Physiological Reviews* 89,

relatively regular body movements, is especially viewed as making it possible to study interpersonal cognition in its natural state for the first time.[32] As two-body neuroscience researcher Guillaume Dumas speculates, these new investigations of interbrain relationships may even be a breakthrough in tackling the infamous "hard problem" of consciousness by developing a notion of mental states that is not bound to one brain but can be measured as "collective consciousness."[33] Clearly, the stakes in hyperscanning research are exceptionally high.

By promising new insights into the neural basis of interpersonal interaction dynamics in more natural settings, researchers in brain-to-brain synchronization respond to concerns about the ecological validity of social interaction research in social neuroscience. Since its rise around 2000, social neuroscience has been critiqued for drawing conclusions about the "social brain" on the basis of experiments that simulate personal interaction in too-artificial laboratory settings, such as the narrow space of the fMRI scanner.[34] The recent enthusiasm about the more naturalistic possibilities of hyperscanning (especially with EEG) is part of a broader discourse in cognitive neuroscience about "real-world" or "real-life" practices: the importance of studying brain/mind functioning in more ecologically valid settings—going out of the lab, into the real world.[35]

Advocating more attention to the "real world," a 2019 paper with the title "Are We Ready for Real-World Neuroscience?" authored by Pawel Matusz, Alexander Huth, Catherine Perrodin, and *Mutual Wave Machine* creator Suzanne Dikker launched the term "real-world neuroscience" to denote a shared impetus behind a number of new approaches in developing a more ecologically valid neuroscience.[36] In the field of perception and social

---

no. 2 (2009); Guillaume Dumas, "Towards a Two-Body Neuroscience," *Communicative & Integrative Biology* 4, no. 3 (2011).

[32] Francisco J. Parada and Alejandra Rossi, "Commentary: Brain-to-Brain Synchrony Tracks Real-World Dynamic Group Interactions in the Classroom and Cognitive Neuroscience: Synchronizing Brains in the Classroom," *Frontiers in Human Neuroscience* 11 (2017): 1; Liu et al., "Interactive Brain Activity."

[33] Dumas, "Towards a Two-Body Neuroscience," 351.

[34] Ruth Leys, "How Did Fear Become a Scientific Object and What Kind of Object Is It?" *Representations* 110, no. 1 (2010); Rose and Abi-Rached, *Neuro*; Simon Cohn, "Making Objective Facts from Intimate Relations: The Case of Neuroscience and Its Entanglements with Volunteers," *History of the Human Sciences* 21, no. 4 (2008).

[35] See Simone G. Shamay-Tsoory and Avi Mendelsohn, "Real-Life Neuroscience: An Ecological Approach to Brain and Behavior Research," *Perspectives on Psychological Science* 14, no. 5 (2019).

[36] The paper resulted from a 2017 "real-world neuroscience" panel at the Cognitive Neuroscience Society conference. Matusz et al., "Are We Ready for Real-World Neuroscience?"

behavior research, Alexander Huth explains, this means a shift from the "twentieth-century mindset" of controlled laboratory experiments toward a twenty-first-century practice of "less-controlled experiments using ethologically relevant, natural behavior."[37]

Real-world neuroscience, as proposed in 2019, studies living subjects with brains "in the wild" by proverbially moving "out of the lab." Exiting the laboratory here means devising new types of less controlled experiments or constructing new (public, artistic, experimental) labs (a practice that is always complementary to more established laboratory experimentation, the authors underline). This move has recently become possible, as Huth and Dikker explain, because of new computational modeling (premised on computational power, data gathering, and data storage) that can start to grapple with the complexity of the many factors affecting a less controlled experiment. Researchers interested in real-world neuroscience base their research on data-driven analysis (often described as "exploratory data analysis").[38] Researching "real-world" actions with data-driven approaches is helpful, as Huth explains, because neuroscience is at a loss in complex situations (such as social interaction or navigating an environment): "Except in very few cases, we don't know which hypotheses to test. The space of hypotheses is too big, and we know too little about the system. This is why we need real-world/natural experiments: to show us the general shape of the hypotheses space efficiently, instead of shooting in the dark with controlled experiments."[39] This "real-world" approach to behavioral and social neuroscience requires assembling large datasets (such as measuring many museum visitors with the *Mutual Wave Machine* experiment) that may, as Suzanne Dikker puts it, "inspire and inform subsequent laboratory experimentation" by potentially revealing novel correlations that spur new hypotheses.[40]

This "real-world neuroscience" is characteristic of a broader data-driven approach to social neuroscience following a data-centric logic by which researchers produce, as Sabina Leonelli puts it, "vast quantities of data in

---

[37] Alexander Huth in Matusz et al., "Are We Ready for Real-World Neuroscience?" 330.

[38] Ralph Adolphs, "Investigating the Cognitive Neuroscience of Social Behavior," *Neuropsychologia* 41, no. 2 (2003): 122; Danilo Bzdok and B. T. Thomas Yeo, "Inference in the Age of Big Data: Future Perspectives on Neuroscience," *NeuroImage* 155 (2017); Yves Frégnac, "Big Data and the Industrialization of Neuroscience: A Safe Roadmap for Understanding the Brain?" *Science* 358, no. 6362 (2017); Ralph Adolphs et al., "Data-Driven Approaches in the Investigation of Social Perception," *Philosophical Transactions of the Royal Society B: Biological Sciences* 371, no. 1693 (2016).

[39] Alexander Huth in Matusz et al., "Are We Ready for Real-World Neuroscience?" 335. As Matusz puts it, an "exploration-confirmation scientific investigation cycle." Ibid.

[40] Suzanne Dikker in ibid., 331.

the hope that they might yield unexpected insights."[41] This approach has critics. In 2017, for example, a group of reputable neuroscientists led by John Krakauer criticized behavioral neuroscience in a much-cited article titled "Neuroscience Needs Behavior: Correcting a Reductionist Bias" over what they described as a flawed epistemic prioritization of data gathering over hypothesis building.[42] Current studies were not invested enough, they argued, in developing conceptual frameworks that hypothesize the relation between neural data and behavior through "careful dissection of behavior into its component parts or subroutines," a practice that should be "epistemologically prior" to neurophysiological measuring.[43] Instead, the authors noted, behavioral neuroscientists (including social neuroscientists) were too preoccupied with acquiring big datasets through technologically demanding procedures and with managing the deluge of data they themselves produce.[44] Yet, without "well-characterized behavior and theories," the authors warn, "brains and behavior will be like two ships passing in the night."[45] *Mutual Wave Machine* is located in this contested domain of data-driven real-world neuroscience and navigates such vexed issues of reductionism in behavioral neuroscience in implicit and evocative ways.

If real-world neuroscience proposes a move out of the lab into the real world, *Mutual Wave Machine* connects this to another move "out of the lab, into public space."[46] In their proposal for crowdsourcing neuroscience that enables acquiring the bigger and more diverse datasets necessary to perform real-world neuroscience, the researchers connect this approach to notions of public engagement with science, even speaking of "citizen science."[47] In doing so, *Mutual Wave Machine* joins a number of recent neuroscience projects that have invoked the power of the crowd in conjunction with notions of "citizen science." One example that has received a great deal of attention in recent

---

[41] Sabina Leonelli, *Data-Centric Biology: A Philosophical Study* (Chicago: University of Chicago Press, 2016), 3. Whether data-driven social neuroscience has moved from an "exploratory" mode of experimentation to a "gathering mode" is a pertinent question for further analysis. See Ulrich Krohs, "Convenience Experimentation," *Studies in History and Philosophy of Science Part C: Studies in History and Philosophy of Biological and Biomedical Sciences* 43, no. 1 (2012).

[42] John W. Krakauer et al., "Neuroscience Needs Behavior: Correcting a Reductionist Bias," *Neuron* 93, no. 3 (2017).

[43] The danger is a "granularity mismatch between levels that prevents substantive alignment between different levels of description." Ibid., 481, 88.

[44] Ibid., 481. See John Krakauer, interview by Ana Gerschenfeld, November 23, 2017.

[45] Hypothesizing about behavior should come epistemologically *prior* to neuroscience research. Krakauer et al., "Neuroscience Needs Behavior," 484.

[46] Dikker, *Crowdsourcing Neuroscience,* @ 0:43.

[47] Suzanne Dikker et al., "Crowdsourcing Neuroscience: Inter-Brain Coupling during Face-to-Face Interactions Outside the Laboratory," *NeuroImage* 227 (2021), 1.

years is the rise of "serious" online (neuroscience) games such as *EyeWire* and *Mozak*, which have gamified image recognition tasks (determining the outlines of a neuron in a microscopic image, for example).

In professional neuroscience publications, such crowdsourcing games receive enthusiastic welcomes. In a 2016 paper in *Neuron*, Jane Roskams and Zoran Popović describe current neurobiology as a big data challenge that is best tackled by new technological infrastructures and aided by a "new generation of expert citizen neuroscientists."[48] When researchers speak about serious games as a way to give "power to the people" and to create a "World Cup of Neuroscience" that allows everyone to "contribute to brain research," they echo the language by which game makers hope to mobilize volunteers to participate in online citizen science projects. Analyzing such mobilization narratives and structures, Dick Kasperowski and Thomas Hillman point to the way crowdsourcing games rhetorically craft a hybrid profile for the citizen scientist: a gamer who is both a distributed epistemic subject that contributes to a small piece of the bigger puzzle, and an individual discoverer who helps scientists with analytical tasks.[49]

In the wake of enthusiastic reporting about citizen neuroscientists analyzing data, *Mutual Wave Machine* belongs to a different category of crowdsourced science based on public experiments or gaming apps (such as *The Great Brain Experiment*) employed to gather large sets of participants' *cognitive data*.[50] By invoking the notion of "citizen science" in the context of cognitive data gathering, the idea of contributing or participating now suggests a different, hybrid epistemic subject.[51] Here the citizen participant is envisioned to both voluntarily provide cognitive (and other) data and engage with science on display. The subject of this cerebral citizen science is viewed as doubly involved in contributing to the "public good," both by offering information and by becoming an informed citizen. Referring to the latter element, Dikker has emphasized the potential of *Mutual Wave*

---

[48] Jane Roskams and Zoran Popović, "Power to the People: Addressing Big Data Challenges in Neuroscience by Creating a New Cadre of Citizen Neuroscientists," *Neuron* 92, no. 3 (2016): 660, 63.

[49] Dick Kasperowski and Thomas Hillman, "The Epistemic Culture in an Online Citizen Science Project: Programs, Antiprograms and Epistemic Subjects," *Social Studies of Science* 48, no. 4 (2018): 584.

[50] As an example of crowdsourced neuroscience at a Toronto art festival, see Natasha Kovacevic et al., " 'My Virtual Dream': Collective Neurofeedback in an Immersive Art Environment," *Plos One* 10, no. 7 (2015). On the *Great Brain Experiment*, see Harriet R. Brown et al., "Crowdsourcing for Cognitive Science: The Utility of Smartphones," *PLOS One* 9, no. 7 (July 15, 2014): e100662.

[51] On heterogeneous notions of "citizen science" as well as a critical view of the historiography of the term, see Bruno J. Strasser et al., " 'Citizen Science'? Rethinking Science and Public Participation," *Science & Technology Studies* 32, no. 2 (2019).

Machine to critically educate and engage participants through "science in the making." As she explained during a 2019 lecture on "Neuroscience in the Wild," it is important that visitors and participants witness the technical setup of measuring brain activity. By seeing the raw EEG signal during the headset adjustments, "it becomes very intuitive that we're not dealing with a mind-reading device," that is, that "you need do a lot of interpretation before those squiggly lines become meaningful." This is important because "there's a lot of misunderstandings ... about what we can tell from our neuroscience findings."[52] Hence, similar to readers of magazine and newspaper reports on new EEG technologies in the 1930s (as described in Chapter 3), the citizen participant of crowdsourced neuroscience today is interpellated as a critical assessor of the status of the science on display. Yet, just *how* participants interpret what they see is a question that remains actively suspended. What becomes evident from my analysis is that while part of their critical abilities are assumed, they are not quite enabled.

What is left out in *Mutual Wave Machine* and its paratexts is the connection between the two poles of this cerebral citizen science: subject volunteering and critical engagement. While the volunteer subject offers their data as part of a crowdsourced, data-driven approach to real-world social neuroscience, the participant of *Mutual Wave Machine* gains little insight into what it means to be one of thousands of subjects in this study, nor to what ends these data will be used. The project does little to prompt reflection on its fundamental reliance on big data neuroscience, its data gathering, data journeys, data infrastructures, and data interpretation, and the ultimate directionality of big data social neuroscience in finding biomarkers for normal and abnormal social behavior.[53] In fact, this gray zone at the heart of citizen neuroscience speaks to a broader difficulty of how to address "data subjectivity" in contemporary research where the weight has shifted to cross-correlations and analysis after data gathering.[54] While critical data scholars

---

[52] Suzanne Dikker, *CMBC 2019 Workshop: Neuroscience in the Wild*, https://www.youtube.com/watch?v=ISeOWAAglQ02019.

[53] Investigating research into data-centric biology, Sabina Leonelli observes an epistemic frame that separates "content" and "context," and that pays some attention to the space of the laboratory, but almost none to what she calls "data-journeys." Leonelli, *Data-Centric Biology*, 179–80.

[54] See a discussion on data subjectivity in Jacob Metcalf and Kate Crawford, "Where Are Human Subjects in Big Data Research? The Emerging Ethics Divide," *Big Data & Society* 3, no. 1 (2016). And the "data divide" between individuals and data collectors in Mark Andrejevic, "Big Data, Big Questions | the Big Data Divide," *International Journal of Communication* 8, no. 1 (2014): 1674. And specifically in the context of neurodata, see Dara Hallinan et al., "Neurodata and Neuroprivacy: Data Protection Outdated?" *Surveillance and Society* 12, no. 1 (2014).

have demonstrated alternative engagements with data, it is unclear what "citizen agency" or "public engagement" could be available to subjects of crowdsourced neuroscience.[55] Addressing this difficulty, Mike Michael and Deborah Lupton state:

> Publics are, in varying degrees, both the subjects and objects of knowledge, both authors and texts, simultaneously informants, information and informed ... What exactly counts as knowledge or understanding where the data themselves are continually in flux for both lay people and experts?[56]

These issues are not directly addressed in *Mutual Wave Machine*. By performing crowdsourcing as art–science experiment, the authors propose their own kind of engagement, away from discussions of data gathering, toward a positive evocation of public participation and engagement with art–science. Scholars of science and technology studies (STS), as well as of participatory art, have extensively discussed notions of the "participatory" and the "engaged" in relation to governance, legitimization, popularization, and hierarchical power dynamics in imagining publics for science and the arts.[57] In light of these ongoing debates, *Mutual Wave Machine* proves a particularly interesting case because it reenvisions the museum space as a vanguard "real-world" neuroscience laboratory. In this sense, it is not only the public but also the institution that is called to perform a civic (scientific) duty. As Dikker et al. state: "Institutions such as museums, galleries, or any other organization where the public actively engages out of self-motivation, can help facilitate this type of 'citizen science' research, and support the collection of large datasets under scientifically controlled experimental conditions."[58]

Within the walls of this new museum-cum-laboratory, "engagement" gains new meaning: the subject is not just an engaged participant of an immersive artwork, or an engaged reviewer of science, but also an engaged experimental subject. In the context of cognitive science, "task engagement"

---

[55] Helen Kennedy, Thomas Poell, and José van Dijck, "Data and Agency," *Big Data & Society* 2, no. 2 (2015): 6.

[56] Mike Michael and Deborah Lupton, "Toward a Manifesto for the 'Public Understanding of Big Data,'" *Public Understanding of Science* 25, no. 1 (2016): 109.

[57] Georgina Born and Andrew Barry, "Art-Science. From Public Understanding to Public Experiment," *Journal of Cultural Economy* 3, no. 1 (2010); Ian Welsh and Brian Wynne, "Science, Scientism and Imaginaries of Publics in the UK: Passive Objects, Incipient Threats," *Science as Culture* 22, no. 4 (2013); Claire Bishop, "Delegated Performance: Outsourcing Authenticity," *October* 140 (2012).

[58] Dikker et al., "Crowdsourcing Neuroscience," 1.

refers to the levels of motivation and attendance to experimental assignments (a vital issue, since many cognitive science experiments can be tedious and boring, and inattentiveness makes data unreliable). Festivals and museum environments not only offer atmospheres for potentially higher levels of task engagement but also allow scientists access to large pools of self-motivated and engaged subjects[59]—though some might not have the right type of attitude as Dikker notes. At several recording sites, she recounts, "visitors did not treat the experience as a scientific experiment but rather as a curiosity (e.g., taking selfies instead of interacting with each other)."[60] Instead of being task-engaged, some visitors exhibited social interaction behavior typical of museums—selfie-taking—that cannot be contained in a "real-world" laboratory after all.

When Works of Art Become Scientific Papers: Neurocentrism Revisited

Mutual Wave Machine is a flickering, buzzing, and illuminated contraption that envelopes two people in order to performatively meld their mind/brains together for ten minutes at a time. As an immersive and interactive art–science experiment, it persuasively conjures the imaginary media and media imaginaries that have been attached to brainwaves for more than a century. In a world of vibratory energies and ubiquitous media technologies, the human nervous system was invoked as the preeminent site of influence, excitation, and breakdown long before the rise of EEG, as I discussed in previous chapters. Sometimes conceived as a mediating apparatus itself, it was always intimately part of the energetic-media-technological fabric of modern life. This conception is strikingly mirrored in the *Mutual Wave Machine* dome: the installation places its "brainwave partners" amidst multisensory turbulence. Facing each other, you also see brightly flashing images that provide feedback on your performance, hear loud buzzing sounds, are aware of others waiting in line, feel the measuring equipment squeezing your head, sense the technical assistants ready to jump in, know the room is full of spectators, glimpse the luminescence of phone screens, feel your itchy scalp wearing a brain device. Whereas much EEG art results in meditative spaces where colorful blobs augment a desired state of concentration, *Mutual Wave Machine* offers a site of overstimulation, displaying a decidedly modern

[59] On motivation in a crowdsourced neuroscience experiment by Kovacevic et al., "My Virtual Dream."
[60] Dikker et al., "Crowdsourcing Neuroscience," 1.

"real-world" neuroscience. However, although it evokes real-life media intensity, I argue that there is a strange disjunction between the real-world experiences it invokes and the data it ultimately models to compute brain-to-brain synchronization.

What we see during art–science experiments is not necessarily what we get in scientific operationalization. Even if sound and light are integral to the experience of social interaction in the dome, subsequent data processing removes the synchronization effects of audio-visual stimuli from the potentially "social" brain-to-brain synchronization measurements.[61] The rationale behind this filtering is that while two people seeing and hearing the same thing may elicit similar brain activities, such synchronization is not necessarily "social," so stimulus-induced synchrony is viewed as a separate "low-level" measurement of synchrony that may influence actually "social" brain-to-brain synchronization, the latter hypothesized as corresponding to "high-level" measurements (necessary for cognitive inference, for example).[62] Brain-to-brain synchronization, as we now see, is operationalized in highly circumscribed ways, building on prior theories about specific EEG frequencies and particular brain areas connected to social cognition (in previous social-neuroscientific papers).[63] Synchronization's "social" element in *Mutual Wave Machine*'s complex, stimuli-heavy, "real-world" environment is difficult to pin down—and is based on previous research on localizing social cognition in the brains. Ultimately, the media-environmental and mind-melding *form* of this real-time artistic medium has a tendentious relation to the *content* of the scientific paper in which it results.

A similar tension—between the art and the science of synchronization research—emerges when we turn to the notion of feedback enacted by the experiment. While the striking visual feedback imagery on the dome is a structuring element of the installation—telling us how in sync we are with our partner—the 2019 paper based on the experiment reveals that what visitors see is in fact a meaningless measure, a minimal kind of feedback that is mainly geared toward keeping participants motivated.[64] And how could it be otherwise? If the *Mutual Wave Machine*'s aim is to explore what

---

[61] Matusz et al., "Are We Ready for Real-World Neuroscience?" 6; Dikker et al., "Crowdsourcing Neuroscience."

[62] Dikker et al., "Crowdsourcing Neuroscience," 2.

[63] Svenja Matusall, Ina Maria Kaufmann, and Markus Christen, "The Emergence of Social Neuroscience," in *The Oxford Handbook of Social Neuroscience*, ed. Jean Decety and John T. Cacioppo (Oxford: Oxford University Press, 2011); Svenja Matusall, "Social Behavior in the 'Age of Empathy'? A Social Scientist's Perspective on Current Trends in the Behavioral Sciences," *Frontiers in Human Neuroscience* 7 (2013).

[64] Dikker et al., "Crowdsourcing Neuroscience," 3.

brain-to-brain synchronization is (if it is a meaningful measure in relation to social behavior at all), at this stage it would be impossible to give feedback on a fact that is itself in the making. Yet, the suggestion of feedback signals an even more central conundrum in the field of brain-to-brain synchronization.

The presence of feedback in this art–science experiment points to the feedback imaginary that fundamentally structures the field of brain-to-brain synchronization and that of social neuroscience more broadly. Part of a longer history of brain trigger circuits and feedback interfaces (sketched in Chapters 4 and 5), acquiring visual or auditory feedback on brain-to-brain synchronization is the imagined horizon of this research field. Exemplary in this respect is a 2017 paper by psychiatrist Kai Vogeley discussing the promise of two-body neuroscience. He sketches a future of hyperscanning neurofeedback that would allow people "to observe both their own and the interactants brain activity during the ongoing interaction in real time. This would substantially enrich simple face-to-face encounters and would allow us to perform intervention studies, for instance, nonverbal communication training in persons with ASD [autism spectrum disorder]."[65] This feedback medium for "being in sync" is a future vision; other hyperscanning researchers are more prudent in projecting future applications of brain-to-brain synchronization. Yet, brain-to-brain research frequently invokes a potential usefulness to the study of mental disorders such as autism (construed as a social cognition disorder).[66] The fundamental premise and impetus of synchronization research is the idea that brain-to-brain synchronization patterns may become *neuromarkers* to predict social behavior.[67] *Mutual Wave Machine* shows us—evocatively and implicitly—how feedback imaginaries, paired with a vision of localizing social behavior, are at the core of the brain-to-brain synchronization paradigm.

My analysis of the relations between the form of art–science and the content of brain-to-brain synchronization research flags two tensions. Firstly, the artistic presence of (deceptive) feedback signals carries the implicit promise of finding neuromarkers for synchronization that fundamentally structures this field. Secondly, even though the piece presents social cognition in a remarkably media-intensive "live" and "real-life" installation that seems to meld minds, the "social brain" that

---

[65] Kai Vogeley, "Two Social Brains: Neural Mechanisms of Intersubjectivity," *Philosophical Transactions of the Royal Society B: Biological Sciences* 372, no. 1727 (2017).

[66] Both Suzanne Dikker and Guillaume Dumas, for example, are working on autism-related projects.

[67] Takahiko Koike, Hiroki C. Tanabe, and Norihiro Sadato, "Hyperscanning Neuroimaging Technique to Reveal the 'Two-in-One' System in Social Interactions," *Neuroscience Research* 90 (2015): 29.

is ultimately enacted through this scientific paradigm is based on an individual and localization-oriented social neuroscience. My observations on what is left out of art–science experimentation (or what is only implicitly evoked but epistemologically structural) are supported by the discussion and critique of brain-to-brain synchronization research in the past years. A number of researchers of enacted, extended, embodied, and affective (4EA) perspectives on cognition have not been impressed by the promises of two-body neuroscience.

Writing in 2013, researchers Shaun Gallagher, Daniel Hutto, Jan Slaby, and Jonathan Cole dismissed hyperscanning's radical claims as a "normal science" of neural correlates dressed in new garb, for example.[68] Brain-to-brain research ultimately harks back to the same "classical computational models, representationalism, localization of function" that single out "measurable brain activation as the most relevant *explanans*."[69] Instead, the authors argue, not the brain (or two or more brains) should be the explanatory unit of interaction, but "a dynamic relation between *organisms*, which include brains."[70] From the perspective of 4EA researchers, cognitive science gives the brain too foundational a priority—even in brain-to-brain studies—and thus strengthens a neuro-centric reductionism. Countering these objections, hyperscanning researcher Leonard Schilbach and colleagues stress that if social neuroscientists refrain from the "neo-phrenological attempt to isolate brain regions," then measuring brain-to-brain activity is in fact a valid part of investigating the nature of social cognition.[71] What is most fundamental about the hyperscanning approach—crucially allowing it to "go social," they emphasize—is the possibility for "innovative experimental setups to investigate social interaction … in more ecologically valid ways."[72] *Mutual Wave Machine*, as an art-science experiment, partakes in this emphasis on innovative setups for a neuroscience of the "real world." And yet, a deeper dive into the artwork-turned-research demonstrates the limits of two-body neuroscience in capturing the real world it has conjured.

---

[68] Shaun Gallagher et al., "The Brain as Part of an Enactive System," *Behavioral and Brain Sciences* 36, no. 4 (2013): 422. See John Cromby, "Integrating Social Science with Neuroscience: Potentials and Problems," *BioSocieties* 2, no. 2 (2007); Shaun Gallagher, "Decentering the Brain: Embodied Cognition and the Critique of Neurocentrism and Narrow-Minded Philosophy of Mind," *Constructivist Foundations* 14, no. 1 (2018): 15.
[69] Gallagher et al., "The Brain as Part of an Enactive System," 422.
[70] Ibid.
[71] Leonhard Schilbach et al., "A Second-Person Neuroscience in Interaction," *Behavioral and Brain Sciences* 36, no. 4 (2013): 445.
[72] Ibid.

The Allure of Synchronization: Toward a Critical Media History of Being on the Same Wavelength

Notions of synchronization are at the heart of contemporary EEG hyperscanning research. Interpersonal interaction is predominantly framed as *social synchronization*—described in terms of "tuning," "being on the same wavelength," or "getting in sync"—and the neural dynamics underlying these social modes have been similarly hypothesized as a manifestation of attunement: a *brain-to-brain synchronization*. This current focus has not gone unnoticed: in a 2019 article in *Nature Reviews Neuroscience*, Elizabeth Redcay and Leonhard Schilbach wrote that the field of two-body neuroscience is "heavily dominated by a search for synchrony, or mirroring, between brains."[73] Synchronization is today's central relationality underpinning the search for the neural basis of social interaction.

If forms of liveness imbue research into the active brain with a sense of presentness, hereness, nearness, nowness, and directness, the logic and rhetoric of synchronization cuts across these dynamics by offering a simultaneity as well as a togetherness. Understood as an alignment that is more than just temporal, synchronization connects to the twenty-first-century form of liveness described in this chapter: the desire to see brain-to-brain dynamics unfold as they happen, but not in one brain: in two interacting brains. Synchronization pairs with this "real-life" or "real-world" liveness to be achieved through hyperscanning technology and less controlled, naturalistic settings. In this final part of the chapter, I sketch the contours of the contemporary "allure of synchronization" in social neuroscience in a back and forth between synchronization's neuroscientific present and its (brainmedia-historical) past.

A first and central observation on the prevalence of synchronization is its strong connection to notions of harmonization. While there is no shared definition of synchronization in current papers in the field, concepts of alignment and positive accord are abundant. In hyperscanning experiments, being socially "in tune" is often literally postulated as an in-tuneness on a brain-to-brain to level and connected to individual mechanisms for empathy, mutual understanding, and trust.[74] When synchronization experiments are

---

[73] Elizabeth Redcay and Leonhard Schilbach, "Using Second-Person Neuroscience to Elucidate the Mechanisms of Social Interaction," *Nature Reviews Neuroscience* 20, no. 8 (2019): 502–3. For an alternative approach, see Sebastian Wallot et al., "Beyond Synchrony: Joint Action in a Complex Production Task Reveals Beneficial Effects of Decreased Interpersonal Synchrony," *Plos One* 11, no. 12 (2016).

[74] About the lack of a definition for synchronization, see "Beyond Synchrony," 1.

performed in naturalistic settings (such as teacher–pupil or therapist–client interactions), synchrony is predominantly connected to a good rapport between subjects.[75] Such positive effects are perhaps naturally central. As psychologist Sebastian Wallot and colleagues remark, the majority of synchronization experiments are in fact "focused on interpersonal coordination as an end in itself."[76] The dominant presence of synchronization in two-body neuroscience is surprising, as it is not the only phenomenon that *could* be examined through hyperscanning. The dynamics between two people interacting ("interpersonal action coordination") is broader than synchronization alone (think of turn-taking, for example). Moreover, on the neural level, brain-to-brain synchronization is but one of many possible "real-time neural dynamics" that may be involved.[77] One of the reasons that research into behavioral synchronization (such as synchronous tapping, swinging, or mimicking) dominates in this field is perhaps simply because it is technologically easier to execute.[78] But there are other fundamental reasons for synchronization's appeal, both in its connection to a very current "real-world" neuroscience and its resonance with a deeper media-historical genealogy.

Contemporary synchronization discourses and practices are part of a longer media-historical lineage that go back to the 1920s, the era I have identified in this study as that of "liveness" connected to brain research. It is by reading synchronization through the twentieth-century media history of "tuning," "being on the same wavelength," and "being in sync" that I propose we can best understand the interpretative resources that give it particular appeal in the present. Synchronization's association with harmonization or attunement actually goes back far beyond the twentieth century. The *OED* cites a figurative use of "to put in tune" in 1530, denoting: "To bring into a proper or desirable condition; to give a special tone or character (esp. of a good kind) to."[79] Yet, importantly, this musical and string-related usage acquired decisively new meaning with the rise of radio in the 1920s (the *OED* notes a first example in 1926), when "to tune" also meant that individuals

---

[75] Ibid.
[76] Ibid.
[77] Redcay and Schilbach note "interaction dynamics" are broader than behavioral synchronization, and "neural synchrony is just one possible neural signature of dynamic, reciprocal social interaction and may not capture the distinct and complementary roles that are inherent to dyads in everyday interactions." Redcay and Schilbach, "Using Second-Person Neuroscience to Elucidate the Mechanisms of Social Interaction," 502–3. Tao Liu and Matthew Pelowski, "Clarifying the Interaction Types in Two-Person Neuroscience Research," *Frontiers in Human Neuroscience* 8 (2014).
[78] Dumas, "Towards a Two-Body Neuroscience," 350.
[79] "Tune, V.2," *Oxford English Dictionary* (Oxford University Press).

could be "mentally receptive," a state of being that was envisioned as *preceding perception* by the human senses.[80] Analogously, the expression "being on the same wavelength" originates from that same era of radio-technical discourse and implied a type of "mutual understanding."[81] This wavelength analogy of social interaction, first noted by the OED in a 1925 novel about a pair of doppelgangers, specifically emphasized a person-to-person situation.[82] The expression "in sync" emerged with sound cinematography (1929) and later with television (1939). Being in or out of sync in a figurative sense arose during television's heyday, first noted in John Steinbeck's 1961 *Winter of Our Discontent* when the protagonist knows that *something* is going on: "I just feel it. Hair on the back of my neck kind of itches. That's a sure sign. Everybody's a little out of synch."[83]

Located at the intersection of the social and the neural, contemporary conceptions of brain-to-brain synchronization are impacted by these media-technical etymologies. By imagining the social through the (media-)technical, these expressions imbue social interaction with specific interpretative resources that may only be implicit in contemporary research, but are conceptually foundational nonetheless. I have mentioned the dimension of "proper" accord: when synchronization turns into *harmonization*. Moreover, imagining interaction as a type of technical optimization—"tuning"—means that it can be controlled or modulated: synchronization gains *operability*. And finally, through the long-time connection of waves with the realm of the mental, synchronization is often understood as working *affectively*, as preceding conscious perception.

These intersecting meanings attached to synchronization—harmonic, operable, and affective—crisscross the history of EEG research as it has interfaced with the study of social synchronization from the early twentieth century until today. Yet, this longer history of synchronization has been obscured by contemporary two-body neuroscientists, who have sketched their own lineage of the synchronized brain and the social subject. Hyperscanning researchers often only mention one curious, parapsychological antecedent of present-day synchronized EEG recordings: a 1965 study of twins that

---

[80] "3. figurative. To become mentally receptive to, or aware of; to comprehend. Const. as preceding sense." Ibid.
[81] "2. figurative with allusion to radio reception, implying (esp. mutual) understanding; esp. in to be on the same wavelength (as someone else), to understand each other." "Wavelength, N.," *Oxford English Dictionary* (Oxford University Press).
[82] OED marks the first uses as Norman Venner's *The Imperfect Imposter* of 1925. Ibid.
[83] "b. gen. Esp. in *in sync, out of sync*. Also *figurative*." J. Steinbeck, *Winter of Our Discontent* (1961), II. xiv. 278, cited in "Sync, N. And V.," *Oxford English Dictionary* (Oxford University Press).

showed simultaneous alpha rhythms without physical contact.[84] After this first erratic attempt, as neuroscientists Fabio Babiloni and Laura Astolfini tell the story, hyperscanning approaches were "rapidly forgotten in the scientific community, and remained so for about 40 years."[85] By invoking this historical study as a "bizarre" forerunner of present-day two-body neuroscience, researchers aim to distance themselves from a naïve interpretation of synchronization research as a literal "communication channel" between two subjects.[86] Hyperscanning researcher Guillaume Dumas emphasizes this in a poetic video presenting brain-to-brain synchrony to a broader audience: "There is nothing magical here"; it is through the action-perception flow that people form a "coherent system." "Each exchange is an opportunity to overcome our individuality," and it is in this nonmagical yet enthralling way that "we are more connected than we think" (Figure 6.3).[87]

Separating the field of two-body neuroscience from extraordinary or even magical views on extrasensory communication, this demarcation work by contemporary neuroscientists is necessary because the realm of studying nonverbal, prereflective, and "automatic" synchronization behavior—as part of the domain of affect—has a contentious proximity to the field of parapsychology. What this obscures, however, is a much longer history of correlating microtemporal measurements of the active brains of two or more people interacting with the microscopic changes of their interacting bodies, in which notions of the harmonic, the operable, and the affective intersect.

To understand present-day hyperscanning research, we have something to gain from a closer analysis of the controversial (and oft-footnoted, but arguably little-read) work of William Condon and W. D. Ogston. Starting in 1967 at the psychiatric institute of the University of Pittsburgh, they recorded the EEG and muscle activity of interacting subjects while simultaneously studying their gestures and utterances through high-speed cameras that

---

[84] T. D. Duane and Thomas Behrendt, "Extrasensory Electroencephalographic Induction between Identical Twins," *Science* 150, no. 3694 (1965). It is beyond this chapter's scope to sketch the interface between ESP research, EEG measurements, and synchronization research. See also Andrew Pickering's footnotes in Pickering, *The Cybernetic Brain*, 416.

[85] Babiloni and Astolfi, "Social Neuroscience and Hyperscanning Techniques," 78.

[86] "Bizarre," in Riitta Hari et al., "Centrality of Social Interaction in Human Brain Function," *Neuron* 88, no. 1 (2015): 187. "Communication channel," in Babiloni and Astolfi, "Social Neuroscience and Hyperscanning Techniques," 80. Two-body neuroscience has developed its own visual language to represent brain-to-brain synchronization by means of lines drawn between brain areas with similar measures of brain activity (see Figure 6.4).

[87] Dumas, "Towards a Two-Body Neuroscience"; Guillaume Dumas and L. Halard, *Phi (Subtitled)*, https://www.youtube.com/watch?v=PPbH6CMw2bU2012; Alejandro Pérez et al., "Differential Brain-to-Brain Entrainment while Speaking and Listening in Native and Foreign Languages," *Cortex* 111 (2019).

Synchronizing Two Dynamic Brains 219

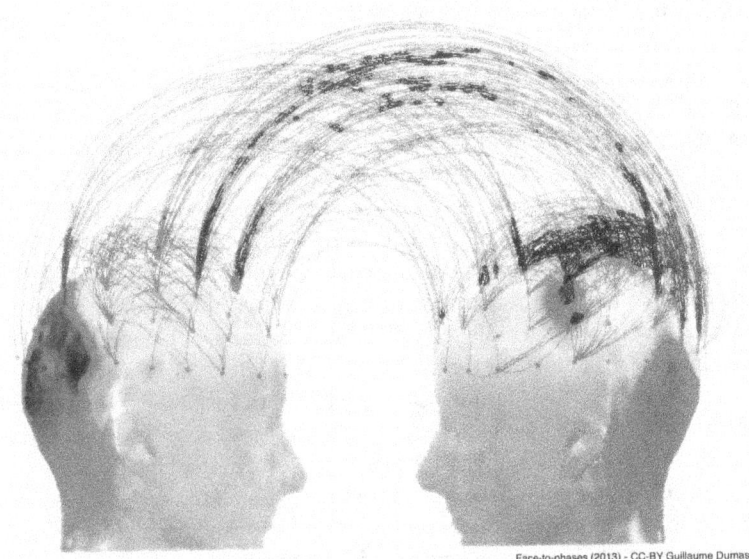

Figure 6.3 Guillaume Dumas, illustration of two-body neuroscience, *Face-to-phases*, 2013 (illustration).

allowed a frame-to-frame comparative microanalysis of these simultaneous records.[88] Synchronization was a structuring feature of both the technical apparatus and the correlative analysis these researchers proposed in their study of social interaction. Condon and Ogston reported on tentative synchronous patterns, "harmonious patterns of change," in both speech, body, motion, EEG, and EMG.[89] A correlative approach between the social and the neural, with a central emphasis on synchronization, was both *proposed* and *produced* by these sync technology approaches. Doing simultaneous measurements had a specific appeal: the possibility of also applying the behavioral segmentation approach (the macro level) "in search for patterns in artifacts at the micro level."[90] In the 1980s, Condon started to speak of behavior as a "wave phenomenon": movements, gestures, and speech were understood as hierarchically organized waveforms exhibiting characteristic

---

[88] William S. Condon and W. D. Ogston, "A Segmentation of Behavior," *Journal of Psychiatric Research* 5, no. 3 (1967); William S. Condon, "Method of Micro-Analysis of Sound Films of Behavior," *Behavior Research Methods & Instrumentation* 2, no. 2 (1970).
[89] Condon, "Method of Micro-Analysis of Sound Films of Behavior," 54.
[90] Condon and Ogston, "A Segmentation of Behavior," 232.

periodicities.⁹¹ Naturally, he argued, such waveforms "suggests that they may be produced by similarly synchronized brain processes."⁹²

Condon's analysis of microscopic biological and behavioral "characteristics" opened his work up to a particular biotypological project. In 1970, he reported on frame-to-frame analyses of films of patients with aphasia, parkinsonism, petit mal, stuttering, schizophrenia, and childhood autism, of which "many were found to display patterns of change that differed in subtle ways from the harmony characteristic of normal behaviour."⁹³ In a similar vein, in the 1980s, he extended his analysis of interactional synchrony and microbehavioral rhythms toward "cultural rhythms," arguing that "those having different cultural rhythms are unable to really 'synch-in' fully with each other."⁹⁴ In this line of reasoning, he contrasted typical rhythms of "black behavior" (who moved with "greater intensities") with that of whites.⁹⁵ While several researchers critiqued Condon's methodological approaches and found they could not replicate his findings, the racist aspects of his biotypological "wave" analysis did not seem to face equal resistance.⁹⁶ Some researchers suggested that scholarship on microbehavioral paralinguistic elements should be more analyzed in relation to the *content* of spoken utterances.⁹⁷ But Condon's work—and other work on microanalysis and interactional synchrony—clearly fitted with a reigning and popular and cultural emphasis on the "hidden messages" of body language, the dimension of the paralinguistic, the realm of the affective.⁹⁸

As the scholarly analysis of "mutual waves" in social brains has a long history, *Mutual Wave Machine* and brain-to-brain synchronization research at large deserve more historical comparative attention. Turning to Condon helps to understand the allure and biopolitical tensions attached

---

[91] William S. Condon, "Communication: Rhythm and Structure," in *Rhythm in Psychological, Linguistic and Musical Processes*, ed. J. R. Evans and M. Clynes (Springfield, IL: Thomas, Charles C., 1986), 67.

[92] Ibid.

[93] Condon, "Method of Micro-Analysis of Sound Films of Behavior," 54.

[94] "Cultural Microrhythms," in *Interaction Rhythms: Periodicity in Communicative Behavior*, ed. Martha Davis (New York: Human Sciences Press, 1982).

[95] Ibid., 66.

[96] J. J. McDowall, "Interactional Synchrony: A Reappraisal," *Journal of Personality and Social Psychology Today* 36, no. 9 (1978).

[97] Condon also compared verbal and nonverbal behavior, but with an emphasis on microscopic elements of nonverbal behavior. William S. Condon, "Sound-Film Microanalysis: A Means for Correlating Brain and Behavior," in *Dyslexia: A Neuroscientific Approach to Clinical Evaluation*, ed. F. H. Duffy and Norman Geschwind (Boston: Little, Brown, 1985).

[98] See Condon and Ogston, "A Segmentation of Behavior," cited in Michael B. McCaskey, "The Hidden Messages Managers Send," *Harvard Business Review*, November 1979.

to this academic field and raises questions about the promise of using synchronization activity as a neural signature for social behavior, as well as the contemporary re-appearance of a cultural emphasis on "visceral literacy," that is, reading elements of social interactional behavior outside the purview of language and symbolic mediation.[99]

It also shows how brain-to-brain synchronization research, sheerly through its shared investigatory focus—synchronization, offers intuitive connections between the very disparate analytical levels in social neuroscience: interpersonal behavior and measuring the electrical activity of nerve cells. Between two people "feeling in sync" and their measured levels of "brain-to-brain synchronization" are countless influencing factors: social hierarchy and etiquette, for example, or setting and atmosphere. By suturing these explanatory levels, synchronization rhetoric fits the rationale that has been fundamental to social neuroscience since its inception: "integrating" social, cognitive, and biological explanations, which social neuroscientist John Cacioppa has outlined as the field's fundamental "doctrine of multilevel analysis."[100] Characteristic of this doctrine is one article explaining the science behind *Mutual Wave Machine*, which takes the reader through various synchronization levels step by step: from synchronizations between oscillating *neurons*, to synchronizations of the activity of *single brains* due to external stimuli (entrainment), to new hyperscanning measurements of *brain-to-brain* synchronization, and finally the phenomenon of *interpersonal* (social) synchronization.[101] By crafting such natural affiliations between levels, the bridges between them can be more easily assumed or projected to future investigation.

The allure of synchronization also has a longer history in the world of art–science experimentation. In 1972, David Rosenboom created a project remarkably similar to *Mutual Wave Machine*. Exhibited at Vancouver Art Gallery, his installation *Vancouver Piece* allowed two EEG-fitted participants to face each other, separated by a two-way mirror (Figure 6.4). If simultaneous

---

[99] Mark Andrejevic, "Reading the Surface: Body Language and Surveillance," *Culture Unbound: Journal of Current Cultural Research* 2, no. 1 (2010).

[100] The "doctrine of multilevel analysis" in John T. Cacioppo and Gary G. Berntson, "Social Psychological Contributions to the Decade of the Brain: Doctrine of Multilevel Analysis," *American Psychologist* 47, no. 8 (1992). See also Kevin Ochsner and Matthew Lieberman, "The Emergence of Social Cognitive Neuroscience," *American Psychologist* 56, no. 9 (2001): 717–34, cited in Matusall et al., "The Emergence of Social Neuroscience," 13. For an example of social neuroscience's bridging "levels" narrative, see Guillaume Dumas, J. A. Scott Kelso, and Jacqueline Nadel, "Tackling the Social Cognition Paradox through Multi-Scale Approaches," *Frontiers in Psychology* 5 (2014).

[101] Dikker et al., "Using Synchrony-Based Neurofeedback in Search of Human Connectedness," 164–8.

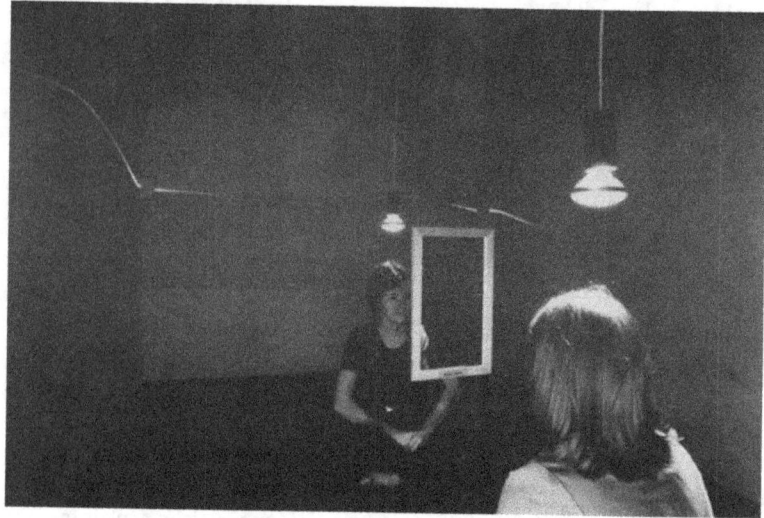

Figure 6.4 Two museum attendees participating in David Rosenboom's Vancouver piece at the Vancouver Art Gallery, 1973 (photograph).

alpha activity was measured, they would see a reflection of their own face seemingly projected on the shoulders of the person in front. "The intended effect," Rosenboom explained, "was to open the participants' consciousness of self to enable them to explore ideas about shared identity."[102] He had developed his ideas on synchronization in tandem with EEG researcher Lester Fehmi of Stony Brook University, who further developed the idea into a patented feedback course that he used in training management executives, people with social anxieties, and meditation practitioners.[103]

I mention these dispersedly networked conceptions of synchronization—from strange alpha-correspondences between twins to Condon's

---

[102] David Rosenboom and Tim Mullen, "More Than One—Artistic Explorations with Multi-Agent BCIS," in *Brain Art: Brain-Computer Interfaces for Artistic Expression*, ed. Anton Nijholt (Cham: Springer, 2019), 122–3.

[103] Rosenboom and Mullen cite a 1971 conference paper with Lester Fehmi (at the conference for Humanistic Psychology in Washington, DC) as the first conceptualization of what was called "contingent" group feedback. Ibid., 120. In the 1970s, Fehmi would further develop his work on feedback, including notions of synchronization, in various biofeedback consulting and training contexts (he tested his methods on management executives, for example, and eventually trademarked his "Open Focus Training," which is still practiced today). Lester G. Fehmi, "EEG Biofeedback, Multichannel Synchrony Training and Attention," in *Expanding Dimensions of Consciousness*, ed. A. Arthur Sugerman and Ralph E. Tarter (New York: Springer, 1978).

biotypological psychiatry research, Rosenboom's art installations, and Fehmi's management focus training—to sketch synchronization's varying historical affiliations to the affective, the operable, and the harmonic. These histories are very closely interconnected from the 1960s to hyperscanning practices today. Rosenboom, who started hypothesizing about using the new microtemporal ERP measurements in musical compositions in the 1970s (as discussed in Chapter 5), recently collaborated with hyperscanning researcher and neurotech entrepreneur Tim Mullen. In 2014 they created *Ringing Minds*, a musical multiperson EEG feedback piece that employs hyperscanning measurements to record multiple participants during a live musical performance and use the changes in the music to give the participants synchronization feedback. Writing about *Ringing Minds* and other multiperson musical pieces, the creators describe synchronization between multiple EEGs as though generated by a "hyperbrain" or "group mind" that produces a "collective neural response" that allows for "effective" practice and "positive results."[104] What those results may be is left undetermined, but they are associatively linked to notions of harmony, creativity, and feeling in sync.

Mullen continues to use hyperscanning technology for his San Diego-based firm Intheon, which offers software packages and computational databases to companies interested in measuring real-time EEG data or doing offline analysis of EEG measurements. The company aims to work, he explained, like an online speech recognition platform: by acquiring more and more EEG data, the detection algorithms for states like "emotion" or "frustration" will get better and better. As he explained in a recent interview, "We need more and more data, out there … in the wild of the world, which is not the lab. So one of the key things is making the world the laboratory."[105]

Artists, then, are instrumental in creating visions for a real-world neuroscience in which "the possible-world-to-come" (to use Shapin's term) is the world as laboratory. Mullen teams up with artists for performances and installations that show intricate visualizations and sonifications of the EEG data in real time and often do so in multiperson, brain-to-brain synchronization settings. Similar to *Mutual Wave Machine*, Mullen's art–science collaborations are key to neuroscientific future-making, creating a vision of a world in which everybody would continuously contribute their EEG (and other bio-)data. A world in which we would instantly see who is on the same wavelength—and who is not.

---

[104] Rosenboom and Mullen, "More Than One—Artistic Explorations with Multi-Agent BCIS," 126, 35, 28, 20.

[105] Tim Mullen, *Neurotech Anytime, Anywhere*, https://www.youtube.com/watch?v=NJmo pBvbALs, Simulation | Transtech, 2018, #248.

Conclusions

In this chapter, I analyzed the art–science experiment *Mutual Wave Machine* to study the new forms of liveness and performances of knowledge that characterize the emerging field of two-body neuroscience. New hyperscanning technologies (simultaneously measuring two brains as they interact) and portable EEG devices have enabled exploring new brain-to-brain measurements of active brains. In this experimental practice, researchers hail the possibility of studying social interaction in less controlled and more naturalistic settings outside of the conventional laboratory, aiming to measure communication "live." This new form of liveness in two-body neuroscience connects to a broader concern in current neuroscientific discussions on the urgency of what is called a "real-world neuroscience" that studies human behavior in a more ecologically valid way.

This new discourse and form of liveness—"real-world" liveness—is fundamentally connected, as I have explained, with social neuroscience's big data-centered approach. *Mutual Wave Machine* signals the search for, and testing of, new forms of "crowdsourcing neuroscience" that allow the harnessing of large amounts of brain activity data necessary to conduct a "real-world" neuroscience. It frames crowdsourcing in a narrative of "citizen science"—out of the lab, into public space. *Mutual Wave Machine* specifically proposes a cerebral citizen science that evokes a hybrid epistemic profile for its participants: both a volunteer experimental subject contributing to an important scientific field and an informed critical reviewer of science. Yet, while the art–science project invites participants to look critically at the "messy" situation of measuring and acquiring data, it does little to engage subjects with the way individual recordings become part of a bigger biotypological project of developing neuromarkers for social interaction behavior. How a distributed epistemic subject of big data cognitive science could be engaged with art–science remains an open question, but I have shown how artistically crowdsourcing neuroscience shifts this discussion to opportunities for museums to "contribute" to "real-world" neuroscience instead.

An immersive art–science installation that gives feedback on participation, *Mutual Wave Machine* allows visitors to experience moment-to-moment interaction anew and conjures a complex experimental environment that foregrounds the multisensory and media-impacted dimension of social contact. Its evocation of feedback, I argued, is a promissory gesture, revealing a "feedback imaginary" at the heart of the scientific program of two-body neuroscience. Moreover, I observe a tension between the art and the science of representing complex social interactions: the invoked multilayered and media-impacted experiences can hardly be modeled within the scientific framework of brain-to-brain analysis. This observation is connected to a broader critique of social neuroscience by a number of scholars who have pointed to reductionist and neuro-centric approaches in current research into interpersonal interaction.

Finally, I analyzed what I call the allure of synchronization by sketching a longer history of synchronization that points to the influence of media-historical conceptions of "being on the same wavelength," "tuning," and "being in sync" employed by researchers correlating brain activity with social activity. This media-impacted conception of synchronization, which originated in the 1920s, imbues the study of social interaction with notions of harmony, operability, and affect. A number of artists and scientists have investigated social interaction as a type of synchronization, and these programs showed the different conceptualizations and research directions, including an objectionable one toward racial "wave" categorizations in the 1980s. My observations here are meant to signal an urgent need, in the light of present-day synchronization research, for more critical historical analyses.

Taking stock of my examination of *Mutual Wave Machine* as a crowdsourcing neuroscience art–science experiment, I have examined it as an exemplary case to unpack epistemic tensions and discourses at the heart of two-body neuroscience. Moreover, I positioned it as part of a broader trend of scientists collaborating with artists (or scientists becoming performers and installation artists themselves), a practice that is at the heart of new forms of "neuroscientific life" preoccupied with "future-making." These new artistic and performance-oriented forms of promissory science are especially prevalent in social neuroscience as it interfaces with the development of EEG devices and BCI software. One promise that *Mutual Wave Machine* enacts is the longstanding projection that EEG records will one day be legible as neural signatures of social interaction and even be used in real-time feedback devices. The other important promise that becomes visible through my analysis is the way that (artistic) crowdsourcing neuroscience experiments propose and produce an image of the "world as laboratory," a vision that underpins the project of "real-world" neuroscience.

Despite my critical view on art–science experiments in this chapter, I think *Mutual Wave Machine* should be regarded as exemplary in one other sense. It positively demonstrates the first steps of how art–science project could discuss science in the making (with *Mutual Wave Machine*, this takes place through scientists' in-class visits as well as a special website to communicate results). Though "engagement" is a contentious notion, as I argued in light of the rhetoric of citizen science, there are critical ways to conduct "neuroscience in public" that should be further explored. All too often, visitors are merely invited to marvel at a technological setup and feel giddy while providing EEG data that enable biodatabases at scales that are incongruous with the intimate interactive experiences on offer. The course of action, I propose, should be to develop not only new forms of "engagement" but also and especially new forms of art–science experimentation.

In the past decade, project-based teams of neuroscientists, sociologists, and anthropologists have set out to experiment with new interdisciplinary configurations in the field of cognitive science. Felicity Callard and Des Fitzgerald have described some first attempts in various European research projects to bring together researchers from the humanities, social sciences, and neurosciences in efforts to "reshape and reimagine the conceptual and empirical contours" of cognitive neuroscientific experiments.[106] We cannot speak of a movement yet; interdisciplinary projects are exceptionally time-consuming and dependent on scarce and often temporary funding schemes. But it is precisely now that such interdisciplinary experiments to understand the brain in action and science in the making could pair up with another scholarly development of the past decade: the shaping of new norms and forms as well as support structures for art–science collaboration. The time for a new direction in art-neuroscience is now.

---

[106] Felicity Callard and Des Fitzgerald, "Entangling the Medical Humanities," in *The Edinburgh Companion to the Critical Medical Humanities*, ed. Anne Whitehead et al. (Edinburgh: Edinburgh University Press, 2016), 44. *Rethinking Interdisciplinarity across the Social Sciences and Neurosciences* (New York: Palgrave Macmillan, 2015).

Conclusion

In *Brainmedia: One Hundred Years of Performing Live Brains, 1920–2020*, I have examined how scientists, science educators, and artists perform knowledge of the brain at work. My main argument is that approaching the history of brain and mind sciences as a history of live brains helps to see the extent of media's imbrication in thinking the brain at work—not only of (medical and experimental) media technologies but also recording and broadcast media. By describing and analyzing assemblages of brains and media in particular historical contexts as brainmedia, I have shown how specific practices *of* and ideas *about* mediation impacted how scientists and science educators conceptualized and demonstrated the active human brain. My five historical case studies of brainmedia assemblages spanning the period of 1920–2020 substantiate my study. I analyzed illuminated brain models from the 1920s until the 1930s, staged brainwave recordings from the 1930s to the 1940s, live brains *on* television and conceptions of brains *as* television in the 1940s and 1950s, EEG feedback circuits and the rise of real-time interfaces around 1970, and "brain-to-brain" art–science experiments between 2013 and 2019.

While previous critical studies of twentieth-century brain and mind sciences have emphasized the rhetoric of transparency and immediacy as having shaped the promise of watching the brain in action, my analysis of brainmedia points to the importance of liveness as a structuring element in enacting and staging the brain at work. I showed how research into cerebral processes and demonstrations of active brains were impacted by different forms of liveness, that is, the different historically situated temporal-spatial configurations offered by different conceptions and practices of mediation. I argue that when brains are rendered "live," they not only become transparent or immediate but also gain dimensions of directness, nearness, hereness, aliveness, liveliness, and nowness. In my analysis of live brains, I move away from a confined view of scientific (brain) image-making. Instead, I establish the urgency of analyzing performing knowledges of the live brain. My study moves between scientists conceptualizing active brains in laboratories and scientific publications, and practices of developing, demonstrating, and exhibiting live brains in public. I thus take an inclusive and recursive approach to circulating knowledges as they are performed, mediated, and configured within and beyond the science establishment.

Ultimately, my account of brainmedia offers an alternative genealogy of the contemporary live brain. My histories of different forms of liveness show how the idea of the active brain depended on different cultural as well as technical spatiotemporalities and spatio-tempo-realities at different points in time over the past century. My analysis reveals how such forms of liveness are connected to emerging new media and concepts of mediation. The verbs in my chapter titles denote the historical assemblages that enabled these new forms of liveness: displaying, demonstrating, broadcasting, interfacing, and synchronizing. Chapter 2 describes how the 1920s and 1930s rise of new display devices (such as light indicators, illuminated circuit diagrams, and signaling systems) in city life were thought to offer new abstractions of active processes. These forms of technical mediation also impacted the way brain activity could be conceptualized and represented: visualized activities could now be viewed as immediate while being mediated, a form of liveness that I call a "logic of direct display." Chapter 3 explains how 1930s demonstrations of brainwave-measuring technologies (in texts, exhibitions, and a Hollywood film) wavered between attributions of "liveliness" (virtuously vivid demonstration strategies) and "aliveness" (the uncanny dimension that viewers were supposed to negotiate wisely). In Chapter 4, an experiment with a live TV broadcast of a brain X-ray shows how the medium of television offered a vision of a networked intimacy between nearby screens and distant machines in the 1950s. At the same time, a number of scientists could perceive brains as TV-like scanning machines vulnerable to disturbance by flickering light (jittery images on a TV screen, for example), thus shaping an image of the brain as having a particular operative accessibility and also susceptibility. In Chapter 5, a new form of liveness manifests around 1970 in (artistic) EEG feedback setups interfacing with the brain, which presented brain activity as part of circuited energetic flows in media environments. At the same time, emerging brain–computer interfaces introduced the technical temporality of "real time" into the brain–machine circuit, enabling microtemporal measurements below human consciousness and new imaginaries of feedback. In the concluding case study, in Chapter 6, I analyze the contemporary discourse on "real-world neuroscience" and its search for more naturalistic social neuroscience experiments of synchronizing brains. Here a new, "real-world" liveness is enacted through an artistic performance of a crowdsourced experiment, buttressing the claim to more lifelike brain measurement.

In each of my case studies, I argued that the forms of liveness acquired in and through these assemblages of brains and media shaped new, historically variable types of "live brains"—that is, visions of brains that are, for example, more intimately connected to media or more open to categorization, circuiting, and training; brain activities that are strangely separate from bodies, or bodies that feel ever more embrained. However, my analysis not only shows how these new visions of brains are produced; it also reveals emerging conceptions of media and mediation: respectively, conceptions of how to interpret abstractions on displays, how to marvel at projected brainwaves, how to approach the TV screen, how to dissolve into a media-technical environment, and how to engage with an artistic media installation. Rather than studying "representation in (neuro)scientific practice," my analysis offers a view of the complexly enmeshed media in hybrid processes of performing the live brain inside and outside the laboratory. The particular forms of liveness integral to histories of live brains described here—direct display, liveliness/aliveness, networked intimacy, operative accessibility, circuited flow, real-time temporality, and real-world liveness—mark crucial junctures in the past century of brainmedia. My methodological proposal for a material-discursive approach to brainmedia also enables future studies of different or alternative forms of liveness. It is with this prospect in mind that I offer my genealogy of contemporary brainmedia: to contribute to a better understanding of the political scientific imaginaries shaped by the live brain today.

Understanding Contemporary Live Brains

Here I return to the first image discussed in this book, the 1920 image of a female writer, a working woman with a working brain sketched inside her head. It opened Chapter 1, which located the emergence of a conception of "live brain" in the early decades of the twentieth century. At a time when media were omnipresent (as were narratives about media technologies' ubiquity), the image of the neurological body became newly paired, I argued, with a conception of a "live brain": a body whose nervous system was understood as intimately imbricated with a network of mediating technologies, and whose active brain was about to be captured in action. The modern human subject—whose nervous system connected to the outside world via the senses—was understood as utterly caught up in, as well as changed by, a media-saturated environment. This particular environmental conception of media—media as milieu—has been an undercurrent of my analysis of brainmedia.

The brainmedia assemblages described in the preceding chapters address media not primarily as (new) media devices but as medial presences in human life. By showing these assemblages of brains and media—of brain activity and active brains—as they were performed in and through specific historical mediations, I show how media have at key points been positioned as a type of nervous middle that exceeds their role as object or technology. Ultimately, these brainmedia are examples of what Jennifer Gabrys has called "atmospheric media": fields of relations shaping a *mediality* that inhabits humanity as its habitat.[1] I have historicized such fields of relations in this study by tracing particular discourses about mediation in relation to scientific brain research and performances of sciences inside and outside the laboratory. The brainmedia assemblages that emerge from my five case studies clearly have such an atmospheric role: as illuminated presences in city life, spheres of broadcasting, artistic media environments, interface assemblages, and (in the last chapter) as computational data infrastructures at the basis of a "real-world neuroscience" that allow potential correlations between social and neural data and feed into future measurements of brain activity.

My historical approach to live brains emphasizes the different ways media have been conceptualized as fields of relations. The alpha-loving brainmedia environments envisioned in the 1970s were based on different approaches and different conceptions of atmospheric media than contemporary performances of EEG measurements. Yet, taken together, these successive descriptions of brainmedia also present a historical backdrop to the intimate imbrication of the (conceptualization of the) sensing brain and technological mediation in the twentieth century. I thus add to the current work of media-theoretical scholars (such as Mark Hansen, Marie-Louise Angerer, Patricia Clough, and N. Katherine Hayles) who have proposed that the present media-technological landscape impacts the human nervous system in ways not only quantitatively but also *qualitatively* different than in previous periods in media history. Hansen, for example, sketches a present in which ubiquitous computational media impinge upon sensory experience and the "sensing brain" at microtemporal, preperceptual levels.[2] By his account, we can no longer speak of media as devices or objects that extend the human sensorium in today's computational media landscape; we should instead interpret our media-saturated present as presenting a new technical mediality, a continuous background presence of informational flows that address the

---

[1] Gabrys cites Régis Debray: "Mediological man does not cohabitate with his technological surroundings, he is inhabited by his habitat; constructed by the niche he has constructed." *Media Manifestos: On the Technological Transmission of Cultural Forms* (London: Verso, 1996), 11, cited in Jennifer Gabrys, "Atmospheres of Communication," in *The Wireless Spectrum: The Politics, Practices, and Poetics of Mobile Media*, ed. Barbara Crow, Michael Longford, and Kim Sawchuk (Toronto: University of Toronto Press, 2010), 53.

[2] "Qualitative" shift and "sensing brain," in Mark Hansen, "Ubiquitous Sensation: Towards an Atmospheric, Collective and Microtemporal Model of Media," in *Throughout: Art and Culture Emerging with Ubiquitous Computing* (Cambridge, MA: MIT Press, 2013), 72; ibid., 67.

sensing brain and affect our experience of (and being in) the world without those mediations becoming perceptible. For Marie-Louise Angerer, this qualitative shift in mediation structures even constitutes a new "affective dispositif," an intertwining of media with power, law, and truth that effects both institutional practices and conditions of subjectivation.[3] In light of these studies, my argumentation in *Brainmedia: One Hundred Years of Performing Live Brains, 1920–2020* contributes to a view of the longer twentieth-century emergence of such an affective dispositif by showing how new temporalities of mediation—new forms of liveness—came into being in relation to active brain research.

I have shown how brainmedia conjoined with forms of liveness and allowed for conceptualizing particular intimacies between media and brains, as well as visions of active brains that were impressionable and could be viewed at work—all before today's omnipresent computational media and before the circulation of blinking functional brain images in the 1990s. Consequently, my brainmedia genealogy offers an alternative path to answer the questions that arise when faced with a media environment operating on levels that are beyond consciousness and that make it hard to envision, as Angerer puts it, a way to "deviate from the great wave" of "all-encompassing modulation."[4] My historical approach to brainmedia as assemblages makes it possible to study the constitution of forms of liveness through the lens of performing knowledges. By carefully considering the ways we are asked to engage with performances of live brains, we might start to see beyond the space of engagement marked out for us.

Engaging Live Brains Today

Over the past years, when people asked me about the content of my book, I would usually describe my project as a history of showing and staging the active brain, or a history of scientists and other researchers employing new media to understand brains. More often than not, the response was an enthusiastic yet puzzled "That's so fascinating." And indeed, *fascination* seems to be the endemic mode of engagement assigned to the public display of images and imaginaries of seeing the brain at work. The most common use of the word fascination today, according to the *Oxford English Dictionary*, is "to attract and 'hold spellbound' by delightful qualities; to charm, enchant."[5]

[3] Marie-Luise Angerer, *Desire after Affect* (London: Rowman & Littlefield, 2015), xv.
[4] Marie-Luise Angerer, drawing on Deleuze in "Postscript: A New Affective Organization," in *Desire after Affect* (London: Rowman & Littlefield, 2015), 130.
[5] "Fascination, N.," *Oxford English Dictionary* (Oxford University Press).

In my introduction, I described how functional brain images, especially fMRI and PET since the Decade of the Brain, were positioned as a type of *eidola*, viewed as exerting undue persuasive powers over the spectator, temporarily short-circuiting the rational thinking of laypeople or even scientists. Fascination signals the fact that narratives about brain science have a proximity to the spectacular, what Jonathan Crary has called "an organization of appearances that are simultaneously enticing, deceptive, distracting, and superficial."[6] We say we are fascinated, as philosopher Ackbar Abbas notes, when we appreciate something but do not have much to say or, alternatively, when there is too much to say. Being fascinated is an "enigmatic experience"; it "captures our attention without at the same time submitting entirely to our understanding."[7] For Abbas, fascination is "neither knowledge nor ignorance," but conjures a "*paracritical* mode of attention" shaping an "enigmatic relation to what we do not know."[8]

In critical accounts of the brain and mind sciences, it is precisely this paracritical mode of attention to the unknown that fascinating public appearances of the brain conjure up, which is viewed as suspect. Scientists use the power of fascination to frame future-oriented endeavors, for example, about the eventual possibility of viewing the mind at work in the brain. Fascination sustains what Nicolas Rose calls the "promissory culture" of technoscience and biotechnology in which brain science partakes, a domain in which groundbreaking transformations are predicted, always "imminent," but "just out of reach."[9] To say that brain images are fascinating means to be dazzled by the sight of something that might be known. Michael Hagner and Cornelius Borck have dubbed this promissory logic in neuroscience a "proleptic structure," emphasizing the present-day conception of neuroscience as based on a promise that is not yet fulfilled, as a science that anticipates "a future of comprehensive understanding."[10] At the same time, neuroscience's proleptic structure is due to not only its promise of revelation, Borck and Hagner contend, but also its ever-present enigma: while gaining more understanding of the brain, researchers at the same time get a better

---

[6] Jonathan Crary, "Spectacle," in *New Keywords. A Revised Vocabulary of Culture and Society*, ed. Tony Bennett, Lawrence Grossberg, and Meaghan Morris (Malden: Blackwell, 2005), 335.
[7] M. A. Abbas, "Dialectic of Deception," *Public Culture* 11, no. 2 (1999): 348.
[8] Ibid.
[9] Nikolas Rose, *The Politics of Life Itself: Biomedicine, Power, and Subjectivity in the Twenty-First Century* (Princeton, NJ: Princeton University Press, 2009), 79.
[10] Michael Hagner and Cornelius Borck, "Brave Neuro Worlds," in *Der Geist Bei Der Arbeit: Historische Untersuchungen Zur Hirnforschung*, ed. Michael Hagner (Göttingen: Wallstein, 2006), 36. Cornelius Borck, "Through the Looking Glass: Past Futures of Brain Research," *Medicine Studies* 1, no. 4 (2009): 330.

grasp of the *difficulties* in understanding the brain, thus framing the issue as a secret to be uncovered. This dialectic between mystery and revelation is crucial to brain discourses, a "chronic anticipation of the solution to one of the last big mysteries of mankind."[11]

With *Brainmedia: One Hundred Years of Performing Live Brains, 1920–2020*, I augment these critiques with a situated historical view of experiencing brains at work. Fascination and its paracritical mode of attention is itself inscribed in historical narratives about engagement and discourses of wonder about the live brain. Earlier I aligned myself with historians of public performances of new (media) technologies and historians of staging science all rallying for no longer approaching the spectacular and the fascinating with a sweeping critique of the "spectacle-as-commodity," but by historicizing experiences of the spectacular and the fascinating instead. My study of live brains as instances of performing knowledges was in pursuit of such a historicized approach to shifting fascinations with the technologically mediated active brain. It has allowed me to study the circulation of knowledge about the active brain between scientific publications and platforms for making science public.

Focusing on such circulation and boundary work has also allowed me to pay significant attention to the way the fascinated layperson—often standing in for any subject with a brain addressed by brain research—was invoked at various historical junctures. In Chapter 3, for example, I show how 1930s accounts of EEG as a new research technology asked readers and viewers to be amazed by a new scientific accomplishment while at the same time not attributing too much mind-reading power to the new technologies with which they were presented. Later, analyzing a contemporary art–science project in the recent field of brain-to-brain synchronization research in Chapter 6, I show how participants of a public scientific experiment are viewed as being educated about the uncertainties of a "science in the making" while at the same time being asked to perform a civic duty in contributing to brain research. These interpellations of nonexperts signal the power of scientific discourse in instituting normative views of *how* one is supposed to engage with science on display.

In the end, my situated view of moments of performing knowledges opens up a better understanding of the modes of engagement, as well as the demarcations of knowledges, inscribed in science demonstrations. My analysis prompts a set of questions about engagement with contemporary practices of neuroscience: What types of publics are evoked? How are publics asked to understand brain science on display? How are they supposed

---

[11] Hagner and Borck, "Brave Neuro Worlds," 36.

to engage with new technological media? I contend that identifying this boundary work in circulating and performing knowledges is a first step toward Angerer's "deviating from the great wave" of the current assembling of brains with computational media. It is recognizing and acknowledging such demarcations in particular assemblages of brains and media, especially when brainmedia are performed in public, that can open up a new view of a situated politics of fascination.

Recently, philosopher and historian of science Isabelle Stengers has urged us to think of engagement in terms of fostering "public intelligence."[12] Here, the "public" is no infantile crowd, nor an attentive body of laypeople eager and capable to participate. Rather, Stengers shows how the current form of scientific life has fostered an atmosphere in which scientists are ever more restricted by the funding opportunities of the knowledge economy and more than ever need to uphold a "fable of 'free' research" and an imaginary of curiosity-driven research of the "mysteries of the world."[13] What is lacking is the work of what Stengers's calls "demanding connoisseurs" who would "hold scientists to the task of taking care when making normative judgments about what does or does not matter, or of presenting their results in a lucid manner that actively situates them in relation to the questions they really can answer, rather than as a response to whatever is the object of a more general interest." Connoisseurs are those who can both appreciate the originality of an idea and pay attention to questions it does not take into account. Public intelligence then means fostering a milieu of connoisseurs dense enough to approach particular scientific propositions with intelligence.

Stengers is suspicious of that which fascinates. The first challenge is for the public "to not let itself be fascinated," by Science with a capital S, "not to be too easily impressed."[14] As my analysis of the entwined histories of forms of liveness and active brains has shown, however, being impressed and fascinated are integral parts of performing knowledges and brainmedia. Perhaps what we need in order to critically engage contemporary live brains is to foster a public intelligence that takes fascination seriously. Stengers speaks of connoisseurs meeting scientists in the spirit of a "shared perplexity."[15] It is in this spirit of serious fascination and shared perplexity that I propose to engage brainmedia today.

---

[12] Isabelle Stengers, *Another Science Is Possible: A Manifesto for Slow Science* (Cambridge, MA: Polity Press, 2018).
[13] Ibid., 6.
[14] Isabelle Stengers and Penelope Deutscher, "Another Look: Relearning to Laugh," *Hypatia* 15, no. 4 (2000): 53.
[15] Isabelle Stengers, *Invention of Modern Science* (Minneapolis: University of Minnesota Press, 2000), 65.

List of Sources of Figures

Figure 1.1. Diagram showing the four chief association centers of the human brain, *c.* 1919 (illustration). In Warren Hilton. *Applied Psychology: Making Your Own World: Being the Second of a Series of Twelve Volumes on the Applications of Psychology to the Problems of Personal and Business Efficiency*. New York: The Literary Digest for the Society of Applied Psychology, 1919: 17.

Figure 1.2. Pierre Marie's diagram for interconnected left hemisphere cortical centers involved in oral and written language, 1888 (illustration). In Pierre Marie. "De L'aphasie (Cécité Verbale, Surdité Verbale, Aphasie Motrice, Agraphie)." *Revue de Médecine* 3 (1888): 693–702. Caption and illustration in Victor W. Henderson. "Alexia and Agraphia." *Neurology* 70, no. 5 (2008): 396.

Figure 2.1. Fritz Kahn, *Die Lichtwarhnehmung*, 1929 (illustration). In Fritz Kahn. *Das Leben Des Menschen: Eine Volkstümliche Anatomie, Biologie, Physiologie Und Entwicklungsgeschichte Des Menschen*. 2 vols. Stuttgart: Kosmos, Gesellschaft der Naturfreunde; Franck'sche Verlagshandlung, 1929. Band iv, plate xxii.

Figure 2.2. Dr. Edith Klemperer and Dr. Robert Exner, luminous brain model from Vienna, *c.* 1931 (photograph). In "Use Glass in Model of Brain." *Popular Science Monthly* 119, no. 6 (December 1931): 32. Used with permission of Popular Science Copyright © 2021. All rights reserved.

Figure 2.3. Edith Klemperer, patent of anatomical model, 1931/4 (drawing). Patent US1951422 a, filed October 29, 1931, and issued March 20, 1934, http://www.google.nl/patents/us1951422.

Figure 2.4. Push-button brain diagram, *c.* 1930 (photograph). In Werner Lincke. "Lehrmeister Licht." *Das Licht. Zeitschrift für praktische Leucht- und Beleuchtungs- Aufgaben* (1930): 124.

Figure 2.5. Constantin Von Economo's plaster models of cytoarchitecture, *c.* 1927 (photograph). In Constantin von Economo. *Cellular Structure of the Human Cerebral Cortex*. Translated by Lazaros Constantinos Triarhou. Basel: Karger, [1927] 2009: 174.

Figure 2.6. Fritz Lang, *Die Spione*, 1928 (film still). Photo/copyright: Friedrich Wilhelm Murnau Stiftung (Wiesbaden).

Figure 2.7. Science's futile attempt to build a perfect mechanical brain, 1934. In *The Salt Lake Tribune*, November 25, 1934, 2. Courtesy The Salt Lake Tribune.

Figure 3.1. "Why Radio May Have Uncovered a Sixth Sense! Science Now Investigating Cases of Broadcast Programs Being Picked Up, Unaided, By the Human Nervous System," (newspaper article). In the Tribune, July 4, 1926, 3. Courtesy the Salt Lake Tribune.

Figure 3.2. Electroencephalography booth in the exhibition *La Biologie: Exposition Internationale de Paris* at the Palais de la Découverte as part of the *Paris International Exposition*, 1937 (photograph). In "Que Savons-Nous Des Ondes Électriques Émises Par Le Cerveau?" *Science et Vie* (September 1937): 218.

Figure 3.3. Edward Dmytryk, *The Devil Commands*, 1941 (film still), https://www.youtube.com/watch?v=lJpwiujHqXI&list=UUqH2YMSzMaGN92Vc3VkhWnQ&index=5563.

Figure 4.1.1. Catching a brain wave, 1954 (television broadcast stills). *Johns Hopkins Science Review*, Du Mont Television Network, broadcasted March 24, 1954. Source: University Archives, Sheridan Libraries, Johns Hopkins University.

Figure 4.1.2. Catching a brain wave, 1954 (television broadcast stills). *Johns Hopkins Science Review*, Du Mont Television Network, broadcasted March 24, 1954. Source: University Archives, Sheridan Libraries, Johns Hopkins University.

Figure 4.2.1. *En direct du Cerveau Humain*, 1956 (television broadcast still). *En direct de*, RTF Télévision/ORTF, broadcasted December 19, 1956, https://www.ina.fr/video/CPF86658437.

Figure 4.2.2. *En direct du Cerveau Humain*, 1956 (television broadcast still). *En direct de*, RTF Télévision/ORTF, broadcasted December 19, 1956, https://www.ina.fr/video/CPF86658437.

Figure 4.2.3. *En direct du Cerveau Humain*, 1956 (television broadcast still). *En direct de*, RTF Télévision/ORTF, broadcasted December 19, 1956, https://www.ina.fr/video/CPF86658437.

Figure 4.2.4. *En direct du Cerveau Humain*, 1956 (television broadcast still). *En direct de*, RTF Télévision/ORTF, broadcasted December 19, 1956, https://www.ina.fr/video/CPF86658437.

Figure 4.3. Illustration of an analogy between a profile-scanning device and the human brain, 1953 (book illustration). In W. Grey Walter. *The Living Brain*. New York: W.W. Norton, [1953] 1963: 110.

Figure 4.4. Illustration of trigger circuit, 1949 (book illustration). In V. J. Walter and W. Grey Walter. "The Central Effects of Rhythmic Sensory

Stimulation." *Electroencephalography and Clinical Neurophysiology* 1, nos. 1–4 (1949): 84. Reprinted from *Electroencephalography and Clinical Neurophysiology*, vol. 1, nos. 1–4, V. J. Walte and W. Grey Walter, "The Central Effects of Rhythmic Sensory Stimulation." Pages 57–86, with permission from Elsevier.

Figure 4.5. Vivian Walter working with EEG-set-up at the Burden Neurological Institute in Bristol, England. Toposcope display visible left from the subject. *c.* 1954 (photograph). In Walter, W. Grey. "The Electrical Activity of the Brain." *Scientific American* 190, no. 6 (1954): 55.

Figure 4.6. Toposcope display, *c.* 1957 (photographs). In W. Grey Walter. "The Brain as a Machine. Section of Pyschiatry [January 8, 1957]." *Proceedings of the Royal Society of Medicine* 50 (1957): 803. Courtesy of the National Library of Medicine.

Figure 5.1. Alvin Lucier (left) and John Cage (right) preparing a performance of *Music for Solo Performer* at the festival "John Cage at Wesleyan," 1988 (photograph). Special Collections & Archives, Wesleyan University. In Volker Straebel and Wilm Thoben. "Alvin Lucier's Music for Solo Performer: Experimental Music beyond Sonification." *Organised Sound* 19, no. 1 (April 2014): 18. Reproduced with permission from Cambridge University Press.

Figure 5.2. Joe Kamiya with EEG feedback setup and subject, *c.* 1968 (photograph). In Joe Kamiya. "Conscious Control of Brain Waves." *Psychology Today* 1 (1968): 58.

Figure 5.3.1. Marvin Karlins and Lewis M. Andrews, *Biofeedback: Turning on the Power of Your Mind.* London: Garnstone Press, 1973.

Figure 5.3.2. Larry Kettelkamp, *A Partnership of Mind and Body, Biofeedback.* New York: Morrow, 1976.

Figure 5.3.3. Barbara B. Brown, *New Mind, New Body; Bio-Feedback: New Directions for the Mind.* New York: Harper & Row, 1974. Courtesy of Penguin Random House LLC (US).

Figure 5.3.4. Anthony A. Zaffuto, *Alphagenics: How to Use Your Brain Waves to Improve Your Life.* Garden City, NY: Doubleday, 1974. Courtesy of Penguin Random House LLC (US).

Figure 5.4. David Rosenboom, *Ecology of the Skin*, 1970 (photograph). In David Rosenboom. "Method for Producing Sounds or Light Flashes with Alpha Brain Waves for Artistic Purposes." *Leonardo* 5, no. 2 (1972): 143. Photo: Peter Moore, New York.

Figure 5.5. Nina Sobell and Michael Trivich, interactive electroencephalographic video drawings, 1972–4 (video stills). In "Interactive Electro-encephalographic Video Drawings," partly taped in the laboratory of Barry Sterman, Veterans Hospital Sepulveda, California. Video produced by Nina Sobel and LBMA Video, 1992, https:// archive.org/details/XFR_2013-08-23_1B_16, courtesy of Nina Sobel.

Figure 5.6. Marc Bjorlund, no title, 1971 (illustration). In *Radical Software* 1 (California Edition) no. 4 (Summer 1971): 4.

Figure 5.7. Richard Lowenberg, *Environetic Synthesis*, 1970 (drawing). In *Radical Software* 2, no. 1 (Winter 1972): 44, courtesy of Richard Lowenberg.

Figure 5.8. Feedback training with alpha train in Barbara Brown's lab, c. 1974 (video still). In *Dialogue on Biofeedback*, https://www.youtube.com/watch?v=3gGQF2ItH8c. Veterans Administration, 1974.

Figure 5.9.1. Edmond Dewan, brain-controlled lamp setup, 1964 (television broadcast still). CBS news item, 1964, https://www.youtube.com/watch?v=nCGcY6sQjcM.

Figure 5.9.2. Edmond Dewan, brain-controlled lamp setup, 1964 (television broadcast still). CBS news item, 1964, https://www.youtube.com/watch?v=nCGcY6sQjcM.

Figure 6.1. Suzanne Dikker and Matthias Oostrik, *Mutual Wave Machine*, 2014 (photography by Oleg Borodin). Installation at Museum of Science and Industry, Moscow, 2014, courtesy of Oleg Borodin.

Figure 6.2. Jean Livet et al., *Brainbow* image of dentate gyrus, 2007 (image on website). Published on Harvard Center for Brain Science website, 2007, http://cbs.fas.harvard.edu/science/connectome-project/brainbow#.

Figure 6.3. Guillaume Dumas, two-body neuroscience synchronization illustration, Fase-to-phases, 2013 (cover illustration). On the cover of a journal with Dumas's article: Alejandro Pérez, Guillaume Dumas, Melek Karadag, and Jon Andoni Duñabeitia. "Differential Brain-to-Brain Entrainment while Speaking and Listening in Native and Foreign Languages." *Cortex* 111 (February 2019): 303–15.

Figure 6.4. Two museum attendees participating in Rosenboom's Vancouver piece at the Vancouver Art Gallery, 1973 (photograph by John Tod Greenaway), first printed in David Rosenboom. "Vancouver Piece." In *Sound Sculpture: A Collection of Essays by Artists Surveying the Techniques Applications and Future Direction of Sound Sculpture*, ed. John Grayson. A.R.C. Publications, 127–31. Copyright 1975 A.R.C. and the individual authors.

Bibliography

Abbas, M. A. "Dialectic of Deception." *Public Culture* 11, no. 2 (1999): 347–63.
Abbott, Alison. "Colours Light Up Brain Structure." *Nature News*, October 31, 2007.
Adolphs, Ralph. "Investigating the Cognitive Neuroscience of Social Behavior." *Neuropsychologia* 41, no. 2 (2003): 119–26.
Adolphs, Ralph, Lauri Nummenmaa, Alexander Todorov, and James V. Haxby. "Data-Driven Approaches in the Investigation of Social Perception." *Philosophical Transactions of the Royal Society B: Biological Sciences* 371, no. 1693 (2016): 1–10.
Adrian, Edgar Douglas. "The Discovery of Berger." In *Handbook of Electroencephalography and Clinical Neurophysiology*, edited by Antoine Rémond, 5–10. Amsterdam: Elsevier, 1971.
Adrian, Edgar Douglas. "The Impulses Produced by Sensory Nerve Endings." *Journal of Physiology* 61, no. 1 (1926): 49–72.
Adrian, E. D., and B. H. C. Matthews. "The Berger Rhythm: Potential Changes from the Occipital Lobes in Man." *Brain* 57 (1934): 355–85.
Albe-Fessard, Denise. "Denise Albe-Fessard." In *The History of Neuroscience in Autobiography, Volume 1*, edited by Larry R. Squire, 2–49. Washington, DC: Society of Neuroscience, 1996.
Albino, R., and G. Burnand. "Conditioning of the Alpha Rhythm in Man." *Journal of Experimental Psychology* 67, no. 6 (1964): 539–44.
Ali, Sabrina S., Michael Lifshitz, and Amir Raz. "Empirical Neuroenchantment: From Reading Minds to Thinking Critically." *Frontiers in Human Neuroscience* 8 (2014): 357.
Ancoli, Sonia, and Joe Kamiya. "Methodological Issues in Alpha Biofeedback Training." *Biofeedback and Self-regulation* 3, no. 2 (1978): 159–83.
Andrejevic, Mark. "Big Data, Big Questions | The Big Data Divide." *International Journal of Communication* 8, no. 1 (2014): 1673–89.
Andrejevic, Mark. "Reading the Surface: Body Language and Surveillance." *Culture Unbound: Journal of Current Cultural Research* 2, no. 1 (2010): 15–36.
Andriopoulos, Stefan. "Psychic Television." *Critical Inquiry* 31, no. 3 (2005): 618–37.
Andriopoulos, Stefan. "The Sleeper Effect: Hypnotism, Mind Control, Terrorism." *Grey Room* 45 (2011): 88–105.
Angerer, Marie-Luise. *Desire after Affect*. London: Rowman & Littlefield International, 2015.

Angerer, Marie-Luise. "Medienkörper: Zur Materialität Des Medialen Und Der Medialität Der Körper." In *Kultur—Medien—Macht: Cultural Studies Und Medienanalyse*, edited by Andreas Hepp and Rainer Winter, 259–70. Wiesbaden: VS Verlag für Sozialwissenschaften/Springer, 1997.

Angerer, Marie-Luise. "Postscript: A New Affective Organization." In *Desire after Affect*, 115–34. London: Rowman & Littlefield International, 2015.

Anon. "Alpha Brain Waves—New Key to Inner Peace?" *Paradiso Fox*, no. 3 (1970): 2–6.

Anon. "Das Leuchtende Gehirn." *Reichspost*, June 23, 1932, 7.

Anon. "Die Schöpferin Des Ersten Wiener Gläsernen Gehirnmodells." *Neues Österreich*, February 6, 1953.

Anon. "Driving Tests Indoors." *The Times*, June 4, 1935, 11.

Anon. "Giant Electrical Sign Has Its Own Elevator." *Popular Mechanics* (1934): 16.

Anon. "A Glass 'Brain' Aids Medical Students." *Maitland Daily Mercury*, March 29, 1932, 8.

Anon. "Kurzkurs 'Das Leuchtende Gehirnmodell.'" *Mitteilungen der Volkshochschule Wien Volksheim* 8, no. 2 (1935): 3–4.

Anon. "Listening-in to the Nerves: Dr. Adrian's Experiments." *The Observer*, March 28, 1926, 16.

Anon. "A Luminous Brain. Nerve Specialist's Invention." *West Australian*, February 12, 1937, 19.

Anon. "Mödosam Onsdag for Fysiologer Och Försöksdjur. Efter Dagens Hundra Föredrag Ger Stockholms Stad Bankett I Gyllene Salen." *Dagens Nyhter*, August 5, 1926, 8.

Anon. "'Movie' of a Brain Is Exhibited Here: Sound Film Shown at Museum of Natural History Records Reactions to Impulses." *New York Times*, April 18, 1937, 42.

Anon. "News and Notes." *British Journal of Photography* (1896): 105.

Anon. "Pattern of the Brain at Work to Be Broadcast by Wsui." *Wilmington Morning News* (1928).

Anon. "Photographic Record Made of a Sleeper's Dream." *Science News-Letter* 27, no. 737 (1935): 333.

Anon. "Psychologisch Congres Te Parijs. Zeshonderd Psychologen Uit Alle Landen Bijeen. Een Hartelijke Samenwerking Van Alle Zijden." *Algemeen Handelsblad*, August 4, 1937, 3.

Anon. "Radio to Solve the Secrets of Telepathy. How Science Seeks to Show That 'Brain Waves' Transmit Thoughts and Find a Way to Talk with the Dead." *Daily Press* (1925).

Anon. "Recording Rhythmic 'Brain-Waves.'" *Literary Digest* (1935): 17.

Anon. "Scientists See 'Dream Walking' with Chart Aid." *Berkeley Daily Gazette*, May 21, 1935, 2.

Anon. "'Seeing' the Brain at Work. Specialists' New Aid to Diagnosis. Important Discovery." *Nottingham Evening Post*, December 1, 1934, 8.

Anon. "Signal Lights Indicate Vacant Theater Seats." *Popular Science*, August (1922): 38.

Anon. "Television Report. Televised Endoscopy. Note about a Programme Broadcast on 14 December 1955 Presented by Jean Painlevé." *Science and Film* (1956): 29–31.

Anon. "Tentoonstelling 'De Mensch.'" *De Telegraaf*, August 31, 1935, 10.

Anon. "Wat Denkt Gij?" *Limburger Koerier*, Maart 1937, 7.

Anon. "Why Radio May Have Uncovered a Sixth Sense! Science Now Investigating Cases of Broadcast Programs Being Picked Up, Unaided by the Human Nervous System." *Indianapolis Star*, July 4, 1926, 64.

Anon. "World's First Luminous Brain Model Made by Woman Doctor." *The Mail*, July 31, 1937, 25.

Anwar, Yasmin. "Scientists Use Brain Imaging to Reveal the Movies in Our Mind." *Berkeley News*, September 22, 2011.

Armagnac, Alden P. "'Human-Eye' Camera Opens New Way to Televison." *Popular Science* (1933): 11–13.

Asaro, Peter. "Working Models and the Synthetic Method: Electronic Brains as Mediators between Neurons and Behavior." *Science Studies* 19, no. 1 (2006): 12–34.

Asendorf, Christoph. *Batteries of Life: On the History of Things and Their Perception in Modernity*. Berkeley: University of California Press, 1993.

Ashley, Robert. *Alvin Lucier—Music with Roots in the Aether, c. 1975*. June 7, 2019. https://www.youtube.com/watch?v=nRa5x6j26Is.

Atkinson, William Walker. *Nuggets of the New Thought: Several Things That Have Helped People*. Chicago: The Psychic Research Company, 1902.

Auslander, Philip. "Digital Liveness: A Historico-Philosophical Perspective." *PAJ: A Journal of Performance and Art* 34, no. 3 (2012): 3–11.

B, J. "Impressions of the Eleventh International Congress of Psychology." *Journal of Consulting Psychology* 2, no. 3 (1938): 65–70.

Babiloni, Fabio, and Laura Astolfi. "Social Neuroscience and Hyperscanning Techniques: Past, Present and Future." *Neuroscience & Biobehavioral Reviews* 44 (2014): 76–93.

Baer, Ulrich. "Photography and Hysteria: Toward a Poetics of the Flash." *Yale Journal of Criticism* 7, no. 1 (1994): 41–77.

Bain, Lisa, Noam I. Keren, Sheena M. Posey Norris, and Clare Stroud. "Brain Reading." In *Neuroforensics: Exploring the Legal Implications of Emerging Neurotechnologies: Proceedings of a Workshop*, edited by Lisa Bain, Noam I. Keren, Sheena M. Posey Norris, and Clare Stroud, 18–20. Washington, DC: National Academies Press, 2018.

Barad, Karen. "Posthumanist Performativity: Toward an Understanding of How Matter Comes to Matter." *Signs* 28, no. 3 (2003): 801–31.

Barbara, Jean-Gaël. "The Fessard's School of Neurophysiology after the Second World War in France: Globalistion and Diversity in Neurophysiological Research (1938–1955)." *Archives Italiennes de Biologie* 149 (2011): 187–95.

Barker, Martin. *Live to Your Local Cinema: The Remarkable Rise of Livecasting.* London: Palgrave Macmillan, 2013.

Barlow, H. B. "Eye Movements during Fixation." *Journal of Physiology* 116, no. 3 (1952): 290–306.

Bateson, Gregory. "The Cybernetics of 'Self': A Theory of Alcoholism." *Psychiatry: Journal for the Study of Interpersonal Processes* 34, no. 1 (1971): 1–18.

Bayle, A. L. J. *Nouvelle Doctrine Des Maladies Mentales.* Paris: Gabon et Compagnie Libraires, 1825.

Bazin, André. "La Télévision: Moyen De Culture, *France-Observateur* 297 (19 January 1956)." In *André Bazin's New Media*, edited by Dudley Andrew, 126–30. Oakland: University of California Press, 2014.

Beatty, Jackson. "Effects of Initial Alpha Wave Abundance and Operant Training Procedures on Occipital Alpha and Beta Wave Activity." *Psychonomic Science* 23, no. 3 (1971): 197–9.

Beaulieu, Anne. "Images Are Not the (Only) Truth: Brain Mapping, Visual Knowledge, and Iconoclasm." *Science, Technology & Human Values* 27, no. 1 (2002): 53–86.

Beaulieu, Anne. "The Space Inside the Skull: Digital Representations, Brain Mapping and Cognitive Neuroscience in the Decade of the Brain." PhD dissertation, University of Amsterdam, 2000.

Bellamy, Edward. *Doctor Heidenhoff's Process.* London: W. Reeves, 1890.

Benjamin, Walter. "The Work of Art in the Age of Its Technological Reproducibility [First Version]." *Grey Room*, no. 39 ([1935] 2010): 11–38.

Bennet, Cynthia Denise. "Science Service and the Origins of Science Journalism, 1919–1950." PhD dissertation, Iowa State University, 2013.

Bensaude-Vincent, Bernadette. "A Historical Perspective on Science and Its 'Others.'" *Isis* 100, no. 2 (2009): 359–68.

Bensaude-Vincent, Bernadette. "In the Name of Science." In *Science in the Twentieth Century*, edited by John Krige and Dominique Pestre, 319–38. Amsterdam: Harwood Academic, 1997.

Benschop, Ruth, and Douwe Draaisma. "In Pursuit of Precision: The Calibration of Minds and Machines in Late Nineteenth-Century Psychology." *Annals of Science* 57, no. 1 (2000): 1–25.

Benshoff, Robert L. "Self-Regulation as an Aid to Human Effectiveness and Biocybernetics Technology and Behavior," 1–40. PhD dissertation, San Diego State University, 1976.

Benson, Don. "Neurone Cluster Grope." *Radical Software* 1, no. 2 (1970).

Benton, Arthur. "Bergson and Freud on Aphasia: A Comparison." In *Bergson and Modern Thought*, edited by Pete A. Y. Gunter and Andrew C. Papanicolaou, 175–86. London: Routledge, 2016.

Berenstein, Rhona J. "Acting Live TV Performance, Intimacy, and Immediacy." In *Reality Squared: Televisual Discourse on the Real*, edited by James Friedman, 25–49. New Brunswick, NJ: Rutgers University Press, 2002.

Berger, Hans. *Über Die Lokalisation Im Großhirn: Rede Gehalten Bei Der Akademischen Preisverteilung Zu Jena Am 18. Juni 1927 [Mit Einer Chronik Der Universität Für Das Jahr 1926/27]*. Jena: G. Fischer, 1927.

Berger, Milton M., Barry Sherman, Janet Spalding, and Robert Westlake. "The Use of Videotape with Psychotherapy Groups in a Community Mental Health Service Program." *International Journal of Group Psychotherapy* 18, no. 4 (1968): 504–15.

Besser, Stephan. "From the Neuron to the World and Back: The Poetics of the Neuromolecular Gaze in Bart Koubaa's 'Het Gebied Van Nevski' and James Cameron's 'Avatar.'" *Journal of Dutch Literature* 4, no. 2 (2013): 43–67.

Bevilacqua, Dana, Ido Davidesco, Lu Wan, Kim Chaloner, Jess Rowland, Mingzhou Ding, David Poeppel, and Suzanne Dikker. "Brain-to-Brain Synchrony and Learning Outcomes Vary by Student-Teacher Dynamics: Evidence from a Real-World Classroom Electroencephalography Study." *Journal of Cognitive Neuroscience* 31, no. 3 (2018): 401–11.

Biro, Matthew. "The New Man as Cyborg: Figures of Technology in Weimar Visual Culture." *New German Critique* 62 (1994): 71–110.

Bishop, Claire. "Delegated Performance: Outsourcing Authenticity." *October* 140 (2012): 91–112.

Bishop, George H. "Electrophysiology of the Brain." In *The Problem of Mental Disorder: A Study Undertaken by the Committee on Psychiatric Investigations, National Research Council*, 120–32. New York: McGraw Hill, 1934.

Bonnardel, R. *La Biologie: Exposition Internationale De Paris 1937*. Paris: Palais de la Découverte, 1937.

Bono, James J. "Making Knowledge: History, Literature, and the Poetics of Science." *Isis* 101, no. 3 (2010): 555–9.

Bono, James J. "Why Metaphor? Toward a Metaphorics of Scientific Practice." In *Science Studies. Probing the Dynamics of Scientific Knowledge*, edited by Sabine Maasen and Matthias Winterhager, 215–34. Bielefeld: Transcript Verlag, 2003.

Borck, Cornelius. "Auf Der Suche Nach Der Verlorenen Kultur: Vom Neuroimaging Über Critical Neuroscience Zu Cultural Neuroscience— Und Zurück Zur Kritik." *Berichte zur Wissenschaftsgeschichte* 41, no. 3 (2018): 238–57.

Borck, Cornelius. "Between Local Cultures and National Styles: Units of Analysis in the History of Electroencephalography." *Comptes Rendus Biologies* 329, nos. 5–6 (2006): 450–9.

Borck, Cornelius. *Brainwaves: A Cultural History of Electroencephalography*. London: Routledge, 2018.

Borck, Cornelius. "Communicating the Modern Body: Fritz Kahn's Popular Images of Human Physiology as an Industrialized World." *Canadian Journal of Communication* 32, no. 3 (2007): 495–520.

Borck, Cornelius. "Die Unhintergehbarkeit Des Bildschirms." In *Mit Dem Auge Denken: Strategien Der Sichtbarmachung in Wissenschaftlichen Und Virtuellen Welten*, edited by Bettina Heintz, Arnold O. Benz, and Jörg Huber. Zürich: Ed. Voldemeer, 2001.

Borck, Cornelius. "Electricity as a Medium of Psychic Life: Electrotechnological Adventures into Psychodiagnosis in Weimar Germany." *Science in Context* 14, no. 4 (2001): 565–90.

Borck, Cornelius. "Electrifying the Brain in the 1920s: Electrical Technology as a Mediator in Brain Research." In *Electric Bodies: Episodes in the History of Medical Electricity*, edited by Paola Bertucci and Giuliano Pancaldi, 239–64. Bologna: Università di Bologna, 2001.

Borck, Cornelius. *Hirnströme: Eine Kulturgeschichte Der Elektroenzephalographie*. Göttingen: Wallstein Verlag, 2005.

Borck, Cornelius. "Media, Technology and the Electric Unconsciousness in the 20th Century." In *L'ère Électrique—The Electric Age*, edited by Olivier Asselin, Silvestra Mariniello, and Andrea Oberhuber, 33–60. Ottawa: University of Ottawa Press, 2011.

Borck, Cornelius. "Through the Looking Glass: Past Futures of Brain Research." *Medicine Studies* 1, no. 4 (2009): 329–38.

Borck, Cornelius. "Toys Are Us: Models and Metaphors in Brain Research." In *Critical Neuroscience: A Handbook of the Social and Cultural Contexts of Neuroscience*, edited by Suparna Choudhury and Jan Slaby, 111–33. Oxford: Wiley Blackwell, 2011.

Borck, Cornelius. "Urbane Gehirne. Zum Bildüberschuss Medientechnischer Hirnwelten Der 1920er Jahre." *Archiv fur Mediengeschichte* 2 (2002): 261–71.

Born, Georgina, and Andrew Barry. "Art-Science. From Public Understanding to Public Experiment." *Journal of Cultural Economy* 3, no. 1 (2010): 103–19.

Bowker, Geof. "How to Be Universal: Some Cybernetic Strategies, 1943–70." *Social Studies of Science* 23, no. 1 (1993): 107–27.

Bowler, Peter J. *Science for All: The Popularization of Science in Early Twentieth-Century Britain*. Chicago: University of Chicago Press, 2009.

Brachet, Charles. "La Découverte Scientifique: Création Continue. Le 'Banc D'essais' De La Machine Humaine Au Palais De La Découverte." *Science et Vie* 241 (1937): 3–11.

Braidotti, Rosi. "The Critical Posthumanities; or, Is Medianatures to Naturecultures as Zoe Is to Bios?" *Cultural Politics* 12, no. 3 (2016): 380–90.

Brain, Robert Michael. *The Pulse of Modernism: Physiological Aesthetics in Fin-De-Siècle Europe*. Washington, DC: University of Washington Press, 2015.

Brenninkmeijer, Jonna. *Neurotechnologies of the Self: Mind, Brain and Subjectivity*. London: Palgrave Macmillan, 2016.

Brown, Barbara B. *New Mind, New Body; Bio-Feedback: New Directions for the Mind*. New York: Harper & Row, 1974.

Brown, Barbara B. "Recognition of Aspects of Consciousness through Association with the EEG Alpha Activity Represented by a Light Signal." *Psychophysiology* 6, no. 4 (1970): 442–52.

Brown, Harriet R., Peter Zeidman, Peter Smittenaar, Rick A. Adams, Fiona McNab, Robb B. Rutledge, and Raymond J. Dolan. "Crowdsourcing for Cognitive Science—The Utility of Smartphones." *Plos One* 9, no. 7 (2014): e100662.

Brusini, Hervé, and Francis James. *Voir La Vérité: Le Journalisme De Télévision*. Paris: Presses Universitaires de France, 1982.

Budzynski, Thomas. "Tuning in on the Twilight Zone." *Psychology Today* 11, no. 43 (1977): 38–44.

Bulwer-Lytton, Edward. "The Haunted and the Haunters, or, the House and the Brain." *Blackwell's Magazine*, August 1859.

Bundzen, P. V. "Autoregulation of Functional State of the Brain: An Investigation Using Photostimulation with Feedback." *Federation Proceedings. Translation Supplement; Selected Translations from Medical-Related Science* 25, no. 4 (1965): 551–4.

Burnham, Jack. "Real Time Systems." *Artforum* 8, no. 1 (1969): 49–55.

Bzdok, Danilo, and B. T. Thomas Yeo. "Inference in the Age of Big Data: Future Perspectives on Neuroscience." *NeuroImage* 155 (2017): 549–64.

Cacioppo, John T., and Gary G. Berntson. "Social Psychological Contributions to the Decade of the Brain: Doctrine of Multilevel Analysis." *American Psychologist* 47, no. 8 (1992): 1019–28.

Calfee, Kennard, Herbert B. Cahan, and Lynn Poole. "Catching a Brain Wave." In *Johns Hopkins Science Review*, 28 min. WAAM Station, DuMont Network, Johns Hopkins University, 1954.

Callard, Felicity, and Des Fitzgerald. "Entangling the Medical Humanities." In *The Edinburgh Companion to the Critical Medical Humanities*, edited by Anne Whitehead, Angela Woods, Sarah J. Atkinson, Jane Macnaughton, and Jennifer Richards, 35–49. Edinburgh: Edinburgh University Press, 2016.

Callard, Felicity, and Des Fitzgerald. *Rethinking Interdisciplinarity across the Social Sciences and Neurosciences*. New York: Palgrave Macmillan, 2015.

Canales, Jimena. "'A Number of Scenes in a Badly Cut Film': Observation in the Age of Strobe." In *Histories of Scientific Observation*, edited by Lorraine Daston and Elizabeth Lunbeck, 230–54. Chicago: University of Chicago Press, 2011.

Canales, Jimena. "Recording Devices." In *A Companion to the History of Science*, edited by Bernard Lightman, 500–14. Chichester: John Wiley, 2016.

Cartwright, Lisa. *Screening the Body: Tracing Medicine's Visual Culture*. Minneapolis: University of Minnesota Press, 1995.
Casper, Stephen T. "History and Neuroscience: An Integrative Legacy." *Isis* 105, no. 1 (2014): 123–32.
Casper, Stephen T., and Delia Gavrus. "Introduction. Technique, Technology, and Therapy in the Brain and Mind Sciences." In *The History of the Brain and Mind Sciences: Technique, Technology, Therapy*, edited by Stephen T. Casper and Delia Gavrus, 1–24. Rochester: University of Rochester Press, 2017.
Cazzamalli, Ferdinando. *Archivio generale di neurologia, psichiatria e psicoanalisi* 4 (1925): 215–33.
Charcot, Jean-Martin. "Faculté De Médecine De Paris: Anatomo-Pathologie Du Système Nerveux." *Progrès Médical*, no. 14 (1879).
Charcot, Jean-Martin. *Tome V. Maladies Des Poumons Et Du Système Vasculaire*. Oeuvres Complètes. Edited by Désiré-Magloire Bourneville. N.p.: Progrès médical, 1888.
Charlton, M. H., and Paul F. A. Hoefer. "Television and Epilepsy." *Archives of Neurology* 11, no. 3 (1964): 239–47.
Choudhury, Suparna, and Jan Slaby. "Introduction." In *Critical Neuroscience: A Handbook of the Social and Cultural Contexts of Neuroscience*, edited by Suparna Choudhury and Jan Slaby, 1–26. Oxford: Wiley Blackwell, 2011.
Chu, N. N. Y. "Brain-Computer Interface Technology and Development: The Emergence of Imprecise Brainwave Headsets in the Commercial World." *IEEE Consumer Electronics Magazine* 4, no. 3 (2015): 34–41.
Clark, J. H. "Adaptive Machines in Psychiatry." In *Progress in Brain Research*, edited by N. Wiener and J. P. Schadé, 224–35. Amsterdam: Elsevier, 1963.
Cobb, Matthew. *The Idea of the Brain: The Past and Future of Neuroscience*. New York: Basic Books, 2020.
Coen, Deborah R. *Vienna in the Age of Uncertainty: Science, Liberalism, and Private Life*. Chicago: University of Chicago Press, 2008.
Cohn, Simon. "Making Objective Facts from Intimate Relations: The Case of Neuroscience and Its Entanglements with Volunteers." *History of the Human Sciences* 21, no. 4 (2008): 86–103.
Colligan, Colette, and Margaret Linley. "Introduction: The Nineteenth-Century Invention of Media." In *Media, Technology, and Literature in the Nineteenth Century: Image, Sound, Touch*, edited by Colette Colligan and Margaret Linley, 1–21. Farnham: Ashgate, 2011.
Collopy, Peter. "The Revolution Will Be Videotaped: Making a Technology of Consciousness in the Long 1960s." PhD dissertation, University of Pennsylvania, 2015.
Collura, T. F. "History and Evolution of Electroencephalographic Instruments and Techniques." *Journal of Clinical Neurophysiology* 10, no. 4 (1993): 476–504.

Condon, William S. "Communication: Rhythm and Structure." In *Rhythm in Psychological, Linguistic and Musical Processes*, edited by J. R. Evans and M. Clynes, 55–78. Springfield, IL: Thomas, Charles C., 1986.

Condon, William S. "Cultural Microrhythms." In *Interaction Rhythms: Periodicity in Communicative Behavior*, edited by Martha Davis, 53–77. New York: Human Sciences Press, 1982.

Condon, William S. "Method of Micro-Analysis of Sound Films of Behavior." *Behavior Research Methods & Instrumentation* 2, no. 2 (1970): 51–4.

Condon, William S. "Sound-Film Microanalysis: A Means for Correlating Brain and Behavior." In *Dyslexia: A Neuroscientific Approach to Clinical Evaluation*, edited by F. H. Duffy and Norman Geschwind, 123–56. Boston: Little, Brown, 1985.

Condon, William S., and W. D. Ogston. "A Segmentation of Behavior." *Journal of Psychiatric Research* 5, no. 3 (1967): 221–35.

Cook, James W. *The Arts of Deception: Playing with Fraud in the Age of Barnum*. Cambridge, MA: Harvard University Press, 2001.

Cooter, Roger. "Neural Veils and the Will to Historical Critique: Why Historians of Science Need to Take the Neuro-Turn Seriously." *Isis* 105, no. 1 (2014): 145–54.

Cooter, Roger, and Stephen Pumfrey. "Separate Spheres and Public Places: Reflections on the History of Science Popularization and Science in Popular Culture." *History of Science* 32, no. 3 (1994): 237–67.

Cotterill, Rodney. *Enchanted Looms: Conscious Networks in Brains and Computers*. Cambridge: Cambridge University Press, 2000.

Couldry, Nick. "Liveness, 'Reality,' and the Mediated Habitus from Television to the Mobile Phone." *Communication Review* 7, no. 4 (2004): 353–61.

Craik, Kenneth J. W. *The Nature of Explanation*. London: Cambridge University Press, [1943] 1967.

Cramer, Florian. "What Is Interface Aesthetics, or What Could It Be (Not)?" In *Interface Critism: Aesthetics beyond Buttons*, edited by Christian Ulrik Andersen and Søren Bro Pold, 117–29. Aarhus: Aarhus University Press, 2011.

Crary, Jonathan. "Spectacle." In *New Keywords. A Revised Vocabulary of Culture and Society*, edited by Tony Bennett, Lawrence Grossberg, and Meaghan Morris, 335–37. Malden: Blackwell, 2005.

Crawford, Matthew B. "The Limits of Neurotalk." *New Atlantis* 19 (2008): 65–78.

Cressman, Dale L. "News in Lights: The Times Square Zipper and Newspaper Signs in an Age of Technological Enthusiasm." *Journalism History* 43, no. 4 (2018): 198–208.

Cromby, John. "Integrating Social Science with Neuroscience: Potentials and Problems." *BioSocieties* 2, no. 2 (2007): 149–69.

Crowley, Karlyn. *Feminism's New Age: Gender, Appropriation, and the Afterlife of Essentialism*. Albany, NY: SUNY Press, 2011.

Crown, Peter. Email communication, December 8, 2018.

Daly, Judith. "Close-Coupled Man/Machine Systems Research (Biocybernetics) Completion Report." Cybernetics Technology Division Program, DARPA, 1981.

Darrow, Chester W. "Psychological and Psychophysiological Significance of the Electroencephalogram." *Psychological Review* 54, no. 3 (1947): 157–68.

Daston, Lorraine J., and Peter Louis Galison. *Objectivity*. Cambridge, MA: MIT Press, 2007.

Daston, Lorraine, and Katharine Park. *Wonders and the Order of Nature, 1150–1750*. New York: Zone Books, 1998.

Daum, Andreas W. "Varieties of Popular Science and the Transformations of Public Knowledge: Some Historical Reflections." *Isis* 100, no. 2 (2009): 319–32.

Debray, Regis. *Media Manifestos: On the Technological Transmission of Cultural Forms*. London: Verso, 1996.

Dennett, Daniel. *Consciousness Explained*. Boston: Little, Brown, 1991.

Dewan, Edmond M. "Communication by Electroencephalography. Experiment at the Stanley Cobb Laboratories at Massachusetts General Hospital." *Air Force Cambridge Research Laboratories, United States Air Force* (1964): 1–6.

Didi-Huberman, Georges. *Invention De L'hystérie: Charcot Et L'iconographie Photographique De La Salpêtrière*. Paris: Macula, 1982.

Dietz, David. "Science and the American Press." *Science* 85, no. 2196 (1937): 107–12.

Dikker, Suzanne. *CMBC 2019 Workshop: Neuroscience in the Wild*. 2019. https://www.youtube.com/watch?v=ISeOWAAglQ0

Dikker, Suzanne. *Crowdsourcing Neuroscience*. 2014. https://www.youtube.com/watch?v=pyIs6Syvf_k. Interview with exhibition footage and brain activity animation.

Dikker, Suzanne. *Mutual Wave Machine*. 2013. http://www.suzannedikker.net/mutualwavemachine#background.

Dikker, Suzanne, Karina Kirnos, and Siena Oristaglio. "Neuroscience Exclusives. Digital Mai Exclusives Package." Marina Abramovic Institute, 2014. https://www.kickstarter.com/projects/maihudson/marina-abramovic-institute-the-founders/posts/764797.

Dikker, Suzanne, Georgios Michalareas, Matthias Oostrik, Amalia Serafimaki, Hasibe Melda Kahraman, Marijn E. Struiksma, and David Poeppel. "Crowdsourcing Neuroscience: Inter-Brain Coupling during Face-to-Face Interactions Outside the Laboratory." *NeuroImage* 227 (2021): 117436.

Dikker, Suzanne, Lauren J. Silbert, Uri Hasson, and Jason D. Zevin. "On the Same Wavelength: Predictable Language Enhances Speaker–Listener Brain-to-Brain Synchrony in Posterior Superior Temporal Gyrus." *Journal of Neuroscience* 34, no. 18 (2014): 6267–72.

Dikker, Suzanne, Suzan Tunca, and Sean Montgomery. "Using Synchrony-Based Neurofeedback in Search of Human Connectedness." In *Brain*

Art: Brain-Computer Interfaces for Artistic Expression, edited by Anton Nijholt, 161–206. Cham: Springer, 2019.

Dikker, Suzanne, Lu Wan, Ido Davidesco, Lisa Kaggen, Matthias Oostrik, James McClintock, Jess Rowland, et al. "Brain-to-Brain Synchrony Tracks Real-World Dynamic Group Interactions in the Classroom." *Current Biology* 27, no. 9 (2017): 1375–80.

Dodge, H. W., R. G. Bickford, A. A. Bailey, C. B. Holman, M. C. Petersen, and C. W. Sem-Jacobsen. "Technics and Potentialities of Intracranial Electrography." *Postgraduate Medicine* 15, no. 4 (1954): 291–300.

Donchin, Emanuel. "Appendix: The Relationship between P300 and the CNV. A Correspondence Conducted by Dr. Emanuel Donchin." In *The Responsive Brain*, edited by John Russell Knott and William Cheyne MacCallum. *The Proceedings of the Third International Congress of Event-Related Slow Potentials of the Brain. Bristol, England 12–18 August 1973*, 222–34. Bristol: John Wright and Sons, 1976.

Donchin, Emanuel. "Event-Related Brain Potentials: A Tool in the Study of Human Information Processing." In *Evoked Brain Potentials and Behavior. Proceedings of a Conference on Evoked Brain Potentials and Behavior Held at the Downstate Medical Center, State University of New York, Brooklyn, New York, May 19–21, 1977*, edited by Henri Begleiter, 13–88. New York: Plenum Press, 1979.

Draaisma, Douwe. "An Enchanted Loom." In *Metaphors of Memory: A History of Ideas about the Mind*, 185–210. Cambridge: Cambridge University Press, 2000.

Draaisma, Douwe. *Metaphors of Memory: A History of Ideas about the Mind*. Cambridge: Cambridge University Press, 2000.

Dror, Otniel E. "Techniques of the Brain and the Paradox of Emotions, 1880–1930." *Science in Context* 14, no. 4 (2001): 643–60.

Drucker, J. "Humanities Approaches to Interface Theory." *Culture Machine* 12 (2011): 1–20.

Druckery, Timothy. "Foreword." In *Deep Time of the Media: Toward an Archaeology of Hearing and Seeing by Technical Means*, edited by Siegfried Zielinski, vii–xi. Cambridge, MA: MIT Press, 2006.

Duane, T. D., and Thomas Behrendt. "Extrasensory Electroencephalographic Induction between Identical Twins." *Science* 150, no. 3694 (1965): 367.

Dumas, Guillaume. "Towards a Two-Body Neuroscience." *Communicative & Integrative Biology* 4, no. 3 (2011): 349–52.

Dumas, Guillaume, and L. Halard. *Phi (Subtitled)*. 2012. https://www.youtube.com/watch?v=PPbH6CMw2bU.

Dumas, Guillaume, Jacqueline Nadel, Robert Soussignan, Jacques Martinerie, and Line Garnero. "Inter-Brain Synchronization during Social Interaction." *Plos One* 5, no. 8 (2010): e12166.

Dumas, Guillaume, J. A. Scott Kelso, and Jacqueline Nadel. "Tackling the Social Cognition Paradox through Multi-Scale Approaches." *Frontiers in Psychology* 5 (2014): 882.

Dumit, Joseph. "How (Not) to Do Things with Brain Images." In *Representation in Scientific Practice Revisited*, edited by Catelijne Coopmans, Janet Vertesi, Michael E. Lynch, and Steve Woolgar, 291–313. Cambridge, MA: MIT Press, 2014.

Dumit, Joseph. "Objective Brains, Prejudicial Images." *Science in Context* 12, no. 1 (1999): 173–201.

Dumit, Joseph. *Picturing Personhood: Brain Scans and Biomedical Identity*. Princeton, NJ: Princeton University Press, 2004.

Dumit, Joseph. "Plastic Diagrams: Circuits in the Brain and How They Got There." In *Plasticity and Pathology: On the Formation of the Neural Subject*, edited by David Bates and Nima Bassiri, 219–67. New York: Fordham University Press, 2016.

Duncan Jr., Starkey. "Nonverbal Communication." *Psychological Bulletin* 72, no. 2 (1969): 118–37.

Dupree, Mary Helen, and Sean B. Franzel. "Introduction: Performing Knowledge, 1750–1850." In *Performing Knowledge, 1750–1850*, edited by Mary Helen Dupree and Sean B. Franzel, 1–24. Berlin: Walter de Gruyter GmbH, 2015.

Dupuy, Jean-Pierre. *On the Origins of Cognitive Science: The Mechanization of the Mind*. Cambridge, MA: MIT Press, 2009.

During, Simon. *Modern Enchantments*. Cambridge, MA: Harvard University Press, 2009.

Durup, G., and A. Fessard. "I. L'électrencéphalogramme De L'homme. Observations Psycho-Physiologiques Relatives À L'action Des Stimuli Visuels Et Auditifs." *L'année psychologique* 36, no. 1 (1935): 1–32.

Eckert Jr., J. P. "Continuous Variable Input and Output Devices." In *The Moore School Lectures; Theory and Techniques for Design of Electronic Digital Computers*, edited by Martin Campbell-Kelly and Michael R. Williams. The Charles Babbage Institute Reprint Series for the History of Computing, 393–423. Cambridge, MA: MIT Press, 1946. Reprint 1985.

Economo, Constantin Von. "Some New Methods for Studying Brains of Exceptional People (Encephalometry and Braincasts) (Presentation with Models and Demonstration at the New York Academy of Medicine, Section on Neurology, December 3, 1929)." *Journal of Nervous and Mental Disease* 72, no. 2 (1930): 125–34.

Edwards, Paul N. *The Closed World: Computers and the Politics of Discourse in Cold War America*. Cambridge, MA: MIT Press, 1997.

Eidelman, Jacqueline, and Odile Welfele. "Enseignement Supérieur Et Universités; Palais De La Découverte (1900–1978)." Archives Nationales, 1990.

Elder, Rachel. "Secrecy & Safety: A Cultural History of Seizures in Mid-Twentieth Century America." PhD dissertation, University of Pennsylvania, 2015.

Engell, Lorenz. "Fernsehen Mit Unbekannten. Uberlegungen Zur Experimentellen Television." In *Fernsehexperimente: Stationen Eines Mediums*, edited by Michael Grisko and Lorenz Engell, 15–46. Berlin: Kulturverlag Kadmos, 2009.

Enns, A., and S. Trower, eds. *Vibratory Modernism*. London: Palgrave Macmillan, 2013.

Enns, Anthony. "Psychic Radio: Sound Technologies, Ether Bodies and Spiritual Vibrations." *Senses and Society* 3, no. 2 (2008): 137–52.

Enns, Anthony. "Vibratory Photography." In *Vibratory Modernism*, edited by A. Enns and S. Trower, 177–97. London: Palgrave Macmillan, 2013.

Eribon, Didier. *Michel Foucault*. London: Faber & Faber, 1993.

Erlanger, Joseph, and Herbert Spencer Gasser. *Electrical Signs of Nervous Activity*. Philadelphia: University of Pennsylvania Press, 1937.

Ernst, Wolfgang. *Chronopoetics: The Temporal Being and Operativity of Technological Media*. London: Rowman & Littlefield, 2016.

Estrin, Thelma. "Neurophysiological Research Using a Remote Time-Shared Computer." In *Data Acquisition and Processing in Biology and Medicine: Proceedings of the 1966 Rochester Conference*, edited by Kurt Enslein, 117–35. Rochester, NY: Elsevier, 1968.

Estrin, Thelma. "On-Line Electroencephalographic Digital Computing System." *Electroencephalography and Clinical Neurophysiology* 19, no. 5 (1965): 524–6.

Exner, Robert. "Das Gehirnmodell." *Annalen des Naturhistorischen Museums in Wien* 59 (1953): 49–53.

Exner, Robert. "Das Leuchtende Gehirnmodell." *Psychiatrisch-Neurologische Wochenschrift* 35 (1933): 501–2.

Exner, Sigmund. *Untersuchungen Über Die Localisation Der Functionen in Der Grosshirnrinde Des Menschen*. Wien: W. Braumuller, 1881.

Ezios, Ralph. "Implications of Physiological Feedback Training." *Radical Software* 1, no. 4 (1971): 2–4.

Fahrion, Steven, and Patricia Norris. "Biofeedback & Self-Regulation." *Subtle Energies & Energy Medicine* 10, no. 1 (1999).

Falkner, Leonard. "The Sky Is His Blackboard." *American Magazine*, March 1931, 78–9.

Farah, Martha J. "Brain Images, Babies, and Bathwater: Critiquing Critiques of Functional Neuroimaging." *Hastings Center Report* 44, no. s2 (2014): S19–30.

Fehmi, Lester G. "EEG Biofeedback, Multichannel Synchrony Training and Attention." In *Expanding Dimensions of Consciousness*, edited by A. Arthur Sugerman and Ralph E. Tarter, 155–82. New York: Springer, 1978.

Felt, Ulrike. "Science and Its Public: Popularization of Science in Vienna 1900–1938." In *3th International Conference on Public Communication of Science and Technology*, Montreal, Canada, 10–13 April 1994. https://pcst.co/archive/pdf/Felt_PCST1994.pdf.

Fessard, A. "Les Ondes Bioélectriques. De La Décharge Du Poisson-Torpille Aux Oscillations Electriques Du Cerveau Humain." In *La Biologie: Exposition Internationale De Paris 1937*, edited by R. Bonnardel, 57–70. Paris: Palais de la Découverte, 1937.

Feuer, Jane. "The Concept of Live Television: Ontology as Ideology." In *Regarding Television: Critical Approaches—An Anthology*, edited by E. Ann Kaplan. American Film Institute Monograph Series, 12–22. Frederick, MD: University Publications of America, 1983.

Fine, Cordelia. *Delusions of Gender: How Our Minds, Society, and Neurosexism Create Difference*. New York: W. W. Norton, 2011.

Finnegan, Diarmid A. "Lectures." In *A Companion to the History of Science*, edited by Bernard Lightman, 414–27. Chichester: John Wiley, 2016.

Fitsch, Hannah. *Dem Gehirn beim Denken zusehen?: Sicht- und Sagbarkeiten in der funktionellen Magnetresonanztomographie*. Bielefeld: Transcript Verlag, 2014.

Fitzgerald, Des, Svenja Matusall, Joshua Skewes, and Andreas Roepstorff. "What's So Critical about Critical Neuroscience? Rethinking Experiment, Enacting Critique." *Frontiers in Human Neuroscience* 8 (2014): 1–12.

Foucault, Michel. *The Birth of the Clinic: An Archeology of Medical Perception*. London: Routledge, 1976.

Foucault, Michel. *Le Pouvoir Psychiatrique: Cours Au Collège De France, 1973–1974*. Paris: Hautes Études. EHESS/Gallimard/Seuil, 2003.

Foucault, Michel. *Psychiatric Power: Lectures at the Collège De France, 1973–1974*. New York: St. Martins' Press, 2008.

Foucault, Michel. *Technologies of the Self: A Seminar with Michel Foucault*. Amherst: University of Massachusetts Press, 1988.

Foucault, Michel. *Ethics: Subjectivity and Truth*. Edited by Paul Rabinow. New York: New Press, 1997.

Frank, Robert G. "Instruments, Nerve Action, and the All-or-None Principle." *Osiris* 9 (1994): 208–35.

Frazzetto, G., and S. Anker. "Neuroculture." *Nature Reviews Neuroscience* 10, no. 11 (2009): 815–21.

Frégnac, Yves. "Big Data and the Industrialization of Neuroscience: A Safe Roadmap for Understanding the Brain?" *Science* 358, no. 6362 (2017): 470–7.

Gabrys, Jennifer. "Atmospheres of Communication." In *The Wireless Spectrum: The Politics, Practices, and Poetics of Mobile Media*, edited by Barbara Crow, Michael Longford, and Kim Sawchuk, 46–59. Toronto: University of Toronto Press, 2010.

Galili, Doron. "Television from Afar: Arnheim's Understanding of Media." In *Arnheim for Film and Media Studies*, edited by Scott Higgins and Doron Galili, 195–211. New York: Routledge, 2011.

Galison, P. "The Ontology of the Enemy: Norbert Wiener and the Cybernetic Vision." *Critical Inquiry* 21, no. 1 (1994): 228–66.

Gallagher, Shaun. "Decentering the Brain: Embodied Cognition and the Critique of Neurocentrism and Narrow-Minded Philosophy of Mind." *Constructivist Foundations* 14, no. 1 (2018): 8–21.

Gallagher, Shaun, Daniel D. Hutto, Jan Slaby, and Jonathan Cole. "The Brain as Part of an Enactive System." *Behavioral and Brain Sciences* 36, no. 4 (2013): 421–2.

Gardner, John. "A History of Deep Brain Stimulation: Technological Innovation and the Role of Clinical Assessment Tools." *Social Studies of Science* 43, no. 5 (2013): 707–28.

Gardner, Paula, and Britt Wray. "From Lab to Living Room: Transhumanist Imaginaries of Consumer Brain Wave Monitors." *ADA. Journal of Gender, New Media and Technology*, no. 3 (2013). https://adanewmedia.org/2013/11/issue3-gardnerwray/.

Garson, Justin. "The Birth of Information in the Brain: Edgar Adrian and the Vacuum Tube." *Science in Context* 28, no. 1 (2015): 31–52.

Gaspart, Alfred W. "L'oscillation Cellulaire Possède-T-Elle Une Vertu Thérapeutique?" *L'Homme Libre*, November 4, 1931, 2.

Gastaut, H. A., C. Jus, F. Morrell, W. Storm Van Leeuwen, S. Dongier, R. Naquet, H. Regis, et al. "Étude Topographique Des Réactions Électroencéphalographiques Conditionnées Chez L'homme: Essai D'interprétation Neurophysiologique." *Electroencephalography and Clinical Neurophysiology* 9, no. 1 (1957): 1–34.

Gastaut, H., H. Regis, and F. Bostem. "Attacks Provoked by Television, and Their Mechanism." *Epilepsia* 3, no. 3 (1962): 438–45.

Gawrylewki, Andrea, Jennifer Leman, and Liz Tormes. "The Brain in Images: Top Entries in the Art of Neuroscience." *Scientific American* (2019).

Gerard, Ralph W. "Brain Waves." *Scientific Monthly* 44, no. 1 (1937): 48–56.

Gerard, Ralph W. "Some of the Problems Concerning Digital Notions in the Central Nervous System." In *Cybernetics—Kybernetik. The Macy-Conferences 1946-1953. 2: Dokumente Und Reflexionen*, edited by Claus Pias, 171–202. Zurich: Diaphanes Verlag, 2004.

Gere, Charlie. *Art, Time and Technology*. Oxford: Berg, 2006.

Gerschenfeld, Ana. "John Krakauer: 'We're in the Grips of a Totalizing Belief in Data and Techniques.'" *AR Magazine*, November 23, 2017.

Ghaziri, Jimmy, and Robert T. Thibault. "Neurofeedback: An inside Perspective." In *Casting Light on the Dark Side of Brain Imaging*, edited by Amir Raz and Robert T. Thibault, 113–16. Cambridge, MA: Academic Press, 2019.

Gibbs, Frederic, and Erna Gibbs. *Atlas of Electroencephalography*. Cambridge, MA: Addison-Wesley Press, 1941.

Gibbs, Frederic, and Erna Gibbs. *Atlas of Electroencephalography*. Cambridge, MA: Cummings, 1941.
Gieryn, Thomas. "Boundary-Work and the Demarcation of Science from Non-Science: Strains and Interests in Professional Ideologies of Scientists." *American Sociological Review* 48, no. 6 (1983): 781–95.
Giese, Fritz. *Psychotechnik*. Breslau: Ferdinand Hirt, 1928.
Gigerenzer, Gerd. "From Tools to Theories: A Heuristic of Discovery in Cognitive Psychology." *Psychological Review* 98, no. 2 (1991): 254–67.
Gijswijt-Hofstra, Marijke, and Roy Porter. *Cultures of Neurasthenia from Beard to the First World War*. Amsterdam: Rodopi, 2001.
Gillespie, R. D. "The Present Status of the Concepts of Nervous and Mental Energy." *British Journal of Psychology. General Section* 15, no. 3 (1925): 266–79.
Gilman, Sander L. "The Image of the Hysteric." In *Hysteria beyond Freud*, edited by Sander L. Gilman, 345–436. Berkeley: University of California Press, 1993.
Gilman, Sander L. *Seeing the Insane*. Lincoln: University of Nebraska Press, 1996.
Gitelman, Lisa. *Always Already New: Media, History, and the Data of Culture*. Cambridge, MA: MIT Press, 2006.
Glazer, Charles H. *Dialogue on Biofeedback*. 1974. https://www.youtube.com/watch?v=3gGQF2ItH8c.
Goldensohn, E. S. "Simultaneous Recording of EEG and Clinical Seizures Using Kinescope." *Electroencephalography and Clinical Neurophysiology* 21 (1966): 623.
Goschler, Juliana. "Metaphors in Cognitive and Neurosciences: Which Impact Have Metaphors on Scientific Theories and Models?" *Metaphorik* 12 (2007): 7–20. http://www.metaphorik.de/12/goschler.pdf.
Grass, Albert M., and Frederic A. Gibbs. "A Fourier Transform of the Electroencephalogram." *Journal of Neurophysiology* 1, no. 6 (1938): 521–6.
Green, Elmer, Alyce M. Green, and E. Dale Walters. "Self-Regulation of Internal States." *Proceedings of the International Congress of Cybernetics 1969* (1970): 1299–316.
Green, Elmer, Alyce M. Green, and E. Dale Walters. "Voluntary Control of Internal States: Psychological and Physiological." *Journal of Transpersonal Psychology* 2, no. 1 (1970): 1–26.
Grossberg, John M. "Brain Wave Feedback Experiments and the Concept of Mental Mechanisms." *Journal of Behavior Therapy and Experimental Psychiatry* 3, no. 4 (1972): 245–51.
Grudin, Jonathan. "Three Faces of Human-Computer Interaction." *IEEE Annals of the History of Computing* 27, no. 4 (2005): 46–62.
Grusin, Richard. "Radical Mediation." *Critical Inquiry* 42, no. 1 (2015): 124–48.

Guenther, Katja M. *Localization and Its Discontents: A Genealogy of Psychoanalysis and the Neuro Disciplines*. Chicago: University of Chicago Press, 2015.

Guillory, John. "Genesis of the Media Concept." *Critical Inquiry* 36, no. 2 (2010): 321–62.

Gunning, Tom. "Re-Newing Old Technologies: Astonishment, Second Nature, and the Uncanny in Technology from the Previous Turn-of-the-Century." In *Rethinking Media Change: The Aesthetics of Transition*, edited by David Thorburn and Henry Jenkins, 39–60. Cambridge, MA: MIT Press, 2003.

Gunning, Tom. "Uncanny Reflections, Modern Illusions: Sighting the Modern Optical Uncanny." In *Uncanny Modernity; Cultural Theories, Modern Anxieties*, edited by John Jervis, 68–90. London: Palgrave Macmillan, 2008.

Gunning, Tom. "The World as Object Lesson: Cinema Audiences, Visual Culture and the St. Louis World's Fair, 1904." *Film History* 6, no. 4 (1994): 422–44.

Hadler, Florian, and Daniel Irrgang. "Instant Sensemaking, Immersion and Invisibility. Notes on the Genealogy of Interface Paradigms." *Punctum. International Journal of Semiotics* 1, no. 1 (2015): 7–25.

Hagner, Michael. "Bilder Der Kybernetik: Diagramm Und Anthropologie, Schaltung Und Nervensystem." In *Konstruierte Sichtbarkeiten: Wissenschafts- Und Technikbilder Seit Der Frühen Neuzeit*, edited by Martina Hessler, 383–404. Munchen: Wilhelm Fink Verlag, 2006.

Hagner, Michael. "Das Kybernetische Gehirn." In *Geniale Gehirne: Zur Geschichte Der Elitegehirnforschung*, 288–96. Göttingen: Wallstein Verlag, 2004.

Hagner, Michael. *Der Geist Bei Der Arbeit: Historische Untersuchungen Zur Hirnforschung*. Göttingen: Wallstein, 2006.

Hagner, Michael. "Der Geist Bei Der Arbeit. Überlegungen Zur Visuellen Repräsentation Cerebraler Prozesse." In *Anatomien Medizinischen Wissens: Medizin- Macht- Moleküle*, edited by Cornelius Borck and Susan M. DiGiacomo, 259–86. Frankfurt am Main: Fischer-Taschenbuch-Verlag, 1996.

Hagner, Michael. *Geniale Gehirne: Zur Geschichte Der Elitegehirnforschung*. Göttingen: Wallstein Verlag, 2004.

Hagner, Michael. "Lokalisation, Function, Cytoarchitektonik. Wege Zur Modellierung Des Gehirns." In *Objekte, Differenzen Und Konjunkturen: Experimentalsysteme Im Historischen Kontext*, edited by Hans-Jörg Rheinberger, Bettina Wahrig-Schmidt, and Michael Hagner, 121–50. Berlin: Akademie Verlag, 1994.

Hagner, Michael. "The Mind at Work: The Visual Representation of Cerebral Processes." In *The Body Within: Art, Medicine and Visualization*, edited by Renée van de Vall and Robert Zwijnenberg, 67–105. Leiden: Brill, 2009.

Hagner, Michael. "Mind Reading, Brain Mirror, Neuroimaging: Insight into the Brain or the Mind." In *Psychology's Territories: Historical and Contemporary*

Perspectives from Different Disciplines, edited by Mitchell Ash and Thomas Sturm, 287–304. London: Routledge, 2007.

Hagner, Michael, and Cornelius Borck. "Brave Neuro Worlds." In *Der Geist Bei Der Arbeit: Historische Untersuchungen Zur Hirnforschung*, edited by Michael Hagner, 7–34. Göttingen: Wallstein, 2006.

Hallinan, Dara, Philip Schütz, Michael Friedewald, and Paul Hert. "Neurodata and Neuroprivacy: Data Protection Outdated?" *Surveillance and Society* 12, no. 1 (2014): 55–72.

Hamery, Roxane. *Jean Painlevé, Le Cinéma Au Cœur De La Vie*. Rennes: Presses Universitaires de Rennes, 2013.

Hanes, R. M. "A Scale of Subjective Brightness." *Journal of Experimental Psychology* 39, no. 4 (1949): 438–52.

Hansen, Mark. "Ubiquitous Sensation: Towards an Atmospheric, Collective and Microtemporal Model of Media." In *Throughout: Art and Culture Emerging with Ubiquitous Computing*, 63–88. Cambridge, MA: MIT Press, 2013.

Haraway, Donna Jeanne. *The Companion Species Manifesto: Dogs, People, and Significant Otherness*, Vol. 1, Chicago: Prickly Paradigm Press, 2003.

Haraway, Donna Jeanne. *Modest−Witness@Second−Millennium. Femaleman−Meets−Oncomouse: Feminism and Technoscience*. New York: Routledge, 1997.

Harbou, Thea von. *Metropolis*. Fankfurt/M: Verlag Ullstein GmbH, [1926] 1984.

Hari, Riitta, Linda Henriksson, Sanna Malinen, and Lauri Parkkonen. "Centrality of Social Interaction in Human Brain Function." *Neuron* 88, no. 1 (2015): 181–93.

Hari, Riitta, and Miiamaaria V. Kujala. "Brain Basis of Human Social Interaction: From Concepts to Brain Imaging." *Physiological Reviews* 89, no. 2 (2009): 453–79.

Harrell, Thomas W. "Proceedings of the Forty-Second Annual Meeting of the Western Psychological Association." *American Psychologist* 17, no. 9 (1962): 597–606.

Harrington, Anne. "The Brain and the Behavioral Sciences." In *The Cambridge History of Science: Volume 6, The Modern Biological and Earth Sciences*, edited by Peter J. Bowler and John V. Pickstone, 504–23. Cambridge: Cambridge University Press, 2009.

Harrington, Anne. *The Cure Within: A History of Mind-Body Medicine*. New York: W. W. Norton, 2009.

Harrington, Anne. "Kurt Goldstein's Neurology of Healing and Wholeness: A Weimar Story." In *Greater Than the Parts: Holism in Biomedicine*, 25–45. Oxford: Oxford University Press, 1998.

Harrington, Anne. *Medicine, Mind, and the Double Brain: A Study in Nineteenth-Century Thought*. Princeton, NJ: Princeton University Press, 1987.

Harrington, Anne. "Metaphoric Connections: Holistic Science in the Shadow of the Third Reich." *Social Research* 62, no. 2 (1995): 357–85.

Harrington, Anne. *Reenchanted Science: Holism in German Culture from Wilhelm II to Hitler*. Princeton, NJ: Princeton University Press, 1999.
Hart, Joseph T. "Autocontrol of EEG Alpha." *Psychophysiology* 4, no. 4 (1968): 506.
Harter, M. Russell. "Excitability Cycles and Cortical Scanning: A Review of Two Hypotheses of Central Intermittency in Perception." *Psychological Bulletin* 68, no. 1 (1967): 47–58.
Hartmann, Dirk. "Neurophysiology and Freedom of the Will." *Poiesis & Praxis* 2 (2001): 275–84.
Hartzell, Emily, and Nina Sobell. "Sculpting in Time and Space: Interactive Work." *Leonardo* 34, no. 2 (2001): 101–7.
Haueis, Philip, and Jan Slaby. "Connectomes as Constitutiveley Epistemic Objects: Critical Perspectives on Modeling in Current Neuroanatomy." *Progress in Brain Research* 233 (2017): 149–77.
Haupt, Sabine. "'Traumkino.' Die Visualisierung Von Gedanken: Zur Intermedialität Von Neurologie, Optischen Medien Und Literatur." In *Das Unsichtbare Sehen*, edited by Sabine Haupt and Ulrich Stadler. Edition Voldemeer, 87–125. Vienna: Springer, 2006.
Hayles, N. Katherine. *How We Became Posthuman: Virtual Bodies in Cybernetics, Literature, and Informatics*. Chicago: University of Chicago Press, 1999.
Hayles, N. Katherine. *How We Think: Digital Media and Contemporary Technogenesis*. Chicago: University of Chicago Press, 2012.
Hayles, N. Katherine. "Komplexe Zeitstrukturen Lebendiger Und Technischer Wesen." In *Die Technologische Bedingung. Beiträge Zur Beschreibung Der Technischen Welt*, edited by Erich Hörl, 193–228. Frankfurt am Main: Suhrkamp, 2011.
Hayles, N. Katherine. "Unfinished Work: From Cyborg to Cognisphere." *Theory, Culture & Society* 23, nos. 7–8 (2006): 159–66.
Hayles, N. Katherine. *Unthought: The Power of the Cognitive Nonconscious*. Chicago: University of Chicago Press, 2017.
Hayward, Rhodri. "'Our Friends Electric': Mechanical Models of Mind in Postwar Britain." In *Psychology in Britain: Historical Essays and Personal Reflections*, edited by G. C. Bunn, A. D. Lovie, and Graham Richards, 290–307. Leicester: British Psychological Society, 2001.
Head, Henry. "Aphasia: An Historical Review: The Hughlings Jackson Lecture for 1920." *Proceedings of the Royal Society of Medicine; Section of Neurology* 14 (1921): 390–411.
Helmreich, Stefan. "Potential Energy and the Body Electric: Cardiac Waves, Brain Waves, and the Making of Quantities into Qualities." *Current Anthropology* 54, no. S7 (2013): S139–48.
Henahan, Donal. "Music Draws Strains Direct from Brains." *New York Times*, November 25, 1970, 24.
Henderson, Victor W. "Alexia and Agraphia." *Neurology* 70, no. 5 (2008): 391–400.

Hickman, Robert J. "A Neurocybernetic Approach to Man-Machine Communication." *Proceedings of the San Diego Biomedical Symposium* 16, no. 1 (1977): 475–85.

Higgins, Hannah, and Douglas Kahn. *Mainframe Experimentalism: Early Computing and the Foundations of the Digital Arts.* Berkeley: University of California Press, 2012.

Hilgartner, Stephen. "The Dominant View of Popularization: Conceptual Problems, Political Uses." *Social Studies of Science* 20, no. 3 (1990): 519–39.

Hilgartner, Stephen. *Science on Stage: Expert Advice as Public Drama.* Stanford, CA: Stanford University Press, 2000.

Hilton, Warren. *Applied Psychology: Initiative Psychic Energy. Being the Sixth of a Series of Twelve Volumes on the Applications of Psychology to the Problems of Personal and Business Efficiency.* New York: The Literary Digest for the Society of Applied Psychology, 1919.

Hilton, Warren. *Applied Psychology: Making Your Own World: Being the Second of a Series of Twelve Volumes on the Applications of Psychology to the Problems of Personal and Business Efficiency.* New York: The Literary Digest for the Society of Applied Psychology, 1920.

Hilton, Warren. *Applied Psychology: Mind Mechanism. Being the Eighth of a Series of Twelve Volumes on the Applications of Psychology to the Problems of Personal and Business Efficiency.* New York: The Literary Digest for the Society of Applied Psychology, 1920.

Hilton, Warren. *Applied Psychology: Psychology and Achievement. Being the First of a Series of Twelve Volumes on the Applications of Psychology to the Problems of Personal and Business Efficiency.* New York: The Literary Digest for the Society of Applied Psychology, 1919.

Hoffmann, Christoph. "Helmholtz's Apparatuses. Telegraphy as Working Model of Nerve Physiology." *Philosophia Scientiæ* 7, no. 1 (2003): 129–49.

Hookway, Branden. *Interface.* Cambridge, MA: MIT Press, 2014.

Huhtamo, Erkki. "Monumental Attractions: Toward an Archaeology of Public Media Interfaces." In *Interface Critism: Aesthetics beyond Buttons*, edited by Christian Ulrik Andersen and Soren Bro Pold, 19–41. Aarhus: Aarhus University Press, 2011.

Huhtamo, Erkki, and Jussi Parikka. "Introduction: An Archaeology of Media Archaeology." In *Media Archaeology: Approaches, Applications, and Implications*, edited by Erkki Huhtamo and Jussi Parikka, 1–21. Berkeley: University of California Press, 2011.

Jacyna, L. S. *Lost Words: Narratives of Language and the Brain, 1825–1926.* Princeton, NJ: Princeton University Press, 2009.

Jacyna, L. S. "Questions of Identity: Science, Aesthetics, and Henry's Head." In *Greater Than the Parts: Holism in Biomedicine, 1920–1950*, edited by George Weisz and Christopher Lawrence, 211–33. Oxford: Oxford University Press, 1998.

Jakob, A. "Die Lokalisation Im Grosshirn." *Klinische Wochenschrift* 10, no. 44 (1931): 2025–30.

Jastrow, Robert. *Enchanted Loom: The Mind in the Universe*. New York: Simon & Schuster Trade, 1981.

Jefferson, Geoffrey. "The Mind of Mechanical Man." *British Medical Journal* 1, no. 4616 (1949): 1105–10.

Joselit, David. *Feedback: Televison against Democracy*. Cambridge, MA: MIT Press, 2007.

Joseph, Branden W. "Biomusic." *Grey Room* 45 (2011): 128–50.

Josephson-Storm, Jason A. *The Myth of Disenchantment: Magic, Modernity, and the Birth of the Human Sciences*. Chicago: University of Chicago Press, 2017.

Jülich, Solveig. "Media as Modern Magic: Early X-Ray Imaging and Cinematography in Sweden." *Early Popular Visual Culture* 6, no. 1 (2008): 19–33.

Kaempffert, Waldemar. "The Week in Science: The New 'Electrical Thinking'; Activity of the Brain Recorded on a Tape by the Delicate Electroencephalogram." *New York Times*, April 21, 1935, 16.

Kaerlein, Timo. "Presence in a Pocket. Phantasms of Immediacy in Japanese Mobile Telepresence Robotics." *communication+1* 1, no. 1 (2012): 1–24.

Kahn, Douglas. *Earth Sound Earth Signal. Energies and Earth Magnitude in the Arts*. Berkeley: University of California Press, 2013.

Kahn, Fritz. "Wie Arbeitet Das Gehirn?" *UHU* 11 (1928): 34–40.

Kaiser, David, and W. Patrick McCray, eds. *Groovy Science: Knowledge, Innovation, and American Counterculture*. Chicago: University of Chicago Press, 2016.

Kaizen, William. "Steps to an Ecology of Communication: Radical Software, Dan Graham, and the Legacy of Gregory Bateson." *Art Journal* 67, no. 3 (2008): 86–106.

Kamiya, Joe. "Conscious Control of Brain Waves." *Psychology Today* 1 (1968): 56–60.

Kamiya, Joe. "The First Communications about Operant Conditioning of the EEG." *Journal of Neurotherapy* 15, no. 1 (2011): 65–73.

Kamp, A., C. W. Sem-Jacobsen, W. Storm van Leeuwen, and L. H. van der Tweel. "Cortical Responses to Modulated Light in the Human Subject." *Acta Physiologica Scandinavica* 48, no. 1 (1960): 1–12.

Kantor, Robert E., and Dean Brown. "On-Line Computer Augmentation of Bio-Feedback Processes." *International Journal of Bio-Medical Computing* 1, no. 4 (1970): 265–75.

Kasperowski, Dick, and Thomas Hillman. "The Epistemic Culture in an Online Citizen Science Project: Programs, Antiprograms and Epistemic Subjects." *Social Studies of Science* 48, no. 4 (2018): 564–88.

Kay, L. E. "From Logical Neurons to Poetic Embodiments of Mind: Warren S. Mcculloch's Project in Neuroscience." *Science in Context* 14, no. 4 (2001): 591–614.

Keilbach, Judith, and Markus Stauff. "When Old Media Never Stopped Being New. Television's History as an Ongoing Experiment." In *After the Break: Television Theory Today*, edited by Marijke de Valck and Jan Teurlings, 79–98. Amsterdam: Amsterdam University Press, 2013.

Kember, Sarah. "Doing Technoscience as ('New') Media." In *Media and Cultural Theory*, edited by James Curran and David Morley, 235–49. London: Routledge, 2007.

Kennedy, Helen, Thomas Poell, and José van Dijck. "Data and Agency." *Big Data & Society* 2, no. 2 (2015): 1–7.

Kevles, Bettyann H. *Naked to the Bone: Medical Imaging in the Twentieth Century*. New Brunswick, NJ: Rutgers University Press, 1998.

Killen, Andreas. *Berlin Electropolis: Shock, Nerves, and German Modernity*. Berkeley: University of California Press, 2006.

Killen, Andreas. "Homo Pavlovius: Cinema, Conditioning, and the Cold War Subject." *Grey Room* 45 (2011): 42–59.

Kirsch, Robert. "Biofeedback: In the Beginning Was Alpha." *Los Angeles Times*, August 18, 1974, 64.

Klapetek, J. "Photogenic Epileptic Seizures Provoked by Television." *Electroencephalography and Clinical Neurophysiology* 11, no. 4 (1959): 809.

Klemperer, Edith. "Anatomical Model." Patent US1951422 A, filed October 29, 1931, and issued March 20, 1934.

Klemperer, Edith. "Demonstration: Das Gehirnmodell, Ein Plastischer Beleuchteter Unterrichtsbehelf Zur Darstellung Der Einzelnen Funktionen." *Zentralblatt fur die gesamte Neurologie und Psychiatrie* 61 (1932): 499.

Kline, Ronald R. *The Cybernetics Moment: Or Why We Call Our Age the Information Age*. Baltimore, MD: Johns Hopkins University Press, 2015.

Kluitenberg, Eric. "On the Archeology of Imaginary Media." In *Media Archaeology: Approaches, Applications, and Implications*, edited by Erkki Huhtamo and Jussi Parikka, 48–69. Berkeley: University of California Press, 2011.

Knott, John Russell, and William Cheyne McCallum. "An Experimental Approach to Brain Slow Potentials." In *The Responsive Brain*, edited by John Russell Knott and William Cheyne McCallum. *The Proceedings of the Third International Congress of Event-Related Slow Potentials of the Brain. Bristol, England 12–18 August 1973*, 211–21. Bristol: John Wright and Sons, 1976.

Knott, John Russell, and William Cheyne MacCallum. *The Responsive Brain: Proceedings of the Third International Congress of Event-Related Slow Potentials of the Brain. Bristol, England 12–18 August 1973*. Bristol: John Wright and Sons, 1976.

Koike, Takahiko, Hiroki C. Tanabe, and Norihiro Sadato. "Hyperscanning Neuroimaging Technique to Reveal the 'Two-in-One' System in Social Interactions." *Neuroscience Research* 90 (2015): 25–32.

Kornfeld, Donald S., and Lawrence C. Kolb. "The Use of Closed-Circuit Television in the Teaching of Psychiatry." *Journal of Nervous and Mental Disease* 138, no. 5 (1964): 452.

Kornhuber, Hans H., and Lüder Deecke. "Hirnpotentialänderungen Bei Willkürbewegungen Und Passiven Bewegungen Des Menschen: Bereitschaftspotential Und Reafferente Potentiale." *Pflüger's Archiv für die gesamte Physiologie des Menschen und der Tiere* 284, no. 1 (1965): 1–17.

Kornmüller, Alois. "Signalisierung Der Langsamen Wellen Des EEG Im Sauerstoffmangel." MPG-A III/16/41, 1945.

Kovacevic, Natasha, Petra Ritter, William Tays, Sylvain Moreno, and Anthony Randal McIntosh. "'My Virtual Dream': Collective Neurofeedback in an Immersive Art Environment." *Plos One* 10, no. 7 (2015): e0130129.

Kracauer, Siegfried. "Kosmos Der Wissenschaften-Konglomerat Der Kunste." *Das Werk; Schweizer Monatschrift für Architektur, Freie Kunst, Angewandte Kunst. Offizielles Organ des Bundes Schweizer Architekten BSA und des Schweizerischen Werkbundes SWB* 25, no. 1 (1938): 21–4.

Kracauer, Siegfried. *The Salaried Masses: Duty and Distraction in Weimar Germany*. London: Verso, [1930] 1998.

Krakauer, John W., Asif A. Ghazanfar, Alex Gomez-Marin, Malcolm A. MacIver, and David Poeppel. "Neuroscience Needs Behavior: Correcting a Reductionist Bias." *Neuron* 93, no. 3 (2017): 480–90.

Krämer, Sybille. "Was Hat 'Performativität' Und 'Medialität' Miteinander Zu Tun? Plädoyer Für Eine in Der 'Aisthetisierung' Gründende Konzeption Des Performativen. Zu Einführung in Diesen Band." In *Performativität Und Medialität*, edited by Sybille Krämer, 13–32. München: Fink, 2004.

Krohs, Ulrich. "Convenience Experimentation." *Studies in History and Philosophy of Science Part C: Studies in History and Philosophy of Biological and Biomedical Sciences* 43, no. 1 (2012): 52–7.

Kutas, M., G. McCarthy, and E. Donchin. "Augmenting Mental Chronometry: The P300 as a Measure of Stimulus Evaluation Time." *Science* 197, no. 4305 (1977): 792–5.

Kwint, Marius, and Richard Wingate. "Curating the Brain." *Interdisciplinary Science Reviews* 38, no. 3 (2013): 195–9.

Labadié, Jean. "Au Congrès Des 'Ondes Courtes.'" *L'Illustration*, no. 4930 (1937): 565.

Labadié, Jean. "Que Savons-Nous Des Ondes Électriques Émises Par Le Cerveau?" *Science et Vie* (1937): 217–24.

LaFollette, Marcel C. *Science on American Television: A History*. Chicago: University of Chicago Press, 2013.

LaFollette, Marcel C. "A Survey of Science Content in U.S. Television Broadcasting, 1940s through 1950s: The Exploratory Years." *Science Communication* 24, no. 1 (2002): 34–71.

Lallemand, François. *Recherches Anatomico-Pathologiques Sur L'encéphale Et Ces Dépendances*. Paris: Imprimerie de Baudouin Frères, 1820.

Lancel, Karen, and Hermen Maat. "E.E.G. Kiss." https://www.lancelmaat.nl/work/e.e.g-kiss/.

Lasareff, P. "The Theory of Nervous Activity." *Science* 59, no. 1530 (1924): 369–71.

Lashley, K. S. "Studies of Cerebral Function in Learning. IV. Vicarious Function in Destruction of the Visual Areas." *American Journal of Psychology* 59, no. 1 (1922): 44–71.

Lashley, Karl Spencer. "Basic Neural Mechanisms in Behavior." *Psychological Review* 37, no. 1 (1930): 1–24.

Lashley, Karl Spencer. *Brain Mechanisms and Intelligence: A Quantitative Study of Injuries to the Brain*. Chicago, IL: University of Chicago Press, [1929] 1989.

Lashley, Karl Spencer. "Functional Determinants of Cerebral Localization." *Archives of Neurology & Psychiatry* 38, no. 2 (1937): 371–87.

Lashley, Karl Spencer. "Integrative Functions of the Cerebral Cortex." *Physiological Reviews* 13, no. 1 (1933): 1–42.

Latour, Bruno. "From Fabrication to Reality. Pasteur and His Lactic Acid Ferment." In *Pandora's Hope: Essays on the Reality of Science Studies*, edited by Bruno Latour, 113–44. Cambridge, MA: Harvard University Press, 1999.

Latour, Bruno. *The Pasteurization of France*. Cambridge, MA: Harvard University Press, 1993.

Latour, Bruno. *Science in Action: How to Follow Scientists and Engineers through Society*, rev. ed. Cambridge, MA: Harvard University Press, 1988.

Latzko, Johann, and Otto Plechl. "Leuchtschaltbild." Bbc Ag Oesterr, 1928.

Latzko, Johann, and Otto Plechl. "Living Diagram." Bbc Brown Boveri & Cie, 1930.

Laurence, William. "Brain Records for Diagnosis." *New York Times*, May 15, 1936, 10.

Laurence, William. "Electricity in the Brain Records a Picture of Action of Thought." *New York Times*, April 14, 1935, 1, 32.

Law, John. *Aircraft Stories: Decentering the Object in Technoscience*. Durham, NC: Duke University Press, 2002.

Lawrence, Jodi. *Alpha Brain Waves*. New York: Avon Books, 1972.

Lee, Pamela M. *Chronophobia: On Time in the Art of the 1960s*. Cambridge, MA: MIT Press, 2004.

Leff, H. Stephen. "A Case Study of Scientists' Opinions about the Regulations of Their Work: Opinions of the Members of the Society for Psychophysiological Research about the Regulation of Biofeedback Research and Technology." *Psychophysiology* 10, no. 5 (1973): 536–43.

Lennox, W. G., E. L. Gibbs, and F. A. Gibbs. "Inheritance of Cerebral Dysrhythmia and Epilepsy." *Archives of Neurology & Psychiatry* 44, no. 6 (1940): 1155–83.

Lenoir, Timothy. "Helmholtz and the Materialities of Communication." *Osiris* 9 (1994): 184–207.
Leonelli, Sabina. *Data-Centric Biology: A Philosophical Study.* Chicago: University of Chicago Press, 2016.
Leys, Ruth. "How Did Fear Become a Scientific Object and What Kind of Object Is It?" *Representations* 110, no. 1 (2010): 66–104.
Lincke, Werner. "Lehrmeister Licht." *Das Licht. Zeitschrift für praktische Leucht- und Beleuchtungs- Aufgaben* (1930): 123–7.
Lindsay, Grace. *Models of the Mind: How Physics, Engineering and Mathematics Have Shaped Our Understanding of the Brain.* London: Bloomsbury Sigma, 2021.
Littlefield, Melissa M. *The Lying Brain: Lie Detection in Science and Science Fiction.* Ann Arbor: University of Michigan Press, 2011.
Littlefield, Melissa M. "'A Mind Plague on Both Your Houses': Imagining the Impact of the Neuro-Turn on the Neurosciences." In *The Human Sciences after the Decade of the Brain*, edited by Jon Leefmann and Elisabeth Hildt, 198–213. London: Academic Press, 2017.
Liu, Difei, Shen Liu, Xiaoming Liu, Chong Zhang, Aosika Li, Chenggong Jin, Yijun Chen, Hangwei Wang, and Xiaochu Zhang. "Interactive Brain Activity: Review and Progress on EEG-Based Hyperscanning in Social Interactions." *Frontiers in Psychology* 9, 1862 (2018): 1–11.
Liu, Tao, and Matthew Pelowski. "Clarifying the Interaction Types in Two-Person Neuroscience Research." *Frontiers in Human Neuroscience* 8 (2014).
Lowenberg, Richard, and Peter Crown. "Environetic Synthesis. Techno-Sensory Interface Projects." Typoscript, 1971.
Luce, Gay, and Erik Peper. "Mind over Body, Mind over Mind." *New York Times*, September 12, 1971, 34–5.
Lucier, Alvin. "Music for Solo Performer (1965) for Enormously Amplified Brain Waves and Percussion." In *Chambers: Interviews with the Composer by Douglas Simon*, 69. Middletown: Wesleyan University Press, 1980.
Luck, Steven J. *An Introduction to the Event-Related Potential Technique.* Cambridge, MA: MIT Press, 2014.
Luckhurst, Roger. *The Invention of Telepathy, 1870–1901.* Oxford: Oxford University Press, 2002.
Lutz, Tom. "Varieties of Medical Experience: Doctors and Patients, Psyche and Soma in America." In *Cultures of Neurasthenia from Beard to the First World War*, edited by Marijke Gijswijt-Hofstra and Roy Porter, 51–76. Amsterdam: Rodopi, 2001.
Lysen, Flora. "The Brain Observatory and the Imaginary Media of Memory Research." In *Memory in the Twenty-First Century: New Critical Perspectives from the Arts, Humanities, and Sciences*, edited by Sebastian Groes, 57–62. London: Palgrave Macmillan, 2016.

Lysen, Flora. "Grey Matter and Colored Wax." In *Textures of the Anthropocene: Grain, Vapor, Ray*, edited by Katrin Klingan, Ashkan Sepahvand, and Bernd M. Scherer, 73–83. Cambridge, MA: MIT Press, 2015.

Lysen, Flora. "The Interface Is the (Art)Work: EEG-Feedback, Circuited Selves and the Rise of Real-Time Brainmedia (1964–1977)." In *Brain Art: Brain-Computer Interfaces for Artistic Expression*, edited by Anton Nijholt, 33–63. Cham: Springer, 2019.

Lysen, Flora. "It Blinks, It Thinks? Luminous Brains and a Visual Culture of Electric Display, Circa 1930." *Nuncius* 32, no. 2 (2017): 412–39.

Lysen, Flora. "Kissing and Staring in Times of Neuromania: The Social Brain in Art-Science Experiments." In *Artful Ways of Knowing, Dialogues between Artistic Research and Science & Technology Studies*, edited by Trevor Pinch, Henk Borgdorff, and Peter Peters. Routledge Advances in Art and Visual Studies, 167–83. London: Routledge, 2019.

Lysen, Flora. "Neurofutures of Love." *Baltan Quaterly* (2016): 4–5.

Lysen, Flora. "Technokittens and Brainbow Mice. Negotiating New Neural Imaginaries." In *Speculations on Anonymous Materials; Nature after Nature; Inhuman [Fridericianum, Kassel, September 29, 2013–June 14, 2015]*, edited by Susanne Pfeffer, 167–83. London: Koenig Books, 2018.

MacCurdy, John T. *Common Principles in Psychology and Physiology*. Cambridge: Cambridge University Press, 1928.

MacCurdy, John T. "The General Nature of Association Processes within the Central Nervous System 1." *British Journal of Psychology. General Section* 22, no. 2 (1931): 136–49.

MacKay, D. M. "Some Experiments on the Perception of Patterns Modulated at the Alpha Frequency." *Electroencephalography and Clinical Neurophysiology* 5, no. 4 (1953): 559–62.

MacKay, Donald. "Towards an Information-Flow Model of Cerebral Organisation." *Advancement of Science* 12, no. 1 (1956): 392–5.

MacVarish, Jan, Ellie Lee, and Pam Lowe. "Understanding the Rise of 'Neuroparenting.'" In *We Need to Talk about Family: Essays on Neoliberalism, the Family and Popular Culture*, edited by Roberta Garrett, Tracey Jensen, and Angie Voela, 95–116. Newcastle: Cambridge Scholars Publishing, 2016.

Marie, Pierre. "De L'aphasie En Général Et De L'agraphie En Particulier, D'après L'enseignement De M. Le Professeur Charcot." *Progrès Médical* 7, no. 5, deuxième serie (1888): 81–4.

Marie, Pierre. "Revue Général De L'aphasie (Cécité Verbale, Surdité Verbale, Aphasie Motrice, Agraphie)." *Revue de Médecine* 3 (1883): 693–702.

Marshall, Jonathan W. *Performing Neurology: The Dramaturgy of Dr Jean-Martin Charcot*. New York: Springer, 2016.

Martin, James. *Design of Real-Time Computer Systems*. Englewood Cliffs, NJ: Prentice Hall, 1967.

Maslow, A. H. "Toward a Humanistic Biology." *American Psychologist* 24, no. 8 (1969): 724–35.
Matthews, Laura. "Scientists Turn Brain Images into Youtube Videos." *International Business Times*, November 24, 2011.
Matusall, Svenja. "Social Behavior in the 'Age of Empathy'? A Social Scientist's Perspective on Current Trends in the Behavioral Sciences." *Frontiers in Human Neuroscience* 7 (2013).
Matusall, Svenja, Ina Maria Kaufmann, and Markus Christen. "The Emergence of Social Neuroscience." In *The Oxford Handbook of Social Neuroscience*, edited by Jean Decety and John T. Cacioppo, 9–27. Oxford: Oxford University Press, 2011.
Matusz, Pawel J., Suzanne Dikker, Alexander G. Huth, and Catherine Perrodin. "Are We Ready for Real-World Neuroscience?" *Journal of Cognitive Neuroscience* 31, no. 3 (2019): 327–38.
McCabe, David P., and Alan D. Castel. "Seeing Is Believing: The Effect of Brain Images on Judgments of Scientific Reasoning." *Cognition* 107, no. 1 (2008): 343–52.
McCaskey, Michael B. "The Hidden Messages Managers Send." *Harvard Business Review*, November 1979, 135–48.
McDowall, J. J. "Interactional Synchrony: A Reappraisal." *Journal of Personality and Social Psychology Today* 36, no. 9 (1978): 963–75.
McGee, Micki. *Self-Help, Inc.: Makeover Culture in American Life*. Oxford: Oxford University Press, 2005.
McLeary, Erin, and Elizabeth Toon. "Here Man Learns about Himself." *American Journal of Public Health* 102, no. 7 (2012): e27–36.
McLuhan, Marshall. *Understanding Media: The Extensions of Man*. Cambridge, MA: MIT Press, [1964] 1994.
McNeil, Maureen, Michael Arribas-Ayllon, Joan Haran, and Adrian Mackenzie. "Conceptualizing Imaginaries of Science, Technology and Society." In *The Handbook of Science and Technology Studies*, edited by Ulrike Felt, Rayvon Fouché, Clark A. Miller, and Laurel Smith-Doerr, 435–64. Cambridge, MA: MIT Press, 2016.
McWhirter, E. M. S. "The Centralisation of Control of Power Networks." *Electrical Communication* 13, no. 3 (1935): 255–73.
Meloni, Maurizio. "Philosophical Implications of Neuroscience: The Space for a Critique." *Subjectivity* 4, no. 3 (2011): 298–322.
Metcalf, Jacob, and Kate Crawford. "Where Are Human Subjects in Big Data Research? The Emerging Ethics Divide." *Big Data & Society* 3, no. 1 (2016).
Meynell, Letitia. "The Politics of Pictured Reality: Locating the Object from Nowhere in Fmri." In *Neurofeminism: Issues at the Intersection of Feminist Theory and Cognitive Science*, edited by Robyn Bluhm, Anne Jaap Jacobson, and Heidi Lene Maibom, 11–29. London: Palgrave Macmillan, 2012.

Michael, Mike, and Deborah Lupton. "Toward a Manifesto for the 'Public Understanding of Big Data.'" *Public Understanding of Science* 25, no. 1 (2016): 104–16.

Milkulski, Antoni Jan, and Eufemjusz Józef Herman. "Die Hirnpulsation Des Mensen Auf Grund Experimenteller Untersuchungen." *Zeitschrift für Neurologie und Psychiatrie* 90 (1924): 469–520.

Miller, Neal E., and Barry R. Dworkin. "Visceral Learning: Recent Difficulties with Curarized Rats and Significant Problems for Human Research." In *Cardiovascular Psychophysiology*, edited by Paul Obrist, A. H. Black, Jasper Brener, and Leo V. DiCara, 312–31. London: Routlege, 2017.

Millett, David. "Wiring the Brain: From the Excitable Cortex to the EEG, 1870–1940." PhD dissertation, University of Chicago, 2001.

Mitchell, Edward. "The Soul Spectroscope." *The Sun*, December 19, 1875, 5.

Modic, Dolores, and Maryann P. Feldman. "Mapping the Human Brain: Comparing the US and EU Grand Challenges." *Science and Public Policy* 44, no. 3 (2017): 440–9.

Monea, Alexander, and Jeremy Packer. "Media Genealogy: Technological and Historical Engagements of Power—Introduction." *International Journal of Communication* 10 (2016): 3141–59.

Moore, Rachel O. *Savage Theory: Cinema as Modern Magic*. Durham, NC: Duke University Press, 2000.

Morus, Iwan Rhys. "'More the Aspect of Magic Than Anything Natural': The Philosophy of Demonstration." In *Science in the Marketplace: Nineteenth-Century Sites and Experiences*, edited by Aileen Fyfe and Bernard V. Lightman, 336–70. Chicago: University of Chicago Press, 2007.

Morus, Iwan Rhys. "'The Nervous System of Britain': Space, Time and the Electric Telegraph in the Victorian Age." *British Journal for the History of Science* 33, no. 4 (2000): 455–75.

Morus, Iwan Rhys. "Placing Performance." *Isis* 101, no. 4 (2010): 775–8.

Morus, Iwan Rhys. "Worlds of Wonder: Sensation and the Victorian Scientific Performance." *Isis* 101, no. 4 (2010): 806–16.

Mullen, Tim. *Neurotech Anytime, Anywhere*. 2018. https://www.youtube.com/watch?v=NJmopBvbALs.

Mulvey, Laura. *Death 24x a Second: Stillness and the Moving Image*. London: Reaktion Books, 2006.

Natale, Simone. "Specters of the Mind: Ghosts, Illusion, and Exposure in Paul Leni's the Cat and the Canary." In *Cinematic Ghosts: Haunting and Spectrality from Silent Cinema to the Digital Era*, edited by Murray Leeder, 59–75. London: Bloomsbury, 2015.

Natale, Simone, and Gabriele Balbi. "Media and the Imaginary in History: The Role of the Fantastic in Different Stages of Media Change." *Media History* 20, no. 2 (2014): 203–18.

Nikolow, Sybilla, ed. *Erkenne Dich Selbst! Strategien Der Sichtbarmachung Des Körpers Im 20. Jahrhundert*. Köln: Böhlau, 2015.

Nikolow, Sybilla. "'Erkenne Und Prüfe Dich Selbst!' in Einer Ausstellung 1938 in Berlin. Körperleistungsmessungen Als Objektbezogene Vermittlungspraxis Und Biopolitische Kontrollmaßnahme." In *Erkenne Dich Selbst! Strategien Der Sichtbarmachung Des Körpers Im 20. Jahrhundert*, edited by Sybilla Nikolow, 227–68. Köln: Böhlau, 2015.

Nishimoto, Shinji, An T. Vu, Thomas Naselaris, Yuval Benjamini, Bin Yu, and Jack L. Gallant. "Reconstructing Visual Experiences from Brain Activity Evoked by Natural Movies." *Current Biology* 21, no. 19 (2011): 1641–6.

Noakes, Richard. "Thoughts and Spirits by Wireless: Imagining and Building Psychic Telegraphs in America and Britain, circa 1900–1930." *History and Technology* 32, no. 2 (2016): 137–58.

Noir, J. "Les avantages et les dangers de la vulgarisation." *Le Concours Médical* (1938): 717–18.

Nowlis, David P., and Joe Kamiya. "The Control of Electroencephalographic Alpha Rhythms through Auditory Feedback and the Associated Mental Activity." *Psychophysiology* 6, no. 4 (1970): 476–84.

Nowlis, Vincent. "Mood: Behavior and Experience." In *Feelings and Emotions: The Loyola Symposium*, edited by Magda B. Arnold, 261–74. New York: Academic Press, 1970.

Nunn, Christopher Mark. "Biofeedback with Cerebral Evoked Potentials and Perceptual Fine-Tuning in Humans." *Journal of Experimental Aesthetics* 1, no. 1 (1977): 11–100.

Nye, David E. *American Technological Sublime*. Cambridge, MA: MIT Press, 1996.

Nyhart, Lynn K. "Historiography of the History of Science." In *A Companion to the History of Science*, edited by Bernard Lightman, 7–22. Chichester: John Wiley, 2016.

Ochsner, Kevin, and Matthew Lieberman. "The Emergence of Social Cognitive Neuroscience." *American Psychologist* 56, no. 9 (2001): 717–34.

Ortega, Francisco, and Fernando Vidal, eds. *Neurocultures: Glimpses into an Expanding Universe*. Frankfurt am Main: Peter Lang, 2011.

Östling, Johan, Erling Sandmo, David Larsson Heidenblad, Anna Nilsson Hammar, and Kari H. Nordberg. "The History of Knowledge and the Circulation of Knowledge: An Introduction." In *Circulation of Knowledge: Explorations in the History of Knowledge*, edited by Johan Östling, Erling Sandmo, David Larsson Heidenblad, Anna Nilsson Hammar, and Kari H. Nordberg, 9–36. Lund: Nordic Academic Press, 2018.

Otis, Laura. "The Metaphoric Circuit: Organic and Technological Communication in the Nineteenth Century." *Journal of the History of Ideas* 63, no. 1 (2002): 105–28.

Otis, Laura. *Networking: Communicating with Bodies and Machines in the Nineteenth Century*. Ann Arbor, MI: University of Michigan Press, 2001.

Oxford English Dictionary. "Fascination, N." Oxford University Press.

Oxford English Dictionary. "Live, Adj.1, N., and Adv." Oxford University Press.
Oxford English Dictionary. "Neurometer, N." Oxford University Press.
Oxford English Dictionary. "Sync, N. And V." Oxford University Press.
Oxford English Dictionary. "Tune, V.2." Oxford University Press.
Oxford English Dictionary. "Wavelength, N." Oxford University Press.
Pantelakis, S. N., B. D. Bower, and H. Douglas Jones. "Convulsions and Television Viewing." *British Medical Journal* 2, no. 5305 (1962): 633–8.
Parada, Francisco J., and Alejandra Rossi. "Commentary: Brain-to-Brain Synchrony Tracks Real-World Dynamic Group Interactions in the Classroom and Cognitive Neuroscience: Synchronizing Brains in the Classroom." *Frontiers in Human Neuroscience* 11 (2017): 554.
Parikka, Jussi. "Imaginary Media: Mapping Weird Objects." In *What Is Media Archaeology*, edited by Jussi Parikka, 41–62. Cambridge, MA: Polity Press, 2012.
Parikka, Jussi. *Insect Media: An Archaeology of Animals and Technology*. Minneapolis: University of Minnesota Press, 2010.
Parikka, Jussi. "New Materialism as Media Theory: Medianatures and Dirty Matter." *Communication and Critical/Cultural Studies* 9, no. 1 (2012): 95–100.
Parikka, Jussi, ed. *What Is Media Archaeology?* Cambridge, MA: Polity Press, 2012.
Park, Lisa. "Eunoia." 2013. http://www.thelisapark.com/#/eunoia/.
Pavlov, Ivan Petrovich. *Die Höchste Nerventätigkeit (Das Verhalten) Von Tieren: Eine Zwanzigjährige Prüfung Der Objektiven Forschung; Bedingte Reflexe. Sammlung Von Artikeln, Berichten, Vorlesungen Und Reden*. München: Verlag J. F. Bergmann, 1926.
Penel, Charles. *1937: The Inauguration of the Palais De La Découverte*. 2003. http://videotheque.cnrs.fr/doc=1114?langue=EN.
Peper, Erik. "Localized EEG Alpha Feedback Training: A Possible Technique for Mapping Subjective, Conscious, and Behavioral Experiences." *Kybernetik* 11, no. 3 (1972): 166–9.
Pérez, Alejandro, Guillaume Dumas, Melek Karadag, and Jon Andoni Duñabeitia. "Differential Brain-to-Brain Entrainment while Speaking and Listening in Native and Foreign Languages." *Cortex* 111 (2019): 303–15.
Peters, John Durham. "Broadcasting and Schizophrenia." *Media, Culture & Society* 32, no. 1 (2010): 123–40.
Petsche, H. "From Graphein to Topos: Past and Future of Brain Mapping." In *Topographic Brain Mapping of EEG and Evoked Potentials*, edited by Konrad Maurer, 11–18. Berlin: Springer, 1989.
Pfeiffer, John E. *The Human Brain*. New York: Harper, 1955.
Pias, Claus. "The Age of Cybernetics." In *Cybernetics: The Macy Conferences 1946–1953*, edited by Claus Pias, 11–27. Zurich: Diaphanes, 2016.
Pias, Claus. "Elektronenhirn Und Verbotene Zone. Zur Kybernetische Ökonomie Des Digitalen." In *Analog, Digital: Opposition Oder Kontinuum?*

Zur Theorie Und Geschichte Einer Unterscheidung, edited by Jens Schröter and Alexander Böhnke, 295–310. Bielefeld: Transcript, 2004.

Pias, Claus. "Zeit Der Kybernetik." In *Cybernetics—Kybernetik. The Macy-Conferences 1946-1953. 2: Dokumente Und Reflexionen*, edited by Claus Pias, 9–42. Zürich: Diaphanes Verlag, 2004.

Pickering, Andrew. *The Cybernetic Brain: Sketches of Another Future*. Chicago: University of Chicago Press, 2010.

Pickering, Andrew. *The Mangle of Practice: Time, Agency, and Science*. Chicago: University of Chicago Press, 1995.

Pickering, Andrew. "Psychiatry, Synthetic Brains and Cybernetics in the Work of W. Ross Ashby." *International Journal of General Systems* 38, no. 2 (2009): 213–30.

Pickersgill, Martyn, Sarah Cunningham-Burley, and Paul Martin. "Constituting Neurologic Subjects: Neuroscience, Subjectivity and the Mundane Significance of the Brain." *Subjectivity* 4, no. 3 (2011): 346–65.

Piéron, Henri, and Ignace Meyerson, eds. *Onzième Congrès International De Psychologie: Paris, 25-31 Juillet 1937: Rapports Et Comptes Rendus*. Nendeln: Kraus Reprint, 1974.

Pitts, Walter, and Warren S. McCullogh. "How We Know Universals: The Perception of Auditory and Visual Forms." *Bulletin of Mathematical Biophysics* 9, no. 3 (1947): 127–47.

Plotnick, Rachel. "Force, Flatness and Touch without Feeling: Thinking Historically about Haptics and Buttons." *New Media & Society* 19, no. 10 (2017): 1632–52.

Plotnick, Rachel. "Touch of a Button: Long-Distance Transmission, Communication, and Control at World's Fairs." *Critical Studies in Media Communication* 30, no. 1 (2013): 52–68.

Pond, D. A. "The EEG Society: Maudsley Hospital, London April 26th, 1952. Society Proceedings." *Electroencephalography and Clinical Neurophysiology* 4, no. 3 (1952): 371–4.

Poole, Lynn. "A Visit to Our Studio." In *John Hopkins Science Review*. USA: DuMont Television Network, 1952, 28:37.

Poole, S. "Your Brain on Pseudoscience: The Rise of Popular Neurobollocks." *New Statesman*, September 6, 2012.

Pötzl, Otto. *Die Aphasielehre Vom Standpunkte Der Klinischen Psychiatrie*. Leipzig: F. Deuticke, 1928.

Pötzl, Otto. "Über Die Rückbildung Einer Reinen Wortblindheit." *Zeitschrift für die gesamte Neurologie und Psychiatrie* 52, no. 1 (1919): 241–72.

Pritchard, Walter S. "Psychophysiology of P300." *Psychological Bulletin* 89, no. 3 (1981): 506–40.

Prpa, Mirjana, and Philippe Pasquier. "Brain-Computer Interfaces in Contemporary Art: A State of the Art and Taxonomy." In *Brain Art: Brain-Computer Interfaces for Artistic Expression*, edited by Anton Nijholt, 65–115. Cham: Springer, 2019.

Rabinbach, Anson. *The Human Motor: Energy, Fatigue, and the Origins of Modernity*. New York: Basic Books, 1990.

Rabinovitz, Lauren. *Electric Dreamland: Amusement Parks, Movies, and American Modernity*. New York: Columbia University Press, 2012.

Rader, Karen A., and Victoria E. M. Cain. *Life on Display: Revolutionizing U.S. Museums of Science and Natural History in the Twentieth Century*. Chicago: University of Chicago Press, 2014.

Redcay, Elizabeth, and Leonhard Schilbach. "Using Second-Person Neuroscience to Elucidate the Mechanisms of Social Interaction." *Nature Reviews Neuroscience* 20, no. 8 (2019): 495–505.

Reicher, Karl. "Kinematografie in Der Neurologie." *Verhandlungen der Gesellschaft Deutscher Naturforscher und Ärtze* (1908): 235–6.

Rheinberger, Hans-Georg. *Toward a History of Epistemic Things: Synthesizing Proteins in the Test Tube*. Stanford, CA: Stanford University Press, 1997.

Rich, Willard. *Brain-Waves and Death*. New York: C. Scribner's Sons, 1940.

Robbins, Jim. *A Symphony in the Brain: The Evolution of the New Brain Wave Biofeedback*. New York: Grove Press, 2008.

Rocca, Mattia Della, and Claudio Pogliano. "Different Histories from 20th Century Neuroscience." *Nuncius* 32, no. 2 (2017): 253–60.

Roepstorff, Andreas. "Brains in Scanners: An Umwelt of Cognitive Neuroscience." *Semiotica* 134, no. 1/4 (2001): 747–65.

Roepstorff, Andreas. "Transforming Subjects into Objectivity. An Ethnography of Knowledge in a Brain Imaging Laboratory." *FOLK, Journal of the Danish Ethnographic Society* 44 (2002): 145–70.

Rose, Nikolas. *Inventing Our Selves: Psychology, Power, and Personhood*. Cambridge: Cambridge University Press, 1998.

Rose, Nikolas. *The Politics of Life Itself: Biomedicine, Power, and Subjectivity in the Twenty-First Century*. Princeton, NJ: Princeton University Press, 2009.

Rose, Nikolas. "Reading the Human Brain How the Mind Became Legible." *Body & Society* 22, no. 2 (2016): 1–38.

Rose, Nikolas, and Joelle Abi-Rached. "Governing through the Brain: Neuropolitics, Neuroscience and Subjectivity." *Cambridge Journal of Anthropology* 32, no. 1 (2014): 3–23.

Rose, Nikolas, and Joelle M. Abi-Rached. *Neuro: The New Brain Sciences and the Management of the Mind*. Princeton, NJ: Princeton University Press, 2013.

Rosenblueth, Arturo, Norbert Wiener, and Julian Bigelow. "Behavior, Purpose and Teleology." *Philosophy of Science* 10, no. 1 (1943): 18–24.

Rosenboom, David. *Biofeedback and the Arts: Results of Early Experiments*. Vancouver: A.R.C. Publications, 1976.

Rosenboom, David. *Extended Musical Interface with the Human Nervous System*. Leonardo Monograph Series. San Francisco, CA: International Society for the Arts, Sciences and Technology, 1997.

Rosenboom, David. "Method for Producing Sounds or Light Flashes with Alpha Brain Waves for Artistic Purposes." *Leonardo* 5, no. 2 (1972): 141–5.

Rosenboom, David, and Tim Mullen. "More Than One—Artistic Explorations with Multi-Agent BCIS." In *Brain Art: Brain-Computer Interfaces for Artistic Expression*, edited by Anton Nijholt, 117–43. Cham: Springer, 2019.

Rosental, Paul-André. "Eugenics and Social Security in France before and after the Vichy Regime." *Journal of Modern European History* 10, no. 4 (2012): 540–62.

Roskams, Jane, and Zoran Popović. "Power to the People: Addressing Big Data Challenges in Neuroscience by Creating a New Cadre of Citizen Neuroscientists." *Neuron* 92, no. 3 (2016): 658–64.

Roskies, Adina L. "Neuroimaging and Inferential Distance: The Perils of Pictures." In *Foundational Issues in Human Brain Mapping*, edited by Stephen José Hanson and Martin Bunzl, 195–216, Cambridge, MA: MIT Press, 2010.

Rösler, Frank. "From Single-Channel Recordings to Brain-Mapping Devices: The Impact of Electroencephalography on Experimental Psychology." *History of Psychology* 8, no. 1 (2005): 95–117.

Ross, Andrew. "New Age Technoculture." In *Cultural Studies*, edited by L. Grossberg, C. Nelson, and P. Treichler. New York: Routledge, 1992.

Ross, Andrew. *Strange Weather: Culture, Science, and Technology in the Age of Limits*. London: Verso, 1991.

Rouse, Joseph. "What Are Cultural Studies of Scientific Knowledge?" *Configurations* 1, no. 1 (1993): 1–22.

S.S. "Not Surprisingly, You Can Now Own Your Own Brain Feedback Device..." *Radical Software* 1, no. 4 (1971): 3.

Saler, Michael. "Modernity and Enchantment: A Historiographic Review." *American Historical Review* 111, no. 3 (2006): 692–716.

Salisbury, Laura, and Andrew Shail, eds. *Neurology and Modernity: A Cultural History of Nervous Systems, 1800–1950*. London: Palgrave Macmillan, 2010.

Sänger, Johanna, Ulman Lindenberger, and Viktor Müller. "Interactive Brains, Social Minds." *Communicative & Integrative Biology* 4, no. 6 (2011): 655–63.

Sappol, Michael. *Body Modern: Fritz Kahn, Scientific Illustration, and the Homuncular Subject*. Minneapolis: University of Minnesota Press, 2017.

Sarasin, Philipp. "Was Ist Wissensgeschichte?" *Internationales Archiv für Sozialgeschichte der deutschen Literatur* 36, no. 1 (2011): 159–72.

Scannell, Paddy. *Television and the Meaning of "Live": An Enquiry into the Human Situation*. Cambridge, MA: Polity Press, 2014.

Schilbach, Leonhard, Bert Timmermans, Vasudevi Reddy, Alan Costall, Gary Bente, Tobias Schlicht, and Kai Vogeley. "A Second-Person Neuroscience in Interaction." *Behavioral and Brain Sciences* 36, no. 4 (2013): 441–62.

Schilbach, Leonhard, Bert Timmermans, Vasudevi Reddy, Alan Costall, Gary Bente, Tobias Schlicht, and Kai Vogeley. "Toward a Second-Person Neuroscience." *Behavioral and Brain Sciences* 36, no. 4 (2013): 393–414.

Schirrmacher, Arne. "Introduction: Communicating Science—National Approaches in Twentieth-Century Europe." *Science in Context* 26, no. 3 (2013): 393–404.

Schirrmacher, Arne. "Nach Der Popularisierung. Zur Relation Von Wissenschaft Und Öffentlichkeit Im 20. Jahrhundert." *Geschichte und Gesellschaft* 34, no. 1 (2008): 73–95.

Schmidgen, H. "Cybernetic Times: Norbert Wiener, John Stroud, and the 'Brain Clock' Hypothesis." *History of the Human Sciences* 33, no. 1 (2020): 80–108.

Schmidgen, Henning. "1900—The Spectatorium: On Biology's Audiovisual Archive." *Grey Room* 43 (2011): 42–65.

Schmidgen, Henning. "Die Geschwindigkeit Von Gefühlen Und Gedanken." *NTM International Journal of History & Ethics of Natural Sciences, Technology & Medicine* 12, no. 2 (2004): 100–15.

Schmidgen, Henning. *Hirn Und Zeit: Die Geschichte Eines Experiments 1800–1950.* Berlin: Matthes & Seitz Verlag, 2013.

Schmidgen, Henning. "Pictures, Preparations, and Living Processes: The Production of Immediate Visual Perception (Anschauung) in Late-19th-Century Physiology." *Journal of the History of Biology* 37, no. 3 (2004): 477–513.

Schmidgen, Henning. "Zeit Als Peripheres Zentrum: Psychologie Und Kybernetik." In *Cybernetics: The Macy Conferences, 1946–1953. Band 2 Cybernetics: Essays and Documents*, edited by Claus Pias, 131–52. Berlin: Diaphanes, 2004.

Schmidt-Brücken, Katharina. *Hirnzirkel: Kreisende Prozesse in Computer Und Gehirn: Zur Neurokybernetischen Vorgeschichte Der Informatik.* Bieleveld: Transcript Verlag, 2014.

Schneider, William H. "After Binet: French Intelligence Testing, 1900–1950." *Journal of the History of the Behavioral Sciences* 28, no. 2 (1992): 111–32.

Schwartz, Mark S., Thomas F. Collura, Joe Kamiya, and Nancy M. Schwartz. "The History and Definitions of Biofeedback and Applied Psychophysiology." In *Biofeedback, Fourth Edition: A Practitioner's Guide*, edited by Mark S. Schwartz and Frank Andrasik, 3–23. New York: Guilford, 2017.

Sconce, Jeffrey. *Haunted Media: Electronic Presence from Telegraphy to Television.* Durham, NC: Duke University Press, 2000.

Sconce, Jeffrey. *The Technical Delusion: Electronics, Power, Insanity.* Durham, NC: Duke University Press, 2019.

Scott Belk, Patrick. "King Features Syndicate." In *Comics through Time: A History of Icons, Idols, and Ideas*, edited by M. Keith Booker, 217–19. Santa Barbara, CA: ABC-CLIO, 2014.

Secord, James A. "Knowledge in Transit." *Isis* 95, no. 4 (2004): 654–72.

Seltzer, Mark. *Bodies and Machines.* New York: Routledge, 1992.

Serlin, David. "Performing Live Surgery on Television and the Internet." In *Imagining Illness: Public Health and Visual Culture*, edited by David Serlin, 223–44. Minneapolis: University of Minnesota Press, 2010.

Shagass, Charles. *Evoked Brain Potentials in Psychiatry.* New York: Plenum Press, 1972.

Shamay-Tsoory, Simone G., and Avi Mendelsohn. "Real-Life Neuroscience: An Ecological Approach to Brain and Behavior Research." *Perspectives on Psychological Science* 14, no. 5 (2019): 841–59.

Shapin, Steven. *The Scientific Life: A Moral History of a Late Modern Vocation.* Chicago: University of Chicago Press, 2008.

Sherrington, Charles Scott. *The Brain and Its Mechanism.* Cambridge: Cambridge University Press, 1934.

Sherrington, Charles Scott. *Man on His Nature.* Cambridge: Cambridge University Press, [1940] 2009 (Gifford Lectures 1937–8).

Shipton, H. W. "An Electronic Trigger Circuit as an Aid to Neurophysiological Research." *Journal of the British Institution of Radio Engineers* 9, no. 10 (1949): 374–83.

Siodmak, Curt. *Donovan's Brain.* New York: Alfred A. Knopf, 1943.

Slaby, Jan. "Steps towards a Critical Neuroscience." *Phenomenology and the Cognitive Sciences* 9, no. 3 (2010): 397–416.

Slee, Richard, and Cornelia Atwood Pratt. *Dr. Berkeley's Discovery.* New York: G. P. Putnam's Sons, 1899.

Sloane, William. *The Edge of Running Water.* London: Hachette, [1939] 2013.

Sluckin, W. *Minds and Machines.* Harmondsworth: Penguin Books, 1954.

Smith, Roger. *Between Mind and Nature: A History of Psychology.* London: Reaktion Books, 2013.

Smith, Roger. *Inhibition: History and Meaning in the Sciences of Mind and Brain.* Los Angeles: University of California Press, 1992.

Smith, Roger. "Physiology and Psychology, or Brain and Mind, in the Age of C. S. Sherrington." In *Psychology in Britain: Historical Essays and Personal Reflections,* edited by G. C. Bunn, A. D. Lovie, and G. D. Richards, 223–42. Leicester: The British Psychological Society, 2001.

Smith, Roger. "Representations of Mind: C. S. Sherrington and Scientific Opinion, c.1930–1950." *Science in Context* 14, no. 4 (2001): 511–39.

Smith, Tiffany Watt. "Of Hats and Scientific Laughter." In *Staging Science,* edited by Martin Willis. Palgrave Studies in Literature, Science and Medicine, 59–82. London: Palgrave Macmillan, 2016.

Smythies, J. R. "The Stroboscope as Providing Empirical Confirmation of the Representative Theory of Perception." *British Journal for the Philosophy of Science* 6, no. 24 (1956): 332–5.

Stermitz, Evelin. "Interview with Nina Sobell." August 22, 2007.

Sokal, Michael M. "The Origins of the Psychological Corporation." *Journal of the History of the Behavioral Sciences* 17, no. 1 (1981): 54–67.

Somaini, Antonio. "Walter Benjamin's Media Theory: The Medium and the Apparat." *Grey Room* 62 (2016): 6–41.

Souques, A. "Quelques Cas D'anarthrie De Pierre Marie. Aperçu Historique Sur La Localisation Du Langage." *Revue Neurologique* 2 (1928): 319–68.

Sprenger, Florian. *Medien Des Immediaten: Elektrizität, Telegraphie, Mcluhan.* Berlin: Kadmos, 2012.

Stadler, Max. "Assembling Life. Models, the Cell, and the Reformations of Biological Science, 1920–1960." PhD dissertation, Imperial College London, University of London, 2010.

Stadler, Max. "Circuits, Algae and Whipped Cream: The Biophysics of Nerve, ca. 1930." In *The History of the Brain and Mind Sciences: Technique, Technology, Therapy*, edited by Stephen T. Casper and Delia Gavrus, 107–35. Rochester: University of Rochester Press, 2017.

Stadler, Max. "The Neuromance of Cerebral History." In *Critical Neuroscience*, edited by Suparna Choudhury and Jan Slaby, 135–58. Hoboken, NJ: Wiley Blackwell, 2011.

Stafford, Jane. "Science Takes a Look at Your Brainstorms." *Arizona Republic*, June 30, 1935, 79.

Star, Susan Leigh. *Regions of the Mind: Brain Research and the Quest for Scientific Certainty*. Stanford: Stanford University Press, 1989.

Staub, Michael E. "The Other Side of the Brain: The Politics of Split-Brain Research in the 1970s–1980s." *History of Psychology* 19, no. 4 (2016): 259–73.

Stelarc. "Spectacle of the Mind." https://vimeo.com/24079719.

Stengers, Isabelle. *Another Science Is Possible: A Manifesto for Slow Science*. Cambridge, MA: Polity Press, 2018.

Stengers, Isabelle. *Invention of Modern Science*. Minneapolis: University of Minnesota Press, 2000.

Stengers, Isabelle, and Penelope Deutscher. "Another Look: Relearning to Laugh." *Hypatia* 15, no. 4 (2000): 41–54.

Stifter, Christian. "'Anschaulichkeit' Als Paradigma. Visuelle Erziehung in Der Fruhen Volksbildung, 1900–1938." *Spurensuche* 14, no. 1–4 (2003): 68–83.

Stone, J. L., and J. R. Hughes. "I. The Gibbs' Boston Years: Early Developments in Epilepsy Research and Electroencephalography at Harvard." *Clinical EEG and Neuroscience* 21, no. 4 (1990): 175–82.

Stoyva, Johann, and Joe Kamiya. "Electrophysiological Studies of Dreaming as the Prototype of a New Strategy in the Study of Consciousness." *Psychological Review* 75, no. 3 (1968): 192–205.

Strasser, Bruno J., Jérôme Baudry, Dana Mahr, Gabriela Sanchez, and Elise Tancoigne. "'Citizen Science'? Rethinking Science and Public Participation." *Science & Technology Studies* 32, no. 2 (2019): 52–76.

Sutton, S., M. Braren, J. Zubin, and E. R. John. "Evoked-Potential Correlates of Stimulus Uncertainty." *Science* 150, no. 3700 (1965): 1187–8.

Tanner, Jakob. "The Visual Culture of Rationalization in the Modern 20th Century Enterprise." *Entreprises et Histoire*, no. 44 (2009): 5–8.

Teale, Edwin. "Amazing Electrical Tests Show What Happens When You Think." *Popular Science Monthly* 128, no. 5 (1936): 11.

Teitelbaum, Richard. "In Tune: Some Early Experiments in Biofeedback Music (1966–74)." In *Biofeedback and the Arts: Results of Early Experiments*, edited by David Rosenboom, 35–65. Vancouver: A.R.C. Publications, 1976.

Thacker, Eugene. *Biomedia*. Minneapolis: University of Minnesota Press, 2004.

Thacker, Eugene. "Dark Media." In *Excommunication: Three Inquiries in Media and Mediation*, edited by Alexander R. Galloway, Eugene Thacker, and Wark McKenzie, 77–149. Chicago: University of Chicago Press, 2013.

Thibault, Robert T., Michael Lifshitz, Niels Birbaumer, and Amir Raz. "Neurofeedback, Self-Regulation, and Brain Imaging: Clinical Science and Fad in the Service of Mental Disorders." *Psychotherapy and Psychosomatics* 84, no. 4 (2015): 193–207.

Thibault, Robert T., Amanda MacPherson, Michael Lifshitz, Raquel R. Roth, and Amir Raz. "Neurofeedback with FMRI: A Critical Systematic Review." *NeuroImage* 172 (2018): 786–807.

Thibault, Robert T., and Amir Raz. "The Psychology of Neurofeedback: Clinical Intervention Even If Applied Placebo." *American Psychologist* 72, no. 7 (2017): 679–88.

Thornton, Davi Johnson. *Brain Culture: Neuroscience and Popular Media*. New Brunswick, NJ: Rutgers University Press, 2011.

Thurschwell, Pamela. *Literature, Technology and Magical Thinking, 1880–1920*. Cambridge: Cambridge University Press, 2001.

Tkaczyk, Victoria. "The Making of Acoustics around 1800, or How to Do Science with Words." In *Performing Knowledge, 1750–1850*, edited by Mary Helen Dupree and Sean B. Franzel, 27–55. Berlin: Walter de Gruyter GmbH, 2015.

Topham, Jonathan R. "Rethinking the History of Science Popularization/Popular Science." In *Popularizing Science and Technology in the European Periphery, 1800–2000*, edited by Agustí Nieto-Galan, Enrique Perdiguero, and Faidra Papanelopoulou, 1–20. Farnham: Ashgate, 2009.

Traetta, Luigi. "Docimology Enters into Psychology: Dagmar Weinberg's Work in French Applied Psychology Laboratories." *International Review of Social Sciences and Humanities* 4, no. 2 (2013): 149–58.

Travis, Lee Edward, and Abraham Gottlober. "Do Brain Waves Have Individuality?" *Science* 84, no. 2189 (1936): 532–3.

Triarhou, Lazaros C. "Women Neuropsychiatrists on Wagner-Jauregg's Staff in Vienna at the Time of the Nobel Award: Ordeal and Fortitude." *History of Psychiatry* 30, no. 4 (2019): 393–408.

UC Berkeley. "Vision Reconstruction." 2011. https://www.youtube.com/watch?v=6FsH7RK1S2E.

Väliaho, Pasi. "Bodies Outside In: On Cinematic Organ Projection." *Parallax* 14, no. 2 (2008): 7–19.

van Dijck, José. *Imagenation: Popular Images of Genetics*. Basingstoke: Macmillan, 1998.

van Dijck, José. *The Transparent Body: A Cultural Analysis of Medical Imaging*. Seattle: University of Washington Press, 2005.

VanRullen, Rufin. "Perceptual Rhythms." In *Stevens' Handbook of Experimental Psychology and Cognitive Neuroscience*, edited by John Serences, 1–44. Chichester: American Cancer Society/John Wiley, 2018.

Vasulka, Woody, and Richard Lowenberg. "Environetic Synthesizer." *Radical Software* 1, no. 2 (1970).

Verdeaux, G., and J. Verdeaux. "Description D'une Technique De Polygraphie." *Electroencephalography and Clinical Neurophysiology* 7, no. 4 (1955): 645–8.

Vidal, Jacques J. "Real-Time Detection of Brain Events in EEG." *Proceedings of the IEEE* 65, no. 5 (1977): 633–41.

Vidal, Jacques J. "Silicon Brains: Whither Neuromimetic Computer Architectures." In *Proceedings IEEE International Conference on Computer Design: VLSI in Computers (ICCD '83)*, 17–20. Washington, DC: IEEE Computer Society Press, 1983.

Vidal, Jacques J. "Toward Direct Brain-Computer Communication." *Annual Review of Biophysics and Bioengineering* 2, no. 1 (1973): 157–80.

Vidal, Jacques J., Marshall D. Buck, Ronald H. Olch, and Ramchandi D. Tulsi. "Biocybernetic Control in Man-Machine Interaction: Final Technical Report 1973-1974," 1–51. Los Angeles: Computer Science Department, School of Engineering and Applied Science, University of California, 1974.

Vidal, Jacques J., L. H. Miller, and G. Devillez. "Adaptive User Interfaces for Distributed Information Management." In *Human-Computer Interaction: Proceedings of the First U.S.A.-Japan Conference on Human-Computer Interaction, Honolulu, Hawaii, August 18–20, 1984*, 77–81. Amsterdam: Elsevier, 1984.

Vogeley, Kai. "Two Social Brains: Neural Mechanisms of Intersubjectivity." *Philosophical Transactions of the Royal Society B: Biological Sciences* 372, no. 1727 (2017): 1–11.

Vogl, Joseph. "Becoming-Media: Galileo's Telescope." *Grey Room* 29 (2007): 14–25.

Waag Society. "Brainwaves during Museum Night." 5 November 2014, https://waag.org/en/blog/brainwaves-during-museum-night (accessed July 27, 2015).

Waite, Mitchell. "Build an Alpha Brain Wave Feedback Monitor." *Popular Electronics* (1973): 40–5.

Wallot, Sebastian, Panagiotis Mitkidis, John J. McGraw, and Andreas Roepstorff. "Beyond Synchrony: Joint Action in a Complex Production Task Reveals Beneficial Effects of Decreased Interpersonal Synchrony." *Plos One* 11, no. 12 (2016): 1–25.

Walsh, E. G. "Experimental Investigation of the Scanning Hypothesis." Paper presented at the The EEG Society, Maudsley Hospital, London, April 26, 1952.

Walsh, E. G. "Visual Reaction Time and the Alpha-Rhythm, an Investigation of a Scanning Hypothesis." *Journal of Physiology* 118, no. 4 (1952): 500–508.

Walter, V. J., and W. Grey Walter. "The Central Effects of Rhythmic Sensory Stimulation." *Electroencephalography and Clinical Neurophysiology* 1, nos. 1–4 (1949): 57–86.

Walter, W. Grey. "Discussion on the Electro-Encephalogram in Organic Cerebral Disease." *Proceedings of the Royal Society of Medicine* 41, no. 4 (1948): 237–42.

Walter, W. Grey. "The Electrical Activity of the Brain." *Scientific American* 190, no. 6 (1954): 54–63.

Walter, W. Grey. "Electrical Signs of Mental Function (Part of Cerebral Activity: Report of a Symposium)." *Advancement of Science* 12, no. 1 (1956): 389–92.

Walter, W. Grey. "Enchanted Loom." Great Britain: BBC Radio, 1948.

Walter, W. Grey. "Features in the Electro-Physiology of Mental Mechanisms." In *Perspectives in Neuropsychiatry: Essays Presented to Professor Frederick Lucien Golla by Past Pupils and Associates*, edited by Frederick Lucien Golla and Derek Richter, 67–78. London: H. K. Lewis, 1950.

Walter, W. Grey. *The Living Brain*. New York: W. W. Norton, [1953] 1963.

Walter, W. Grey. "Pattern-Making and Pattern-Seeking." In *Patterns in Your Head: A Weekly Programme about Work in the World of Science*, 15:00. Great Britain: BBC Home Service, 1951.

Walter, W. Grey. "Spontaneous Oscillatory Systems and Alterations in Stability." In *Neural Physiopathology*, edited by R. G. Grenell, 222–57. New York: Harper & Row, 1962.

Walter, W. Grey. "Studies on Activity of the Brain." In *Cybernetics. Circular Causal and Feedback Mechanism in Biological and Social Systems. Transactions of the Tenth Conference, April 22, 23 and 24, 1953, Princeton, NJ*, edited by Heinz von Foerster, 19–31, New York: Josiah Macy, Jr. Foundation, 1953.

Walter, W. Grey. "Thought and Brain: A Cambridge Experiment." *Spectator Archive*, October 4, 1934, 10.

Walter, W. Grey, R. Cooper, V. J. Aldridge, W. C. McCallum, and A. L. Winter. "Contingent Negative Variation: An Electric Sign of Sensori-Motor Association and Expectancy in the Human Brain." *Nature* 203, no. 4943 (1964): 380–4.

Walter, W. Grey, V. J. Dovey, and H. Shipton. "Analysis of the Electrical Response of the Human Cortex to Photic Stimulation." *Nature* 158, no. 4016 (1946): 540–1.

Walter, W. Grey, and H. W. Shipton. "A New Toposcopic Display System." *Electroencephalography and Clinical Neurophysiology* 3, no. 3 (1951): 281–92.

Walter, W. Grey, and V. J. Walter. "The Electrical Activity of the Brain." *Annual Review of Physiology* 11, no. 1 (1949): 199–230.

Ward, Janet. *Weimar Surfaces: Urban Visual Culture in 1920s Germany*. Berkeley: University of California Press, 2001.

Warner, Maria. *Phantasmagoria*. Oxford: Oxford University Press, 2019.

Weber, Anne-Katrin. "Recording on Film, Transmitting by Signals: The Intermediate Film System and Television's Hybridity in the Interwar Period." *Grey Room* 56 (2014): 6–33.

Weidman, Nadine M. *Constructing Scientific Psychology: Karl Lashley's Mind-Brain Debates*. Cambridge: Cambridge University Press, 1999.

Weingart, Peter. *Die Wissenschaft Der Öffentlichkeit: Essays Zum Verhältnis Von Wissenschaft, Medien Und Öffentlichkeit*. Weilerswist: Velbrück, 2005.

Weisberg, D. S., F. C. Keil, J. Goodstein, E. Rawson, and J. R. Gray. "The Seductive Allure of Neuroscience Explanations." *Journal of Cognitive Neuroscience* 20, no. 3 (2008): 470–7.

Welsh, Ian, and Brian Wynne. "Science, Scientism and Imaginaries of Publics in the UK: Passive Objects, Incipient Threats." *Science as Culture* 22, no. 4 (2013): 540–66.

White, Mimi. "The Attractions of Television. Reconsidering Liveness." In *Mediaspace: Place, Scale and Culture in a Media Age*, edited by Nick Couldry and Anna McCarthy, 75–91. London: Routledge, 2004.

Wiener, Norbert. *Cybernetics: Or Control and Communication in the Animal and the Machine*, 2nd ed. Cambridge, MA: MIT Press, [1948] 1961.

Williams, Raymond. *Keywords: A Vocabulary of Culture and Society*. New York: Oxford University Press, 1976.

Willis, Martin. "On Wonder: Situating the Spectacle in Spiritualism and Performance Magic." In *Popular Exhibitions, Science and Showmanship, 1840–1910*, edited by Jill A. Sullivan, 167–81. New York: Routledge, 2015.

Woodman, Geoffrey F. "A Brief Introduction to the Use of Event-Related Potentials (ERPs) in Studies of Perception and Attention." *Attention, Perception & Psychophysics* 72, no. 8 (2010): 2031–46.

Wünsch, Guido, and Hans Rühle. *Messgeräte Im Industriebetrieb*. Berlin: Julius Springer, 1936.

Yalkut, Jud. "Electronic Zen: The Alternate Video Generation." 1984. http://vasulka.org/archive/Artists10/Yalkut,Jud/ElectronicZen.pdf.

Yuste, Rafael, and George M. Church. "The New Century of the Brain." *Scientific American* 310, no. 3 (2014): 38–45.

Index

Abbas, Ackbar 232
Abi-Rached, Joelle 202
active brain 7, 47, 120, *see also* brain
Adrian, Edgar Douglas 1–2, 85–86, 90, 95, 162
Albert, Eduard 44
aliveness 87, 132, 228
alpha 157
  brainwave activity 157
  conditioning 165
  fad 167–170
  feedback 161–162, 167, 169
  media hype 181
  self-regulation 183
  sellers 169
  train 179
Alpha Bean Lima Brain (1971) 181
Alpha Pacer 169
Alphaphone 169
altered states 161
Amsterdam's EYE Filmmuseum 195
analogies 9–10, 33, 50, 53, 68, 70, 73, 78, 119, 122, 133, 136, 138, 140, 153, *see also* metaphors
analogical loops 119, 136
analogical linking 137
anatomical model 59
anatomoclinical gaze 31
Andriopolous, Stefan 91, 122
Angerer, Marie-Louise 12
animation 61–63
animated brains for modern citizens 61–63
Anschaulichkeit 60
anti-physiognomic iconography of the brain 81
aphasia 37, 220
apparatus of neurological capture 31, 38–42, *see also* Foucault, Michel
applied psychology 32

Applied Psychology: Making Your Own World (1919–1920) 31
artificial lighting 61
art-science
  collaborations 227
  experiment 204, 206–207, 215–218, 228–230
  installations 3, 203, 207, 214
Ashley, Robert 155
Astolfini, Laura 217
atmospheric media 230
Atlas of Encephalography (1941) 85, 110
Auslander, Philip 22
autism spectrum disorder (ASD) 213

Babiloni, Fabio 217
Barad, Karen 18
Bateson, Gregory 174
Bayle, Antoine 38
Bazin, André 125–126
BCIs, *see* brain–computer interfaces
Beard, George 33
behavioral neuroscience 207
Benjamin, Walter 33–34
Bensaude-Vincent, Bernadette 16, 103
Berenstein, Rhona 123
Berger, Hans 90, 95
Berger rhythm 162
Bergson, Henri 43
big data neuroscience 209
Bingham, Walter Van Dyke 32
biocybernetic gadgetry 188
biocybernetics 187
  channel 188
  feedback 190
bioelectrical-media-technical hybrids 53
bioelectrical waves 105

biofeedback 157, 168
  boom 169
  devices 177
  interfaces 186
biomediated body 12
biometrics exhibition 107, 109
Bio-Music 176, *see also* Eaton, Manford
biotypology 109
Birth of the Clinic, The (Foucault) 37–38
bisected brain (split brain) 161
Bishop, George 73, 75
Bjorlund, Marc 175
black box brain 152–153
black-boxed systems 72, 178, 201
blobology 5
Bono, James 11
Borck, Cornelius 6, 35, 41, 51–53, 90–91, 95, 136, 232
Bowker, Geoffrey 135
Bowler, Peter J. 55
Braidotti, Rosi 12
Brain
  activity 2
  anatomy 80
  imaging 4–8, 198, *see also* functional imaging
  cerebral localization 39, 43, 55
  culture 4
  damage 55
  decoding 112
  electrical pulses 85
  histories 25
  lesions 37–44, 56
  mirrors 44, 90, 152
  mind–brain relation 6, 55, 167
  models 3, 19, 49–83
  morphology 63
  prints 134
  processes 44, 62
  rhythms 91
  script 95
  states 6

structure-function relation 67
  training 165
  truths 66
Brain: The Mind as Matter (2012) 202
brain at work 3–4, 6, 20–23, 29–30, 86, 105–106, 151, 227
brain–behavior relation 55
Brainbow 198
brain–computer interfaces (BCIs) 157–158, 185–194
braincurves 45
brainmedia 2–13, 23
  as assemblages 12–13, 17, 27, 79, 129
  defined 15
  performing 18
brain–radio connections 93
brains *on* and *as* television 119–123
  brain "*En direct*" 126–133
  broadcasting science 123–126
  cortical scanning hypothesis 134, 140
  cybernetic living brain 133–137
  television as trigger and mirror 142–147
  toposcope 147–151
brain-to-brain synchronization 14, 197, 199, 204–205, 212–213, 215–217, 220–221
brain vibrations 136, *see also* Hartley, David
brainwaves 11, 109, 119, *see also* electroencephalography (EEG)
  imaginaries 89
  imaginaries and popularizing science 90–96
  measuring technologies 228
  media 91–92
  reading device 86
  supernatural connotations 99
  synchronization 195, 204
  synchrony of 175
Brain Waves installation 2, 8, 20–23
brain–X-ray–television 133

Brandon, Joseph 176
Brazier, Mary 185
Brenninkmeijer, Jonna 171
broadcasting live brains 119–123
broadcast programs 94
Brown, Barbara 167, 179–180
Brown, Dan 165
Burnham, Jack 194

Cacioppa, John 221
Cage, John 156
Callard, Felicity 226
Canales, Jimena 19
Cartwright, Lisa 29
Castel, Alan 8, 27
Catching a Brain Wave (1954) 123
cathode ray oscillograph 23, 71
Cazzamalli, Fernando 93
Century of the Brain 198
cerebral hemispheres 10
cerebral localization 39, 43, 55
cerebroscopes 44, 91
Charcot, Jean-Martin 39, 41, 55
Charcot's photographic
    methods 40
choreography of truth 18
Choudhury, Suparna 24
cinematograph 88
cinematography 87
circuit 74–80, 145
circuited self 172
circulating knowledge 16, 233
citizen science 207–210
clairvoyance 92
Clark, Stephen 25
Clough, Patricia 12
cognitive assemblages 12
cognitive data 208
cognitive functions 6
cognitive psychology 7
cognitive science 6
Cohl, Emile 45
Cole, Jonathan 214
collective consciousness 205

communication
    human-computer 178
    machine-man 189
    science 89, 89, 92, 97
computational metaphor 10
computational modeling 206
Computer and the Brain, The
    (1958) 135
Condon, William 218–219
"constitutive invisibility" of media 80
Cook, James 88
Cooter, Roger 17, 27
correlative practices of brain
    research 37–44
correlative constitution of a
    neurological body 39
cortical scanning hypothesis
    134, 140
Craick, Kenneth 138
Crawford, Matthew 8
critical neuroscience 24–27
crowdsourcing neuroscience 196,
    199–201, 207, 224–225
Crown, Peter 159
cultural imaginaries of brain 6
cultural rhythms 220
cybernetics 121–122, 135
    analogical linking 137
    craze 135, *see also* Kline, Ronald
    living brain 133–137
    modeling 141
    television brain 134

Dale, Henry 1
dark media 114
Daston, Lorraine 21, 108
data-driven approach 209
data gathering 209
data infrastructures 209
data interpretation 209
data journeys 209
data visualizations 149
Decade of the Brain 4, 202
deductive logic 123, 149, 152

Defense Advanced Research Projects Agency (DARPA) 184–185
demonic agency 122
demonstrations 67, 86–87, 89–90, 95–102, 233
  assemblage 18
  of brainwaves 96–97
  electroencephalography (EEG) 98, 101–103, 106, 110, 116, 227
Design for a Brain (1952) 135
Devil Commands, The (1941) 90, 111–115
Dewan, Edmond 157, 159, 180–181
diagnosis 40–42
  psychodiagnosis 107
  (tele-)diagnosis 125
Die Lichtwarhnehmung (Kahn) 54
Die Spione (Lang) 77
Dijck, José van 18
Dikker, Suzanne 195–196, 205–206, 208, 210–211
disenchantment 88
display
  dynamic displays 61, 68
  illuminated displays 61
  logic of direct display 50, 79
Dmytryk, Edward 113
Donchin, Emanuel 188, 190
Draaisma, Douwe 10, 37, 81
Dresden Hygiene exhibition 62
Dumas, Guillaume 205, 218–219
Dumit, Joseph 7, 192
Dupree, Mary 17
During, Simon 88
dynamic brains 49–52, 55–57, 195–200
  analogy of the motograph brain 68–71
  dynamic image 63–68
  dynamic lesions 39, 55
dynamic illumination 79

Eaton, Manford 176
ecological validity of neuroscience research 205

Edison, Thomas 44
EEG, *see* electroencephalography (EEG)
E.E.G. KISS (Lancel & Maat) 201
EEG feedback 155–194, *see also* alpha brain training 165
  circuited self 172–174
  disproportionate excitement 167–170
  groovy science of alpha 158–162
  techno-sensory interface projects 178–182
electrical activity of human brain 1
electrical indicators 80
electrical pulses 85
electro-brains 52–57
electroencephalograms 111
electroencephalography (EEG) 6, 85, 110–111
  booth 106
  demonstrations 96–109
  feedback 162
  frequency analyzer 111
  graphs 89
  headset 2
  hyperscanning 204, 215
  liveliness 103, 110
  Paris International Exposition 86, 102–109
  rhythm 1
  uncanny 92
  white magic 102–109
electroneurophysiology recordings 75
electronic brains 135
electronic mediation 52
Electronic Zen 175
electrophysiology of nervous tissue 73
electropsychophysiology 101
electrotechnical brain 83
electrotechnological displays 87
electrotechnology 52, 95
enacted, extended, embodied, and affective (4EA) perspectives on cognition 214

Index

enchanted loom 49, 70
encephaloscope 44
En direct de 127
En direct du Cerveau Humain
  (1956) 127–133
Engell, Lorenz 131
Environetic Synthesis
  (Lowenberg) 176
epidiascope 1, 42
epilepsy research 143–144, 150
epistemic virtues 21
Ernst, Wolfgang 22–23
Estrin, Thelma 185
Eunoia (Park) 201
event-related potentials (ERP)
  183–191
Exner, Robert 58, 60
experimental scenography 18
experiential physiology 167
extrasensory perception 92
eye-television analogies 138

fascination 231–233
feedback 160, 212, *see also*
  EEG feedback
feedback imaginary 213
Felt, Ulrike 58
Fessard, Alfred 105, 108–109
Fitzgerald, Des 226
flicker 134, 143
Foerster, Otfrid 45
forms of liveness 20–23, *see also*
  aliveness; liveness; liveliness;
  television, liveness
Foucault, Michel 3, 30, 37, 108
Frank, Robert 71–72
Franzel, Sean 17
Freud, Sigmund 43
functional imaging 6–7
functional maps 66

Gabrys, Jennifer 230
Galison, Peter 21, 108
Gallagher, Shaun 214

Gallant, Jack 199
Gastaut, Henri 128, 144
Gavrus, Delia 25
Gehirnwahrheiten (brain truths) 66
genealogy 10, 13, 15, 27, 132, 152,
  216, 228–229, 231
Gerard, Ralph W. 69–70, 72–73, 83
Giant Brains or Machines That Think
  (1949) 135
Giese, Fritz 52
Gigerenzer, Gerd 7
Glass Brain 60
"Glass Man" model 60
Goethe, Johann Wolfgang von 65
Goldstein, Kurt 55
groovy science 158–162
Grossberg, John 167
Grudin, Jonathan 179
Grusin, Richard 12
Guenther, Katja 36, 66
Gunning, Tom 87, 92, 114

Hagner, Michael 5–7, 25, 44, 65–66,
  80–81, 232
Haraway, Donna 11, 18
Harbou, Thea von 77–78
Harrington, Anne 36, 43, 161
Hartley, David 136
Hayles, N. Katherine 12, 23, 137
Head, Henry 55–56
Helmreich, Stefan 160
Hilgartner, Stephen 18
Hillman, Thomas 208
Hilton, Warren 31–34
historical epistemology 21
holism 56, 65
Huhtamo, Erkki 13
Human Brain, The (1955) 135, 142
human–computer interaction
  178–179
human nervous system 30
human sensorium 31
Huth, Alexander 205–206
Hutto, Daniel 214

hyperbrain 223
hyperscanning 204–205, 218, 224
hypnosis 92
hysteria 38

illumination 50, 52, 59–64, 67, 71, 79–80
immediacy 6, 20–21, 40, 47, 71, 75–76, 80, 123, 227
impressiveness in visualizing 60, 67
industrial-scientific entrepreneurship 203
information feedback procedures 164
information-flow systems 141
information-processing
  brain 135
  machines 135
Interactive Brainwave Drawing: EEG Telemetry Environment (Sobell & Trivich) 172
interbrain synchrony 204
interface 178–179
International Congress of Psychology 103
International Exposition, Paris (1937) 103
interwar science communication 95

Kaempffert, Waldemar 101
Kahn, Fritz 53–55, 69, 78, 95
Kaiser, David 162
Kamiya, Joe 163–164, 166, 169
Kantor, Robert 165
Kasperowski, Dick 208
Klemperer, Edith 58–59, 67, 75
Kline, Ronald 122, 135
knowledge in transit 16
Kracauer, Siegfried 76–77, 104, 116
Krakauer, John 207
Krämer, Sybille 17

La Biologie: Exposition Internationale de Paris (1937) 106
Lallemand, Francois 38

Lancel, Karen 201
Lang, Fritz 77–79
La Pensée Artificielle (1956) 135
Lashley, Karl Spencer 55–56, 68–70
Latour, Bruno 18
Law, John 13
Leonelli, Sabina 206
Leuchtschaltbild 76
Lichtlehrmethode 61
Lincke, Werner 61
live brain 29–31, 50, 228
  emergence 3, 47
  operative accessibility 229
  performance 2–4
  real-time temporality 229
  real-world liveness 229–231
liveliness 87, 228
liveness 20–23, *see also* live brain; television, liveness
  forms of 22
Living Brain, The (1953) 122, 135, 152, *see also* Walter, William Grey
living diagram 75–79
Lowenberg, Richard 176
Lucier, Alvin 155–156, 159
luminous brain model (*das Leuchtende Gehirn*) 50, 57–64, 63, 75, 79
Lupton, Deborah 210

Maat, Hermen 201
MacCurdy, John T. 55, 68
Mackay, Donald 141
MacVarish, Jan 8
man–computer
  dialogue 185
  symbiosis 179
man–machine
  communication 185, 188
  everything 179
  system 185
Marie, Pierre 36

Index

Maslow, Abraham 165
material-discursive assemblages 9–12, 18, 123, 135, 137, 152
materialization, apparent 7
material performativity 18
Matthews, Bryan H. C. 1–2, 85–86, 162
Matusz, Pawel 205
Max, Louis 98, 100
McCabe, David 8
McCray, W. Patrick 162
McCullogh, Warren 138
McLuhan, Marshall 6
media archaeology 13
mediabodies 12
media ecology 1974
media environment 198
media genealogy 13, *see also* genealogy
media-historical analysis 13, 15, 200
media imaginaries 14, 211
media marginalism 80
medianaturecultures 12
medianatures 12
media-techno-cultural-biological assemblage 177
media technology 10
mediation 3, 12–14, 19–21, 35, 52, 68, 71–76, 80–81, 115, 135, 230
media ubiquity 43
medico-televisual gaze 131
medium 34
Metropolis (Harbou, Lang) 78
Meloni, Maurizio 7
mental power 33
menticide 122
metaphorical circuits 11
metaphors 9–11
Michael, Mike 210
microanalysis 219–220
micro-temporalities of brain 182–191
Millett, David 1
mind–brain relation 6, 55, 167

mind reading 199, *see also* thought reading
mindreading apparatus 101
Minds and Machines (1954) 135, 140
mind script 91
model-making heuristics 122
modern magic 88
modern sensorium 31–38
Monea, Alexander 15
Mood Ring 169
Moore, Rachel 88
Morus, Ian 87
motograph 70
motograph brain 50, 68–71, 76–77, 79–81, 95
Movie Drome (VanderBeek) 175
Mullen, Tim 223
Music for Solo Performer 155, *see also* Lucier, Alvin
Mutual Wave Machine (Dikker & Oostrik) 195, 197, 199–204, 207–208, 211, 221, 224
mutual waves 220
myographs of dreams 100
mystique populaire 116–117
myth of total transparency 130

naturecultures 12
nerve cell activity 128
nerve exhaustion 33
nerve force 33
nerve recordings 93, *see also* Adrian, Edgar Douglas
nervous breakdowns 142
nervous economy 33
nervous exhaustion 33
nervous system 50, 80–81, 83
nervous transmissions 11
network analogy 51
neural coupling 204
neural networks 10–11
neurasthenia 33, *see also* Beard, George
neurobiologization 26

neurobiological citizens 202
neuro-centric reductionism 214
neurocentrism 26, 202, 211–214
neuroculture 4, 7
neuroenchantment 202
neurofeedback 192
neuro-governmentality 24, 202
neurological capture, apparatus of 31
neurological gaze 31
neurometers 44
neurophotography 91
neurophysiology 72
neuroscience 25
  big data 209
  crowdsourcing 199–201
  ecological validity 205
  social neuroscience 197, 200, 203, 205–207, 209, 213–215, 221, 224–225, 228
  in the wild 200
neuroscientific life, forms of 203
neuroscientism 8
New Mind, New Body (Brown) 168
Nikolow, Sybilla 109
Nunn, Christopher 191
Nye, David E. 71

Ogston, W. D. 218–219
Oostrik, Matthias 195–196
optical uncanny 86, 114–115
Otis, Laura 10–11, 51

Packer, Jeremy 15
Painlevé, Jean 125–126
Palais de la Découverte 103
paradox of transparency 21
Parikka, Jussi 11
Paris International Exhibition, 1937 103, 116
Park, Lisa 201
patterned whole-brain activity 90
Patterns in Your Head (1951) 134
Pavlov, Ivan Petrovich 46
pedagogies of brain awareness 202

perceptual psychology 60
performativity 17–18
performing knowledges 2, 16–19, 27
Perrodin, Catherine 205
person-to-person communication 175
Pfeiffer, John 142–143
photomontage technique 93
physiological television apparatus 73
physiognomic sight 108
Pias, Claus 137
Pickering, Andrew 18, 139, 171
Pickersgill, Martyn 202
Pitts, Walter 138
plasticity 67
pneumoencephalography 128, 132
Poole, Steven 8
Popović, Zoran 208
popular science 16, 103
  communication 89, 92, 97
  education 57
  journals 7, 35
  materialism in 1920s 55
  vulgariser 103
  writers 8, 168
Porter, Roy 33
Pötzl, Otto 64–65
Pratt, Cornelia Atwood 44
promissory neuroscience 232
prosthetic approach 185
Psychiatric Power (Foucault) 39–40, 43
psychic radio 91
psychic television 91
psychodiagnosis 107
psychological warfare 122
psychophysiology 45, 110
psychometrics 107
psychotechnics 60
public art–science installation 199
public engagement 210
public intelligence 234, *see also* Stengers, Isabelle

Pumfrey, Stephen 17
push-button brain diagram 63

Rabinovitz, Lauren 51
Radical Software 174–175, 177
radio sense 93
real time 23, 186
real-time brain 155–158
real-time systems 194
real-world neuroscience 206
Redcay, Elizabeth 215
Reicher, Karl 45
Rose, Nikolas 202, 232
Rosenboom, David 172–173, 190, 221–222
Roskams, Jane 208

SABRE 194
SAGE 194
Saler, Michael 88
Salisbury, Laura 33
Salpêtrière, La 39–41, 127, 132
Sarasin, Philip 14
Schilbach, Leonard 214–215
schizophrenia 220
Schmidgen, Henning 19
science communication 89, 92, 97, 233
science exhibition 18, 35, 67, 89, 102–103, 109–110, 135
scientification 35
Sconce, Jeffrey 132
Secord, James 16
seductive allure of brain images 5
self
 californian cult of the self 160
 circuited self 172
 EEG feedback as communicating with the self 165–166
 self-improvement 160
 technologies of the self 171, *see also* Foucault, Michel
self-image of television 132
self-regulation of brain activity 165

sequenced illumination 60–63, 67, *see also* animation
Shapin, Steven 203
Sherrington, Charles Scott 49–50, 70, 75, 81
Shipton, Edward 162
skull breaker 38
Slaby, Jan 8, 24, 214
Slee, Richard 44
Sluckin, W. 140–141
Smith, Roger 6, 27, 45, 55
Smythies, John 146
Sobell, Nina 172, 174
social interaction 197, 200, 205–206, 211–212, 214–219, 221, 224–225
social neuroscience 205
Sokal, Michael 32
somatization 40
sound cinematography 217
Spacewar! 185
spatiotemporalities 228
spatio-tempo-realities 228
spectacle 19, 223
Spectacle of Mind (Stelarc) 201
spectators 1, 9, 20, 60, 63, 71, 87–88, 104–105, 115, 211
spiritual wireless 91
Stadler, Max 25, 72, 81
staging science 18, 233
Star, Susan Leigh 36
Staub, Michael 161
Stelarc 201
Stengers, Isabelle 234
Sterman, Barry 172
Stoyva, Johann 164
stroboscope 122
supernatural powers 92
synaptic gaps 10
synchronization 200
 affect 217
 being in sync 216
 being on the same wavelength 216

brain-to-brain 14, 193, 199–201, 204, 212–218, 220, 223, 225, 227, 233
brainwave 195, 204
dynamic brains 215–223
harmonization 216
interpretation 218
notion 215
neuromarkers 213
operability 217
tuning 216

techno-sensory interface projects 178–182
Teitelbaum, Richard 176, 181
Telefile 194
telepathic brainwaves 93
telepathic radiations 91
television 73–75, 80
  analogy 75, 141
  of attractions 124
  brains *on* and *as* television 119–123
  epilepsy 143–144
  le direct 125
  liveness 123–125
  networked intimacy 129
  scanning mechanism 121, 139
  spatial pyrotechnics 125
  transmissions of visualizations 120
  as trigger and mirror 142–147
tele-visual and televisual neurophysiology 72–75
temporalities and temporealities 22
Thacker, Eugene 12, 114
theater of proof 18
Thibault, Robert 192
thinking machines 135
Thornton, Davi 202
thought-controlled device 157
thought reader 95
thought reading
  helmets 91
  mind script 91

neurophotography 91
  recorder 91
tools-to-theory heuristics 7
toposcope 147–151
transparency 20–21, 60, 80, 130, 227
transposition 11
Travis, Lee 96
trigger circuit 145, 162
Trivich, Michael 174
two-body neuroscience 197, 200, 204, 214, 218, 224

uncanny
  optical uncanny 86, 114
  uncanny technologies 92
urban brains 53

Väliaho, Pasi 11
Vancouver Piece (Rosenboom) 222
VanderBeek, Stan 175
van Dijck, José 130
van Leeuwen, Willem Storm 159
Verdeaux, Georges 41
Verdeaux, Jacqueline 41
vibratory photography 91
Vidal, Jacques 191
visceral literacy 221
visual cortex 149
Volkshochschulen 58, 67
Vogeley, Kai 213
Vogl, Joseph 14
Von Economo, Constantin 64–66

Wallot, Sebastian 216
Walter, Vivian 144
Walter, William Grey 122, 133, 138, 143–144, 162
wave phenomenon, behaviour as 219
Weidman, Nadine 32, 56
Weingart, Peter 18
Wernicke, Carl 43
Wertheimer, Max 64

White, Mimi 124
white magic 89
Wiener, Norbert 138
world brain 174
wonder, discourses of 87, 89, 103–104, 109

writer's brain 36

X-ray microscope 55
X-rays 29, 87–88

Zentrenlehre 35

www.ingramcontent.com/pod-product-compliance
Lightning Source LLC
Chambersburg PA
CBHW052153300426
44115CB00011B/1647